Advanced Materials and Manufacturing Processes

Science, Technology, and Management Series

Series Editor: J. Paulo Davim, Professor, Department of Mechanical Engineering, University of Aveiro, Portugal

This book series focuses on special volumes from conferences, workshops, and symposiums, as well as volumes on topics of current interested in all aspects of science, technology, and management. The series will discuss topics such as, mathematics, chemistry, physics, materials science, nanosciences, sustainability science, computational sciences, mechanical engineering, industrial engineering, manufacturing engineering, mechatronics engineering, electrical engineering, systems engineering, biomedical engineering, management sciences, economical science, human resource management, social sciences, engineering education, etc. The books will present principles, models techniques, methodologies, and applications of science, technology and management.

Manufacturing and Industrial Engineering
Theoretical and Advanced Technologies
Edited by Pakaj Agarwal, Lokesh Bajpai, Chandra Pal Singh, Kapil Gupta, and J. Paulo Davim

Multi-Criteria Decision Modelling
Applicational Techniques and Case Studies
Edited by Rahul Sindhwani, Punj Lata Singh, Bhawna Kumar, Varinder Kumar Mittal, and J. Paulo Davim

High-K Materials in Multi-Gate FET Devices
Edited by Shubham Tayal, Parveen Singla, and J. Paulo Davim

Computational Technologies in Materials Science
Edited by Shubham Tayal, Parveen Singla, Ashutosh Nandi, and J. Paulo Davim

Advanced Materials and Manufacturing Processes
Edited by Amar Patnaik, Malay Kumar Banerjee, Ernst Kozeschnik, Albano Cavaleiro, J. Paulo Davim, and Vikas Kukshal

For more information about this series, please visit: www.routledge.com/Science-Technology-and-Management/book-series/CRCSCITECMAN

Advanced Materials and Manufacturing Processes

Edited by

Amar Patnaik
Malay Kumar Banerjee
Ernst Kozeschnik
Albano Cavaleiro
J. Paulo Davim
Vikas Kukshal

CRC Press is an imprint of the
Taylor & Francis Group, an **informa** business

MATLAB® is a trademark of The MathWorks, Inc. and is used with permission. The MathWorks does not warrant the accuracy of the text or exercises in this book. This book's use or discussion of MATLAB® software or related products does not constitute endorsement or sponsorship by The MathWorks of a particular pedagogical approach or particular use of the MATLAB® software.

First edition published 2022
by CRC Press
6000 Broken Sound Parkway NW, Suite 300, Boca Raton, FL 33487–2742

and by CRC Press
2 Park Square, Milton Park, Abingdon, Oxon, OX14 4RN

© 2022 selection and editorial matter, Amar Patnaik, Malay Kumar Banerjee, Ernst Kozeschnik, Albano Cavaleiro, J. Paulo Davim, and Vikas Kukshal; individual chapters, the contributors

CRC Press is an imprint of Taylor & Francis Group, LLC

Reasonable efforts have been made to publish reliable data and information, but the author and publisher cannot assume responsibility for the validity of all materials or the consequences of their use. The authors and publishers have attempted to trace the copyright holders of all material reproduced in this publication and apologize to copyright holders if permission to publish in this form has not been obtained. If any copyright material has not been acknowledged please write and let us know so we may rectify in any future reprint.

Except as permitted under U.S. Copyright Law, no part of this book may be reprinted, reproduced, transmitted, or utilized in any form by any electronic, mechanical, or other means, now known or hereafter invented, including photocopying, microfilming, and recording, or in any information storage or retrieval system, without written permission from the publishers.

For permission to photocopy or use material electronically from this work, access www.copyright.com or contact the Copyright Clearance Center, Inc. (CCC), 222 Rosewood Drive, Danvers, MA 01923, 978–750–8400. For works that are not available on CCC please contact mpkbookspermissions@tandf.co.uk

Trademark notice: Product or corporate names may be trademarks or registered trademarks and are used only for identification and explanation without intent to infringe.

Library of Congress Cataloging-in-Publication Data
Names: Patnaik, Amar, editor.
Title: Advanced materials and manufacturing processes / edited by Amar Patnaik, Malay Kumar Banerjee, Ernst Kozeschnik, Albano Cavaleiro, J. Paulo Davim, Vikas Kukshal.
Other titles: Advanced materials and manufacturing processes (CRC Press)
Description: Boca Raton : CRC Press, 2022. | Series: Science, technology, and management series | Includes bibliographical references and index.
Identifiers: LCCN 2021017600 (print) | LCCN 2021017601 (ebook) | ISBN 9780367553746 (hbk) | ISBN 9780367553760 (pbk) | ISBN 9781003093213 (ebk)
Subjects: LCSH: Manufacturing processes—Periodicals. | Materials—Periodicals.
Classification: LCC TS183 .A385 2022 (print) | LCC TS183 (ebook) | DDC 670—dc23
LC record available at https://lccn.loc.gov/2021017600
LC ebook record available at https://lccn.loc.gov/2021017601

ISBN: 978-0-367-55374-6 (hbk)
ISBN: 978-0-367-55376-0 (pbk)
ISBN: 978-1-003-09321-3 (ebk)

DOI: 10.1201/9781003093213

Typeset in Times
by Apex CoVantage, LLC

Contents

Preface vii
Acknowledgments ix
Editors xi

Chapter 1 Analysis of the Tribological Behavior of Al6061 Alloy–Graphene Oxide Composites 1

Atirek Gaur, Rajendra Kumar Duchaniya, and Upender Pandel

Chapter 2 Multi-Objective Optimization of Steel Turning Variables under Metal Oxide Nanofluid Cooling Conditions 11

Anup A. Junankar, Yashpal, and Jayant K Purohit

Chapter 3 Modal Analysis of a Wind Turbine Blade Using a Computational Method 23

M.J. Pawar, Amar Patnaik, Vikas Kukshal, Ashiwani Kumar, and Vikash Gautam

Chapter 4 A Comprehensive Review of Ultrasonic Machining: A Tool for Machining Brittle Materials 35

T.G. Mamatha, Mudit K. Bhatnagar, Vansh Malik, Siddharth Srivastava, and Mohit Vishnoi

Chapter 5 Analysis of Mechanical and Sliding Wear Performance of AA6061–SiC/Gr/MD Hybrid Alloy Composite Using a PSI Approach for Rotor Applications 77

Ashiwani Kumar, Amar Patnaik, Mukesh Kumar, Vikas Kukshal, M.J. Pawar, and Vikash Gautam

Chapter 6 Fabrication and Characterization of Metallic Biomaterials in Medical Applications: A Review 95

Ganesh Kumar Sharma, Vikas Kukshal, Deepika Shekhawat, and Amar Patnaik

Chapter 7 Microstructural and Tribological Behavior of 5083-TiB$_2$ Cast Composites Fabricated by a Flux-Assisted Synthesis Technique 107

Vikash Gautam, M.J. Pawar, Amar Patnaik, Vikas Kukshal, and Ashiwani Kumar

Chapter 8	Study and Prediction of Response Parameters during Oblique Machining for Different Machining Parameters by WEDM of Inconel-HX	119

I.V. Manoj and S. Narendranath

Chapter 9	A Review on Milestones Achieved in the Additive Manufacturing of Functional Components	135

T. Sathies and P. Senthil

Chapter 10	Fatigue Behavior of Particulate-Reinforced Polymer Composites: A Review	155

Vijay Verma, Arun Kumar Pandey, and Chaitanya Sharma

Chapter 11	Characterization of Surface Engineering and Coatings	173

K. Ashish Chandran, A. Inbaoli, C.S. Sujith Kumar, and S. Jayaraj

Chapter 12	Review on the Influence of Retained Austenite on the Mechanical Properties of Carbide-Free Bainite	191

Siddharth Sharma, Ravi Kumar Dwivedi, and Rajan Kumar

Chapter 13	Tribological Behavior of Carbide-Free Bainite in High Silicon Steel: A Critical Review	203

Rajan Kumar, Ravi Kumar Dwivedi, and Siraj Ahmed

Chapter 14	Impact of Nanoparticles in Lube Oil Performance: A Review	213

Anoop Pratap Singh, Ravi Kumar Dwivedi, and Amit Suhane

Chapter 15	Non-Asbestos Organic Brake Pad Friction Composite Materials: A Review	227

Mukesh Kumar and Ashiwani Kumar

Chapter 16	Impact of Water Particles on Fly Ash–Filled E-Glass Fiber–Reinforced Epoxy Composites	257

Pankaj Kumar Gupta, Rahul Sharma, and Gaurav Kumar

Chapter 17	A Study on the Silane Chemistry and Sorption/Solubility Characteristics of Dental Composites in a Wet Oral Environment	269

Sukriti Yadav and Swati Gangwar

Index	281

Preface

Advanced Materials Processing and Manufacturing Processes embraces innovations in the fields of materials, manufacturing processes, and distinguished applications proposed by numerous researchers, highly qualified professionals, and academicians. This book emphasizes the properties and end applications of various engineering materials including metals, polymers, composites, biocomposites, fiber composites, ceramics, and so on, along with their property characterizations and potential applications of these materials. The book focuses on all wider aspects relating to advanced materials, materials manufacturing and processing, optimization and sustainable development, and tribology for industrial applications.

All chapters have been subjected to the peer-review process by researchers working in relevant fields. The chapters were selected based on their quality and their relevance to the title of the book. This book will enable researchers working in the field of advanced materials and manufacturing processes to explore the various areas of research and educate future generations.

Successful completion of this book includes the efforts of many people. It is imperative to acknowledge their contribution in shaping the structure of the book. Hence, all the editors would like to express special gratitude to all the reviewers for their valuable time spent on the reviewing process and completing it on time. Their valuable advice and guidance helped in improving the quality of the chapters selected for publication in the book. Finally, we would like to thank all the authors of the chapters for their timely submission of their chapters during the rigorous review process.

Amar Patnaik
Malay Kumar Banerjee
Ernst Kozeschnik
Albano Cavaleiro
J. Paulo Davim
Vikas Kukshal

Acknowledgments

The present book is the outcome of the efforts of many people. It includes all the authors, reviewers, and the concerned institutes of the editors. First, the editors would like to acknowledge all whose work, research, and support have helped and contributed to this book. We gratefully express gratitude to all the authors of the accepted chapters for transforming their work into the chapters of the book. We are grateful to the authors for their continuous support during the rigorous review process.

We would like to acknowledge all the reviewers associated with the review process of all the book chapters. Their continuous support and timely completion of the review process helped the editors complete the book within the stipulated time. Their valuable advice and guidance helped improve the quality of the chapters selected for publication in the book.

It's very difficult to thank everyone associated with this book. Therefore, we would like to acknowledge all the persons associated with the completion of the book directly or indirectly. And last but not the least, we would like to thank our family members for extending their support in allowing us to successfully complete this book.

Editors

Amar Patnaik is Associate Professor of Mechanical Engineering at Malaviya National Institute of Technology Jaipur, India. Dr. Patnaik has more than 10 years of teaching experience and has taught a broad spectrum of courses at both the undergraduate and graduate levels. He also served in various administrative functions, including Dean International Affairs, and as the coordinator of various projects. He has guided 20 PhDs and several M.Tech theses. He has published more than 200 research articles in reputed journals, contributed five book chapters, edited one book, and filed seven patents. He is also the guest editor of various reputed international and national journals. Dr. Patnaik has delivered more than 30 guest lecturers in different institutions and organizations. He is a life member of the Tribology Society of India, Electron Microscope Society of India, and ISTE.

Malay Kumar Banerjee served as former Ministry of Steel Chair Professor in the Department of Metallurgical and Material Engineering at Malaviya National Institute of Technology Jaipur, India. He has over 43 years of experience in academia and has taught a broad spectrum of courses related to metallurgical and material engineering at both the undergraduate and graduate levels. He has also served in various administrative functions and coordinator of various projects. He received many research grants from DST India, Ministry of Steel India, ISRO, DRDO, UGC, DMRL, AICTE, IIF, CPCB, NMRL, MHRD, and BRNS. He has consulted for numerous national and international research bodies. He has published more than 110 research articles in national and international journals, presented papers at 30 conferences, contributed 13 book chapters in different books, edited three books, and filed three patents.

Ernst Kozeschnik is Head of the Institute of Materials Science and Technology at TU Wien since 2009. He has supervised numerous master's and PhD theses, teaches basic and advanced university courses, and has authored and coauthored more than 140 scientific papers. He is the author of the textbook *Modeling Precipitation Kinetics*. Dr. Kozeschnik is Lead Developer of the MatCalc software package, which is used in many different universities and research centers worldwide. He is Head of the Scientific Advisory Board of MatCalc Engineering GmbH. Dr. Kozeschnik was rewarded a "Christian Doppler Laboratory" in 2007, which is one of the most renowned midterm scientific funds in Austria.

Albano Cavaleiro is Professor in the Mechanical Engineering Department of the Coimbra University, Portugal. He has supervised numerous master's and PhD theses, teaches basic and advanced university courses, and has authored and coauthored more than 100 scientific papers, 60 of them in the *International Journals of the Science Citation Index*. He has participated in more than 60 international events, such as congresses, conferences, workshops, and exhibitions. He has given 15 invited talks in national and international scientific events. He has participated in more than 25 projects (10 international).

J. Paulo Davim is Professor at the Department of Mechanical Engineering of the University of Aveiro, Portugal. He has more than 30 years of teaching and research experience in manufacturing, materials and mechanical and industrial engineering, with a special emphasis in machining and tribology. He has guided large numbers of postdoc and PhD theses for many universities in different countries. He is the editor in chief of several international journals, a guest editor of journals, a book editor, a book series editor, and a scientific advisor for many international journals and conferences. Presently, he is an editorial board member of 30 international journals and acts as a reviewer for more than 100 prestigious Web of Science journals. In addition, he has also published, as an editor (and coeditor), more than 250 books and, as an author (and coauthor), more than 15 books, 100 book chapters, and 500 articles in journals and conferences.

Vikas Kukshal is presently working as Assistant Professor in the Department of Mechanical Engineering, NIT Uttarakhand, India. He has more than 10 years of teaching experience and has taught a broad spectrum of courses at both the undergraduate and postgraduate levels. He has authored and coauthored more than 26 articles in journals and conferences and has contributed seven book chapters. Presently, he is a reviewer of various national and international journals. He is a life member of Tribology Society of India, the Indian Institute of Metals, and the Institution of Engineers. His research area includes material characterization, composite materials, high-entropy materials, simulation, and modeling.

1 Analysis of the Tribological Behavior of AL6061 Alloy–Graphene Oxide Composites

Atirek Gaur, Rajendra Kumar Duchaniya, and Upender Pandel

CONTENTS

1.1 Introduction 1
1.2 Experiment 2
    1.2.1 Materials 2
        1.2.1.1 GO Preparation 2
        1.2.1.2 AL6061 Alloy as a Matrix Alloy 2
1.3 Experimental Procedure 3
    1.3.1 Tribological Analysis 3
    1.3.2 Hardness Test 3
    1.3.3 Air-Jet Erosion 4
1.4 Results and Discussions 4
    1.4.1 Microstructure Evolution 4
    1.4.2 Hardness Testing 4
    1.4.3 Weight-Loss Evolution 5
    1.4.4 Evolution of Coefficient of Friction 6
    1.4.5 Evolution of Wear Rate 6
1.5 Conclusion 8
References 8

1.1 INTRODUCTION

AL6061 alloy and its alloys have a great deal of importance on account of its outstanding properties. These are widely utilized in aviation applications, also in the structure field because of their high quality and corrosion obstruction [1]. These metal matrix composites (MMCs) were widely utilized for various applications, for example, airplanes, space transports, and automobiles because of their improved properties [2–5]. These attributes incorporate lightweight, high inflexibility, high strength, and are highly resistive to corrosion cracks [6–8]. Few data are accessible with respect to the appraisal of strong molecule disintegration wear of MMCs. The reduction in the point of impact from 90°

to 20° has brought about a critical increase in erosive wear misfortune and flexibility [7–8]. Their extraordinary properties are the explanation behind the need to grow more and more new composite materials [9–10]. The unreinforced graphene oxide (GO) and its composites with copper (Cu), magnesium (Mg), silicon carbide (SiC) [11–13]. Carbon allotropes like GO, carbon nanotubes, and all other two-dimensional (2D) carbon nanomaterials have excellent electrical and thermal properties [13–17]. GO created by a Hummer's method progression for regular, excellent higher graphitic synthetic concoctions were utilized, which keeps up the GO layers [18]. First, oxidized oxygen and, afterward, GO are diminished through reddening the equivalent to hydrogen [19–22]. An electro-less route exploited on the GO sheets for solid interfacial holding [23–24]. An ensuing advance was utilized where graphene nanocomposite was presented. These works feature the significance of exploring AL6061 as an alloy and AL6061 alloy and GO composites. Few data are accessible with respect to the tribological studies of an AL6061 alloy and GO composite. This work also centers on the use of GO particulates as a fortification for growing excellent MMCs and describing its sliding wear nature.

1.2 EXPERIMENT

1.2.1 MATERIALS

For making the GO using Hummer's method, various chemicals and powder were used as raw material like pure graphite powder 98 wt%, sulfuric acid (H_2SO_4), potassium permanganate ($KMnO_4$), deionized water, sodium nitrate ($NaNO_3$), and an aqueous solution of $NH_3.H_2O$, hydrogen chloride (HCl) in diluted form, 30% of H_2O_2.

1.2.1.1 GO Preparation

The GO was prepared with a modified Hummer's method, having a proportion of 2:1; graphite and sodium nitrite was taken into a beaker with a quantity of H_2SO_4, at 15 °C to 21 °C, to attain the suspension. $KMnO_4$, used for the oxidation, was steadily added into the suspension with permanent stirring. A temperature of 15 °C was maintained for 2.5 hours. A specific measure of deionized water was included in the blend at an agonizingly slow pace when H_2SO_4 blended into the solution. After a time, a mixture of boiling water and 32% H_2O_2 water was included in the blend, and after separate, continuous mixing, the yellow-colored GO suspension came about [24].

1.2.1.2 AL6061 Alloy as a Matrix Alloy

AL6061 alloy at a purity level of 96.4% in form of ingots was purchased from the industry and procured in the material testing lab in the metallurgy department in Malaviya National Institute of Technology (MNIT) for use as a matrix alloy. The composition of the AL6061 alloy is given in Table 1.1.

TABLE 1.1
Elemental Composition of AL6061 Alloy as a Substrate

Mg	Cr	Ti	Zn	Fe	Cu	Mn	Si	Al
0.8	0.30	0.15	0.25	0.70	0.35	0.15	0.70	Balance

1.3 EXPERIMENTAL PROCEDURE

AL6061 alloy was hot extruded at 550 °C, and after some time, the GO was mixed into the hot-extruded solution of AL6061 alloy. Then after cooling, the rectangular sections of AL6061 alloy were fabricated. After fabrication, the rectangular sections were weathered in erosion, hardness, and wear testing. To see the proper reduction of GO characterizations, scanning electron microscopy (SEM) was used to check the morphological investigation. The sample was made by stirring aluminum alloy into AL6061 alloy as a matrix. The sample size was $2.5 \times 2.5 \times 1.0$ cm^3 each at initial, but for the wear testing, the sample size was taken in different measurements according to the need of the testing machine.

1.3.1 TRIBOLOGICAL ANALYSIS

Tribological testing was completed via an erosion-wear tribometer (courtesy: tribology lab MNIT Jaipur) as indicated by the ASTM G99 standard. A pin 8 mm in width and 24 mm in elevation of the ousted sealing compound was used for testing samples. Test samples were cleaned and leveled. The basic surface finishing of the steel plate was 1 μm. The duration of the subjective test was 20 minutes whereas the descending speed was 0.310 m/s. Masses were contrasted as of 10–40 N. The coefficient of friction was resolved using frictional load and normal load data.

1.3.2 HARDNESS TEST

Vickers hardness testing was done on the fabricated sample; the test passed at many regions, and then the average of all the values was engaged as the hardness significance of the sample. A load of 100 kgf was used for the testing.

FIGURE 1.1 Pin-on-Disk Tribometer

TABLE 1.2
Testing Parameters Taken for Air-Jet Erosion Testing

S.No	Testing Parameters	
1	Test Temperature	Room Temperature
2	Test Time (minutes)	30
3	Sample Size (mm)	40 × 18 × 10
4	System Pressure (bar)	1.5
5	Stand-off Distance (mm)	20, 30, 40
6	Nozzle Diameter (Inner)	1.5
7	Impact Angle (degree)	30, 45, 60
8	Erodent Feed Rate (g/m)	4
9	Particle Velocity (m/s)	30
10	Erodent Size (μm)	215
11	Erodent Material	Silica Sand

1.3.3 Air-Jet Erosion

By pushing a stream of rough particulates from a little outlet with a known hole width toward the sample, repeated testing was directed for different trials by differing all the constraints as referenced previously. A loss of material every time is accomplished by utilizing the impingement of little grating particles on the test surface. The samples were evaluated prior to and then afterward testing to discover the weight reduction. Rectangular examples (both in combination and composites) of measurements 40 mm × 18 mm × 10 mm were utilized as testing samples (Table 1.2).

1.4 RESULTS AND DISCUSSIONS

1.4.1 Microstructure Evolution

Figure 1.2a shows the optical image (100 μm) of Keller's etched AL6061 alloy, which was used to check the microstructures of the AL6061 alloy, and Figure 1.2b and Figure 1.2c show the SEM morphology of the AL6061 alloy strengthened with GO. During the study of the images, we found that the optical images show the uniform surface finish of the AL6061 alloy after etching; no debris was found. On the other hand, the SEM images show a larger area of the small particle size, which leads to uniformity of the distribution of reinforcement of GO, and it also indicates an improved mechanical strength.

1.4.2 Hardness Testing

Figure 1.3 shows the results of the Vickers hardness testing; it was found an increase in hardness for the AL6061 alloy sample reinforced with GO. Since small grain sizes were achieved during the SEM testing, the dislocation resistance movements will be greater.

Analysis of Tribological Behavior

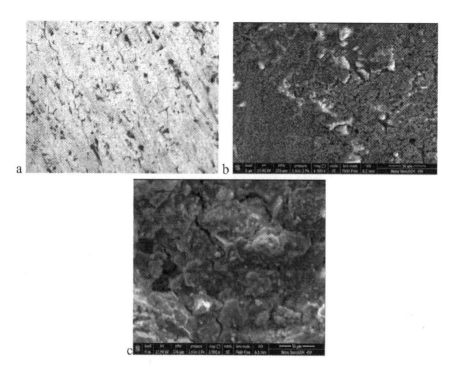

FIGURE 1.2 SEM Images of AL6061 SEM Images of Al6061 with Different GO%

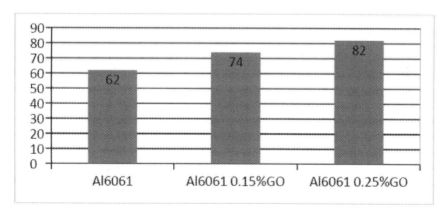

FIGURE 1.3 Microhardness of Bare AL6061 and AL6061 with different wt.% of GO

1.4.3 Weight-Loss Evolution

Various variations were found in the air-jet erosion testing of the bare AL6061 Alloy and the AL6061 alloy reinforced with 0.15% and 0.25% of GO. In Figure 1.4, as the test duration increased, the weight loss due to erosion also increased for the AL6061 alloy and the AL6061 alloy reinforced with 0.15% and 0.25% of GO, respectively.

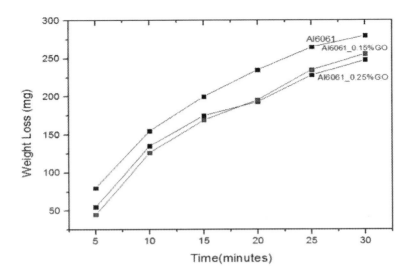

FIGURE 1.4 Weight-Loss Evaluation by Air-Jet Erosion Testing

The effects of increased kinetic energy lead the removal of the surface in large amounts. As compared to bare AL6061 alloy, less erosion of surface material was found for the AL6061 alloy reinforced with GO. Figure 1.4 shows the variations of weight loss during air-jet erosion testing.

1.4.4 Evolution of Coefficient of Friction

For evaluating the hot-extruded AL6061 coefficient of friction for AL6061 alloy reinforced by GO, a drop-in coefficient of friction is found with a growth in weight. A lower load of 10 N to a maximum load of 40 N was used for the testing, and as the experiment went on, we found the results very familiar. Compared to hot-extruded AL6061 alloy, the GO-reinforced AL6061 alloy showed a low coefficient of friction. GO also has the merits of high frictional properties, so it is reinforced in AL6061 alloy. The alloy shows a low coefficient for friction. The 0.15% and 0.25% GO used as a reinforcing form to AL6061 alloy are shown in Figure 1.5.

1.4.5 Evolution of Wear Rate

Figure 1.6 shows that during the evaluation of wear rate for the AL6061 alloy and AL6061 alloy reinforced with GO in various percentages of amounts, many variations were found. Testing was done at a minimum load of 10 N to a maximum load of 40 N at the sliding distance of 0.314 m/s. During the testing, we found that AL6061 alloy reinforced with GO shows a greater amount of wear resistance than bare AL6061 alloy does. The upgrading was achieved due to the superior properties of GO as it has a high resistance to friction and a high resistance to wear when it is used as a reinforcing or coating material to various MCCs.

Analysis of Tribological Behavior

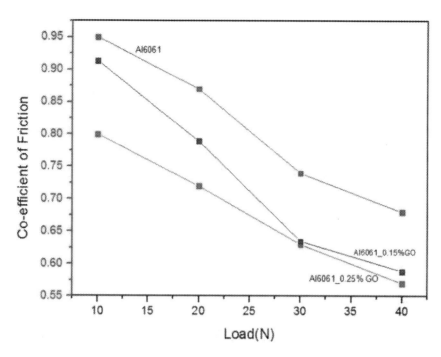

FIGURE 1.5 Hot-Extruded AL6061 Alloy and AL6061 with Different wt.% of GO

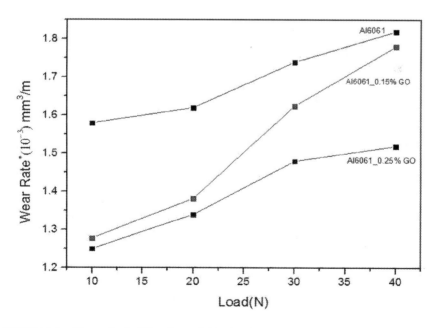

FIGURE 1.6 Change in the Wear Rate of AL6061 Alloy Reinforced with GO

1.5 CONCLUSION

1. GO was made successfully using the novel Hummer method.
2. SEM images and optical images of AL6061 show the presence of GO as the white color portion and microstructure of AL6061.
3. AL6061 alloy reinforced with GO with various contents like 0.15% and 0.25%; 1.0% exhibits a low coefficient of friction.
4. Hardness also increased for reinforced AL6061 alloy. A reduction in wear loss was also observed for the hot-extruded AL6061 alloy reinforced with GO.
5. Wear rates also decreased for the GO-reinforced AL6061 as compared to only hot-extruded AL6061 alloy.
6. During the evaluation of the coefficient of friction, when increasing the load, a reduction was found for AL6061 alloy reinforced with GO having different wt.% amounts.
7. Using GO as a reinforcing material improves bond strength.
8. Through all the testing discussed, it is quite possible that the use of GO as a reinforcing material can improve the overall life for AL6061 as compared to base AL6061 alloy.

As per the discussion, the various tests assures that this composite of AL6061/GO may be used for the automotive sector in many applications in terms of wear and frictional applications.

REFERENCES

1. Kaufman, J. G., ed. *Properties of aluminum alloys: tensile, creep, and fatigue data at high and low temperatures.* ASM International (1999).
2. Veeresh Kumar, G. B., A. R. K. Swamy, and A. Ramesha. "Studies on properties of as-cast AL6061 Alloy-WC-Gr hybrid MMCs." *Journal of Composite Materials* 46, no. 17 (2012): 2111–2122.
3. Veeresh Kumar, G. B., P. S. Shivakumar Gouda, R. Pramod, and C. S. P. Rao. "Synthesis and characterization of TiO2 reinforced AL6061 Alloy composites." *Advanced Composites Letters* 26, no. 1 (2017): 096369351702600104.
4. Suryanarayana, R. C., A. Hiriyannaiah, P. R. Gnanamurthy, and A. Bharadwaj. "Air jet erosion wear behavior of AL6061 Alloy-carbon fiber rod composites." *Tribology Online* 10, no. 1 (2015): 27–34.
5. Miracle, D. B. "Metal matrix composites—from science to technological significance." *Composites Science and Technology* 65, no. 15–16 (2005): 2526–2540.
6. Sinmazçelik, T., S. Fidan, and V. Günay. "Residual mechanical properties of carbon/polyphenylene sulfide composites after solid particle erosion." *Materials & Design* 29, no. 7 (2008): 1419–1426.
7. Mahapatra, S. S., and A. Patnaik. "Study on mechanical and erosion wear behavior of hybrid composites using Taguchi experimental design." *Materials & Design* 30, no. 8 (2009): 2791–2801.
8. MacMillin, B. E., C. D. Roll, and P. Funkenbusch. "Erosion and surface structure development of metal–diamond particulate composites." *Wear* 269, no. 11–12 (2010): 875–883.

9. Li, J.F., L. Zhang, J.K. Xiao, and K.C. Zhou. "The sliding wear behavior of copper-based composites reinforced with graphene nanosheets and graphite." *Transactions of Nonferrous Metals Society of China* (English Ed.) 25 (2015): 3354–3362.
10. Zhang, F.X., Y.Q. Chu, and C.S. Li. "Fabrication and tribological properties of copper matrix solid self-lubricant composites reinforced with Ni/NbSe2 composites." *Materials* 12 (2019).
11. Nautiyal, H., S. Kumari, O.P. Khatri, and R. Tyagi. "Copper matrix composites reinforced by rGO-MoS2 hybrid: strengthening effect to the enhancement of tribological properties." *Composites Part B: Engineering* 173 (2019): 106931.
12. Lian, W., Y. Mai, J. Wang, L. Zhang, C. Liu, and X. Jie. "Fabrication of graphene oxide Ti3AlC2 synergistically reinforced copper matrix composites with enhanced tribological performance." *Ceramics International* 45 (2019): 18592–18598.
13. Abd-Elwahed, M.S., and A.F. Meselhy. "Experimental investigation on the mechanical, structural, and thermal properties of Cu—ZrO2 nanocomposites hybridized by graphene nanoplatelets." *Ceramics International* 1 (2020).
14. Wang, J., L.N. Guo, W.M. Lin, J. Chen, S. Zhang, S. Da Chen, T.T. Zhen, and Y.Y. Zhang. "The effects of graphene content on the corrosion resistance, and electrical, thermal and mechanical properties of graphene/copper composites." *Xinxing Tan Cailiao/New Carbon Materials* 34 (2019): 161–169.
15. Huang, Y., Q. Yao, Y. Qi, Y. Cheng, H. Wang, Q. Li, and Y. Meng. "Wear evolution of monolayer graphene at the macroscale." *Carbon N. Y.* 115 (2017): 600–607.
16. Umanath, K., S.T. Selvamani, and K. Palanikumar. "Friction and wear behavior of AL6061 ALLOY (SiCp+Al2O3p) hybrid composites." *International Journal of Engineering Science and Technology* 3, no. 7 (2011): 5441–5451.
17. Ren, S., P. Rong, and Q. Yu. "Preparations, properties and applications of graphene in functional devices: a concise review." *Ceramics International* 44 (2018): 11940–11955.
18. Amanov, A., S. Sasaki, D.-E. Kim, O. V. Penkov, and Y.-S. Pyun. "Improvement of the tribological properties of AL6061 ALLOY—T6 alloy under dry sliding conditions." *Tribology International* 64 (2013): 24–32.
19. Manivannan, I., S. Ranganathan, S. Gopalakannan, and S. Suresh. "Mechanical properties and tribological behavior of AL6061 ALLOY—SiC—Gr self-lubricating hybrid nanocomposites." *Transactions of the Indian Institute of Metals* 71, no. 8 (2018): 1897–1911.
20. Mishra, A. K., R. Sheokand, and R. K. Srivastava. "Tribological behavior of Al-6061/SiC metal matrix composite by Taguchi's techniques." *International Journal of Scientific and Research Publications* 2, no. 10 (2012): 1–8.
21. Vencl, A., I. Bobic, and B. Stojanovic. "Tribological properties of A356 Al-Si Alloy composites under dry sliding conditions." *Industrial Lubrication and Tribology* (2014).
22. Zhang, D., F. Gao, X. Wei, G. Liu, M. Hua, and P. Li. "Fabrication of textured composite surface and its tribological properties under starved lubrication and dry sliding conditions." *Surface and Coatings Technology* 350 (2018): 313–322.
23. Wu, J., and X. H. Cheng. "The tribological properties of Kevlar pulp reinforced epoxy composites under dry sliding and water-lubricated condition." *Wear* 261, no. 11–12 (2006): 1293–1297.
24. Cao, N., and Y. Zhang. "Study of reduced graphene oxide preparation by Hummers' method and related characterization." *Journal of Nanomaterials* Vol. 2015, Article ID 168125, 5 pages, http://dx.doi.org/10.1155/2015/168125.

2 Multi-Objective Optimization of Steel Turning Variables under Metal Oxide Nanofluid Cooling Conditions

Anup A. Junankar, Yashpal, and Jayant K. Purohit

CONTENTS

2.1 Introduction 11
2.2 Investigational Conditions and Machining Variables 12
2.3 SiO$_2$ Nanofluid Preparation 13
2.4 Process Evaluation 14
2.5 Results and Discussion 18
2.6 Conclusion 20
References 20

2.1 INTRODUCTION

In the machining industry, the performance of a machining process is measured in reference to product surface quality and tool life. For those in the machining industry, enhanced machining performance without compromising the machined component's surface quality and the life of the cutting tool is the main challenge. In the past decade, sustainable outcomes were presented by eminent investigators to resolve the problems associated with machining performance. The alternate cooling techniques for flood cooling during a machining operation were presented. In the prior literature, among all the cooling techniques, minimum quantity lubrication (MQL) is projected to be sustainable and significant. Due to a minimum consumption of cutting fluid in MQL, the exposure of people on the shop floor to microparticles of cutting fluid and harmful gases is minimized. The proper penetration of cutting fluid in a machining zone is achieved with the implementation of MQL.

For upgrading machining performance, a new product, "nanofluid," combined with MQL, delivered noticeable results. When nanometer-sized particles are added to a base fluid, the prepared fluid is known as a "nanofluid." A one- or two-step method is used for the preparation of a nanofluid. Most of the nanoparticles are in

DOI: 10.1201/9781003093213-2

TABLE 2.1
Review of the Utilization of Metal Oxide Nanofluids during a Turning Operation

Ref.	Nanofluid	Surface Roughness	Tool Wear	Cutting Temperature	Cutting Forces
[1]	Al_2O_3	*	*	*	
[2]	Al_2O_3	*	*		*
[3]	Al_2O_3	*	*		*
[4]	Al_2O_3		*		
[5]	Al_2O_3	*	*		
[6]	Al_2O_3	*			
[7]	Al_2O_3	*			
[8]	Al_2O_3	*	*	*	
[9]	Al_2O_3	*			*
[10]	Al_2O_3		*		
[11]	Al_2O_3		*		*
[12]	TiO_2	*	*		*
[13]	TiO_2	*	*	*	
[14]	SiO_2	*	*	*	

the form of metals, oxides, carbonics, and carbides. The metal oxide nanofluid plus MQL combination raised the thermophysical properties of cutting fluid, such as thermal conductivity, the rate of heat transfer, and viscosity. A detailed review of the utilization of metal oxides nanofluid during a turning operation is presented and the influenced turning process response parameters are shown in Table 2.1.

From a detailed literature review, it was observed that very few investigators utilized metal oxide nanofluid during turning operations specifically. This motivated us to conduct the investigation to study the effect of silicon dioxide (SiO_2) nanofluid during a bearing steel turning operation. Surface roughness and machining zone temperature were selected as response variables by focusing on machined component surface quality and tool life. Grey relational analysis was executed to optimize the process variables with two cooling environments—vegetable oil plus MQL and SiO_2 nanofluid plus MQL.

2.2 INVESTIGATIONAL CONDITIONS AND MACHINING VARIABLES

The investigational conditions and machining variables are presented in Table 2.2. The investigation performed on computer numerical control (CNC) turning with an external MQL attachment. The detailed investigational setup is shown in Figure 2.1. Bearing steel having a hardness 60 HRBW was selected as a work material (80 mm length and 25 mm diameter). The surface roughness and machining zone temperature were measured by a roughness tester and an infrared (IR) thermometer, respectively.

Metal Oxide Nanofluid Cooling

TABLE 2.2
Investigational Conditions and Machining Variables

Parameters	Details
Machining Variables	Cutting Speed (CS)
	Feed Rate (FR)
	Depth of Cut (DOC)
Response Variables	Surface Roughness (SR) (um)
	Machining Zone Temperature (MZT) (Deg. Cel.)
Cooling Environment	MQL + Vegetable Oil (5bar, 125ml/hr)
	MQL + SiO$_2$ Nanofluid (5bar, 125ml/hr)
Insert Tool	DNMG 1108E-TMT9125 (Tungalloy Make)

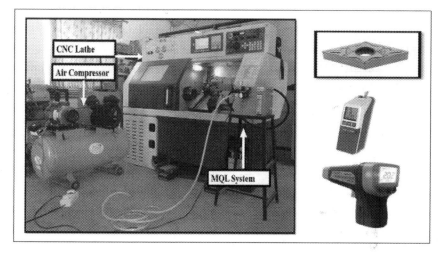

FIGURE 2.1 Investigation Setup with Cutting Tools and Response Variable Measurement Devices

2.3 SIO$_2$ NANOFLUID PREPARATION

The synthesized nanoparticles of SiO$_2$ were purchased, and characterization tests using scanning electron microscopy (SEM) and Energy Dispersive X-Ray Analysis (EDAX) performed at the Sophisticated Analytical Instrumentation Centre (SAIF) at the Indian Institute of Technology (IIT) Madras. The results of the SEM and the EDAX are shown in Figure 2.2. The observed size of SiO$_2$ nanoparticles is in the 500-nm fine category. After characterization, ethylene glycol and butenol were also added the SiO$_2$ nanoparticle blend with vegetable biodegradable oil to evade agglomeration. Ultrasonication and magnetic stirring were performed on prepared SiO$_2$ nanofluid. At last, it was noted that SiO$_2$ nanoparticles scattered properly without settling at the bottom of a glass container.

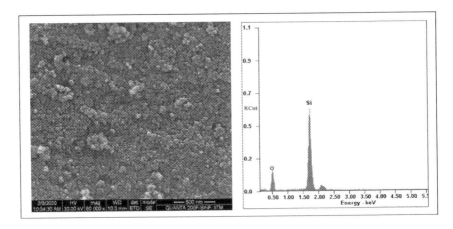

FIGURE 2.2 SEM and EDAX Test Results

2.4 PROCESS EVALUATION

A multi-objective response optimization was performed to optimize the process variables. Grey relational analysis (GRA) was implemented to optimize the turning process variables. Most eminent investigators have utilized GRA to evaluate various machining process performances. For the present investigation, "the smaller, the better" objective function was selected for surface roughness and machining zone temperature. Following are the steps to be followed in the GRA technique:

Step 1. Normalization of response variables in the range 0 to 1.
Step 2. Objective functions accepted for response variables

$$Y_1^*(k) = \frac{\max Y_i^0(k) - Y_i^0(k)}{\max Y_i^0(k) - \min Y_1^0(k)}, \quad (2.1)$$

where $Y_i^0(k)$ is the collected in experimentation and $Y_1^*(k)$ is the sequence afterward the normalization step.

Step 3. Grey relational coefficient (GRC) assessment:

$$GRC, \in (k) = \frac{\Delta_{min} + \varphi \Delta_{max}}{\Delta_{0i}(k) + \varphi \Delta_{max}}, \quad (2.2)$$

where $\Delta_0 i$ is the deviation sequence and $\Delta_{0i} = |Y_0^*(k) - Y_1^*(k)|$, $\varphi = 0.5$, is unique coefficient.

Step 4: Grey relational grade (GRG) calculation:

$$GRG, \alpha_j = \frac{1}{N}\sum_{i=1}^{N} w_k \in (k), \quad (2.3)$$

where $w_k = 1$.

Metal Oxide Nanofluid Cooling

Step 5: Rank calculation—By focusing on a higher value of GRG, optimal variables are determined.

Step 6: Confirmation test—Subsequently, by determining rank value, an additional trial is executed to validate the optimization accuracy. $\gamma_{forecast}$ (Forecasted GRG) as the optimal initial settings is estimated:

$$\gamma_{forecast} = \gamma_m + \sum_{i=1}^{n}(\gamma_i - \gamma_m), \quad (2.4)$$

where γ_m is the GRG mean, γ_i is the GRG mean at the optimal level, and n is the number of machining variables.

Steps 1 to 6 were utilized for the MQL plus vegetable oil and the MQL plus SiO_2 nanofluid cooling environments. For the investigation, an L9 orthogonal array was applied to the design of experiment; the collected data for the response variables are shown in Table 2.3, and the GRG calculations for the response variables as per the GRA methodology are shown in Table 2.4.

Response values were estimated after the rank calculation shown in Table 2.5. The main effect plots for both cooling conditions are shown in Figures 2.3 and 2.4. The results for optimum conditions are presented in Table 2.6. The confirmation test was performed, and the outcomes are shown in Table 2.7.

TABLE 2.3
Collected Response Variables Data

Trial No.	Cooling Environment	Cutting Speed	Feed Rate	Depth of Cut	Surface Roughness	Machining Zone Temperature
01	MQL + Veg. Oil	140	0.2	0.15	1.90	31.6
02	MQL + Veg. Oil	140	0.3	0.25	2.05	32.1
03	MQL + Veg. Oil	140	0.4	0.35	2.13	28.3
04	MQL + Veg. Oil	170	0.2	0.25	1.80	31.5
05	MQL + Veg. Oil	170	0.3	0.35	2.35	31.4
06	MQL + Veg. Oil	170	0.4	0.15	1.85	32.0
07	MQL + Veg. Oil	200	0.5	0.35	2.05	33.5
08	MQL + Veg. Oil	200	0.3	0.15	2.85	29.5
09	MQL + Veg. Oil	200	0.4	0.25	2.40	29.0
10	MQL + SiO_2 NF	140	0.2	0.15	2.00	27.9
11	MQL + SiO_2 NF	140	0.3	0.25	2.63	31.0
12	MQL + SiO_2 NF	140	0.4	0.35	2.60	31.6
13	MQL + SiO_2 NF	170	0.2	0.25	2.12	31.4
14	MQL + SiO_2 NF	170	0.3	0.35	1.60	29.8
15	MQL + SiO_2 NF	170	0.4	0.15	2.33	30.4
16	MQL + SiO_2 NF	200	0.5	0.35	1.95	30.9
17	MQL + SiO_2 NF	200	0.3	0.15	1.96	30.2
18	MQL + SiO_2 NF	200	0.4	0.25	2.20	31.1

TABLE 2.4
GRG Calculation

MQL + Vegetable Oil Cooling Environment

Trial No.	Normalization Surface Roughness (SR)	Normalization Machining Zone Temperature (MZT)	GRC SR	GRC MZT	GRG	Rank
1	0.905	0.365	0.840	0.441	0.640	4
2	0.762	0.269	0.677	0.406	0.542	6
3	0.686	1.000	0.614	1.000	0.807	1
4	1.000	0.385	1.000	0.448	0.724	2
7	0.476	0.404	0.488	0.456	0.472	9
6	0.952	0.288	0.913	0.413	0.663	3
7	0.762	0.000	0.677	0.333	0.505	8
8	0.000	0.769	0.333	0.684	0.509	7
9	0.429	0.865	0.467	0.788	0.627	5

MQL + SiO$_2$ Nanofluid Cooling Environment

Trial No.	Normalization SR	Normalization MZT	GRC SR	GRC MZT	GRG	Rank
10	0.612	1.000	0.563	1.000	0.781	1
11	0.000	0.162	0.333	0.374	0.354	8
12	0.029	0.000	0.340	0.333	0.337	9
13	0.495	0.054	0.498	0.346	0.422	5
14	1.000	0.486	1.000	0.493	0.747	2
15	0.291	0.324	0.414	0.425	0.419	6
16	0.660	0.189	0.595	0.381	0.488	4
17	0.650	0.378	0.589	0.446	0.517	3
18	0.417	0.135	0.462	0.366	0.414	7

TABLE 2.5
Response Values for GRA Techniques

MQL + Vegetable Oil Cooling Environment

Variables	L1	L2	L3	Δ	Rank
CS	0.6630	0.6197	0.5470	0.1160	2
FR	0.6230	0.5077	0.6990	0.1913	1
DOC	0.6040	0.6310	0.5947	0.0363	3

MQL + SiO$_2$ Nanofluid Cooling Environment

Variables	L1	L2	L3	Δ	Rank
CS	0.4907	0.5293	0.4730	0.0563	3
FR	0.5637	0.5393	0.3900	0.1737	2
DOC	0.5723	0.3967	0.5240	0.1757	1

Metal Oxide Nanofluid Cooling

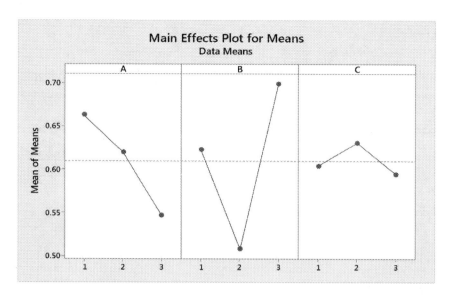

FIGURE 2.3 Main Effect Plot for MQL + Vegetable Oil Cooling Environment

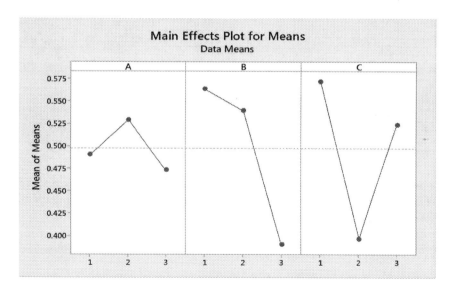

FIGURE 2.4 Main Effect Plot for MQL + SiO$_2$ Nanofluid Cooling Environment

The confirmation test for both cooling environments ensured that the outcomes noticed are remarkable as equal to the outcomes obtained with machining variables combination used previously. In the MQL with vegetable oil cooling environment, the surface roughness decreased from 1.91 μm to 1.88 μm, and the machining zone

TABLE 2.6
Optimum Conditions

Variables	MQL + Vegetable Oil	MQL + SiO$_2$ Nanofluid
CS	140	140
FR	0.4	0.2
DOC	0.25	0.15

TABLE 2.7
Confirmation Test

MQL + Vegetable Oil Cooling Environment

	Initial Setting	Forecast	Experiential
Level	$A_1B_1C_1$	$A_1B_3C_2$	$A_1B_3C_2$
Surface Roughness	1.91		1.88
Machining Zone Temperature	31.6		29.2
GRG	0.640	0.773	0.805
Improved GRG	0.165		

MQL + SiO$_2$ Nanofluid Cooling Environment

	Initial Setting	Forecast	Experiential
Level	$A_3B_2C_1$	$A_1B_1C_1$	$A_1B_1C_1$
Surface Roughness	1.96		1.61
Machining Zone Temperature	30.2		30.0
GRG	0.517	0.670	0.725
Improved GRG	0.208		

temperature was reduced from 31.6 °C to 29.2 °C. Similarly, in the MQL with SiO$_2$ nanofluid cooling environment, surface roughness declined from 1.96 μm to 1.61 μm and machining zone temperature decreased from 30.2 °C to 30.0 °C. Also, the GRG was enhanced to a notable level in both cooling conditions by 0.165 and 0.208, respectively.

2.5 RESULTS AND DISCUSSION

Surface roughness was used to define the surface quality of the machined product. For the current investigation, the lowest surface roughness, 1.60 μm, was observed in the MQL plus SiO$_2$ nanofluid cooling environment. The comparative surface roughness values for both cooling environments are shown in Figure 2.5. Similarly, to enhance the life of the cutting tool insert, the machining zone temperature plays an important role. The lowest machining zone temperature, 27.9 °C, was observed in the MQL plus SiO$_2$ nanofluid cooling environment. The comparative machining zone temperatures for both cooling environments are shown in Figure 2.6. SiO$_2$ nanoparticles possess high surface energy and resulted in the formation of a hydroxyl group

Metal Oxide Nanofluid Cooling

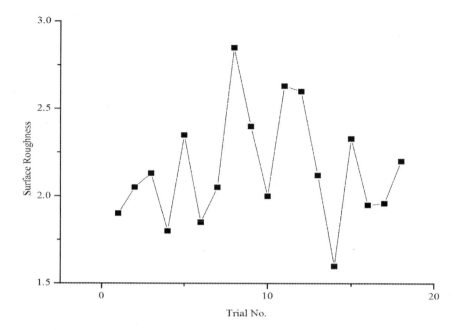

FIGURE 2.5 Comparison of Surface Roughness

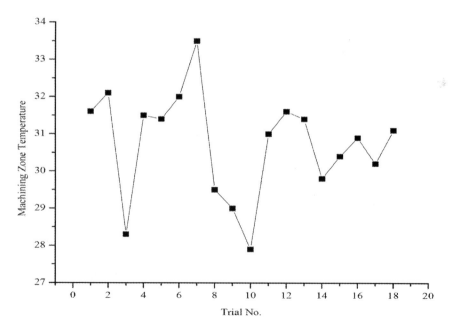

FIGURE 2.6 Comparison Machining Zone Temperature

on the work material surface. At the tool–work material interface, a thin boundary layer was generated and facilitated the decline in friction at the machining zone.

2.6 CONCLUSION

The present investigation focused on bearing steel turning under MQL plus vegetable oil and MQL plus SiO_2 nanofluid environments for the optimization of response variables. The following are the investigation's outcomes:

- For the MQL plus vegetable oil cooling environment, the feed was rate observed as an influencing machining variable, followed by cutting speed and depth of cut.
- For the MQL plus SiO_2 nanofluid cooling condition, the depth of cut was noted as an effective machining variable, followed by feed rate and cutting speed.
- The thin boundary layer of SiO_2 nanofluid reduced the friction at the machining zone, resulting in a significant minimization in surface roughness and the machining zone temperature as compared to the MQL plus vegetable oil cooling condition.
- The MQL with SiO_2 nanofluid efficiently upgraded the performance of the bearing steel turning process as compared to other cooling environments.
- The utilization of biodegradable oil (vegetable oil) as a base fluid during the preparation of SiO_2 nanofluid promotes sustainability.

REFERENCES

1. Vasu, V., Pradeep, R.G., 2011, Effect of minimum quantity lubrication with Al_2O_3 nanoparticles on surface roughness, tool wear and temperature dissipation in machining Inconel 600 alloy. Proc. Inst. Mech. Eng. 225(1), 3–16
2. Khandekar, S., Sankar, M.R., Agnihotri, V., 2012, Nano-cutting fluid for enhancement of metal cutting performance. Mater. Manuf. Proc. 27(9), 963–967
3. Sharma, A.K., Singh, R.K., Dixit, A.R., 2016, Characterization and experimental investigation of Al_2O_3 nanoparticle based cutting fluid in turning of AISI 1040 steel under minimum quantity lubrication (MQL). Mater. Today. Proc. 3, 1899–1906
4. Behera, B.C., Chetan Setti, D., Ghosh, S., 2017, Spreadability studies of metal working fluids on tool surface and its impact on minimum amount cooling and lubrication turning. J. Mater. Process. Technol., 244, 1–16
5. Mahboob Ali, M.A., Azmi, A.I., Khalil, A.N., 2017, Experimental study on minimal nanolubrication with surfactant in the turning of titanium alloys. Int. J. Adv. Manuf. Technol. 92(1–4), 117–127
6. Faheem, A., Husain, T., Hasan, F., Murtaza, Q., 2020, Effect of nanoparticles in cutting fluid for structural machining of Inconel 718. Adv. Mater. Proc. Technol. 1–20, https://doi.org/10.1080/2374068X.2020.1802563
7. Abbas, A.T., Gupta, M.K., Soliman, M.S., Mia, M., Higab, H., Luqman, M., Pimenov, D.Y., 2019, Sustainability assessment associated with surface roughness and power consumption characteristics in nanofluid MQL-assisted turning of AISI 1045 steel. Int. J. Adv. Manuf. Technol. 1–17, https://doi.org/10.1007/s00170-019-04325-6

8. Kumar, R., Sahoo, A.K., Mishra, P.C., Das, R.K., 2019, Influence of Al_2O_3 and TiO_2 nanofluid on hard turning performance. Int. J. Adv. Manuf. Technol. 1–16, https://doi.org/10.1007/s00170-019-04754-3
9. Duc, T.M., Long, T.T., Chien, T.Q., 2020, Performance evaluation of MQL parameters using Al_2O_3 and MoS_2 nanofluids in hard turning 90CrSi steel. Lubricants. 40(7), 1–17, https://doi.org/10.3390/lubricants7050040
10. Ghalme, S., Koinkar, P., Bhalerao, Y., 2020, Effect of aluminium oxide nanoparticles addition into lubricating oil on tribological performance. Tribol. Industry. 1–10
11. Das, A., Patel, S.K., Biswal, B.B., Sahoo, N., Pradhan, A., 2021, Performance evaluation of various cutting fluids using MQL technique in hard turning of AISI 4340 alloy steel. Measurement. 150, 1–28
12. Sharma, A.K., Tiwari, A.K., Dixit, A.R., 2016, Effects of Minimum Quantity Lubrication (MQL) in machining processes using conventional and nanofluid based cutting fluids. J. Cleaner Prod. 127, 1–18
13. Kumar, R., Sahoo, A.K., Mishra, P.C., Das, R.K., 2019, Influence of Al_2O_3 and TiO_2 nanofluid on hard turning performance. Int. J. Adv. Manuf. Technol. 1–16, https://doi.org/10.1007/s00170-019-04754-3
14. Nune, M.M.R., Chaganti, P.K., 2020, Performance evaluation of novel developed biodegradable metal working fluid during turning of AISI 420 material. J. Braz. Soc. Mech. Sci. Eng. 42(319), 1–16

3 Modal Analysis of a Wind Turbine Blade Using a Computational Method

M.J. Pawar, Amar Patnaik, Vikas Kukshal, Ashiwani Kumar, and Vikash Gautam

CONTENTS

3.1 Introduction 23
3.2 Materials and Methods 24
3.3 Natural Frequency and Mode Shapes for a Wind Turbine Blade 27
References 33

3.1 INTRODUCTION

In the 20th and 21st centuries, the requisition for development in the area of wind energy has rapidly increased worldwide. The blades of a wind turbine rotor were considered to be a major section that was subject to failure initially, which resulted in significant damage to the total windmill system. The potential materials considered for blades of wind turbines have to be able to sustain load due to wind forces along with load due to gravitational forces with minimum tip deformation and frequency of vibrations. Considering these requirements, the potential blade material having excellent strength and stiffness properties (as hybrid polymer composite) was considered the finest alternative for manufacturing blades for wind turbines. Hence, special efforts were employed on the modal analysis, with an emphasis on the natural frequency and the deflection of the blade. These are two important considerations when selecting a suitable polymeric material for manufacturing of blades for wind turbines. The work started with the development of a micro-wind turbine installed for educational purposes (at research institutes or environment centers). Quite small numbers of such units were in existence, particularly in urban regions. Earlier, wind turbines with a power capacity of 2.5–20 kW were commissioned for awareness regarding clean energy development. However, recently, work has been initiated with aim of developing micro-sized (less than 1.5-kW power capacity) wind turbines mounted on building roofs [1]. Therefore, wind turbines with low power capacity were specially designed and developed for built surroundings. These were erected on building roofs or on the empty land available near the buildings [2].

DOI: 10.1201/9781003093213-3

Manufacturers and researchers working on the development of the wind turbine initially analyzed the performance of system using computer-aided analysis software. This analysis was very helpful for them when deciding the feasibility of materials. The computer-aided design (CAD) model for the blade of the wind turbine was generated according to the description given by the National Advisory Committee for Aeronautics (NACA). The response of the blade for the wind turbine model was analyzed for different forces due to wind and gravity [3, 4]. It was a challenging task to simulate the modal analysis process using finite element codes. For structural health monitoring and flutter testing of wings excited by force due to wind, Operational Modal Analysis is the best suitable method. Wind tunnel experiments were performed to obtain data. The obtained set of data was used to extract modal parameters from the wing with external excitation [5].

A blade of a wind turbine is usually subjected to a system of coupled forces. Hence, whenever the natural frequency of the system and the natural frequency of the excitation forces match, the state of resonance is reached. The resonance results in high-amplitude vibrations of blade. It is the very primary objective for any researcher to avoid the resonance. For the same prediction of natural frequencies, different modes of vibration were of great importance [6]. Hence, the present work aimed at determining the natural frequencies at different modes during free vibration for a blade with a horizontal axis floating wind turbine of 5 kW capacity. A finite element software—Ansys—was used to predict the natural frequencies and avoid the resonance effect. The results show the natural frequencies for different modes for the series of materials under consideration. Also, the amplitude of vibration for different modes is reported. The studies were available on modal analysis of cantilever members, such as turbine blades and wings of airplanes. The use of finite element analysis software was also described for those members [7, 8].

3.2 MATERIALS AND METHODS

The modal analysis of the wind turbine blade reported in this chapter was carried out in the following steps:

a. Generation of a CAD model for a blade of a wind turbine
b. Allocation of material properties for different potential materials considered for the wind turbine blade
c. Application of boundary conditions for constraints and forces considered for analysis
d. Modal analysis for natural frequency and tip deflection for different modes of vibration

A turbine blade is composed of three basic portions. They are the root section, the mid-portion and the tip portion, as illustrated in Figure 3.1. The structure of a wind turbine blade has an airfoil cross section with a decreasing chord length from blade root to blade tip. A cross section of the hub is cylindrical in shape. This facilitates the blade being attached to the rotor with the help of studs and nuts. The hub assembles to the shaft of the wind turbine [9]. The motive of the finite

Modal Analysis of a Wind Turbine Blade

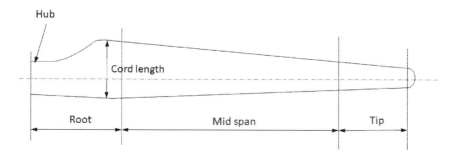

FIGURE 3.1 A Typical Blade Plan and Region Classification

element analysis in the present chapter is to link the modal behavior of a wind turbine blade structure with the mechanical properties of test coupons observed experimentally. The work presented in this chapter considers the wind turbine blade capacity of power generation as 5 kW. For the entire length of the wind turbine blade model, from root to tip, the airfoil profile is generated using NACA 4412 coding. The NACA 4412 code for the airfoil shape elaborated the variation of the chord length and the angle of twist from the root-to-tip section of airfoil. The geometric measurements of blade are total span of blade = 2500 mm, chord length at root = 330 mm, chord length at tip = 130 mm, diameter of blade at hub = 67.5 mm, and root-to-hub length = 300 mm, respectively [3].

Airfoil profile points at 10 equidistance sections along the blade length were calculated according to NACA 4412 profile. The Ansys R14.5 Design Modeler was used for generating the airfoil profile points, which were further joined for getting the airfoil profiles for respective sections. The airfoil profiles generated at all 10 sections were converted into surfaces. The solid geometrical model for the wind turbine blade was generated using the loft command. For the computational analysis, the complete wind turbine blade model was discretized into 41,698 elements and 78,588 nodes by employing an adaptive meshing technique.

Figures 3.2 to 3.4 illustrate the phases of the geometrical model generation using modeling software. Figure 3.2 illustrates the airfoil profile of all 10 sections along the length of the blade. Figure 3.3 shows the model generated from the profile points at the 10 sections. Figure 3.4 shows the discretized model for the blade with mesh patterns. The composite materials under consideration are a hybrid epoxy composite reinforced with glass fiber and waste granite powder. The series of unfilled polymer composites were fabricated varying the glass fiber weight percentage from 10% to 50% (i.e., EG_10, EG_20, EG_30, EG_40, and EG_50). However, composites filled with granite powder were fabricated with reinforcement of variation of weight percentage of granite powder/dust as 8, 16, and 24 maintaining a fiber reinforcement constant at 40 wt.% (EGG_8, EGG_16 and EGG_24). The composite fabrication process and experimentally obtained material properties of the test samples that are allotted to the wind turbine blade geometry as homogeneous isotropic are reported elsewhere [10]. In this analysis, all six degrees of freedom are constrained at the hub portion of the wind turbine blade model. The hub portion was bolted to

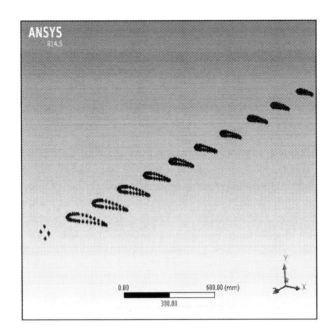

FIGURE 3.2 Stages of Model Generation and Meshing—Profile Points along the Length

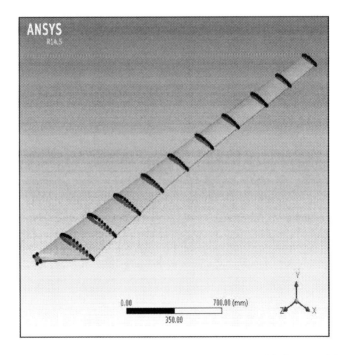

FIGURE 3.3 Stages of Model Generation and Meshing—Model Generated from Profile Points

Modal Analysis of a Wind Turbine Blade

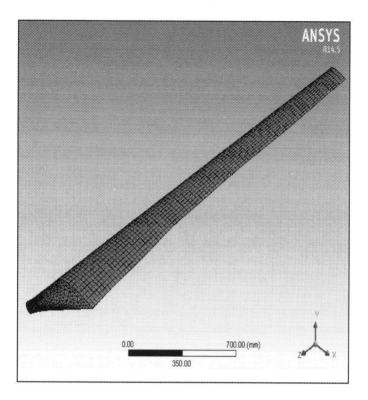

FIGURE 3.4 Stages of Model Generation and Meshing—Mesh Generated for the Model

the rotor. The blade was analyzed using Ansys R 14.5 for the natural frequency and tip deflection for different modes of vibrations under its self-weight. In the current work, the initial six natural frequencies, along with the amplitude of the vibration, are under consideration.

3.3 NATURAL FREQUENCY AND MODE SHAPES FOR A WIND TURBINE BLADE

The mode corresponds to the natural frequency of the vibrating structures, and the mode shape corresponds to the displacement (amplitude) pattern of the vibrating system for a particular mode. Mathematically, the mode indicates an eigenvalue, and the mode shape indicates an eigenvector of the vibrating structures. A continuous system has infinite number natural frequency. Wind turbines are subjected to an unsteady environment that leads to a vibrational response [11]. These structures are composed of slender structures with more length. The structure has elements named the rotor, the blades and the tower. Vibrations generated in the blade result in a frequency of resonance. These criteria are very important and need to be addressed during the preliminary phase of design and manufacturing. Whenever the external excitation frequency of the vibrating structures is in close proximity of any natural frequency, an uninvited resonance state is reached with vibrations of higher amplitude. These high-amplitude vibrations may result in small destructions or even the complete failure of the power generation by the wind turbine or its elements. For a wind turbine rotor blade, the vibration response is a function of the strength, which is a function of the properties of material employed, the design considerations, and the dimension of the blade [12]. Therefore, the prediction and analysis of the natural frequencies and amplitude of vibration in the primary design stage execute significant characteristics in eliminating resonance.

Figures 3.5 to 3.10 show the natural frequency and tip deflection for EG_40 composite for modes 1 through 6, respectively. The fundamental natural frequency for EG_40 is reported as 2.43 Hz, with a blade tip deflection of 27.27 mm. However, higher-order natural frequencies are 9.14 Hz, 10.6 Hz, 21.28 Hz, 39.35 Hz, and 45.83 Hz for mode 2, mode 3, mode 4, mode 5, and mode 6, respectively. The tip deflection recorded for mode 2, mode 3, mode 4, mode 5, and mode 6 are 27.30 mm, 21.35 mm, 28.95 mm, 31.03 mm, and 27.59 mm, respectively. The increase in natural frequency was observed from mode 1 to mode 6. The tip deflection was initially increased for modes 1 and 2; then there was a drastic reduction in tip deflection for mode 3. Furthermore, for modes 4 and 5, tip deflection increased considerably, but for mode 6, the tip deflection reduced marginally. The number of loops generated for the different modes of vibration is also evident.

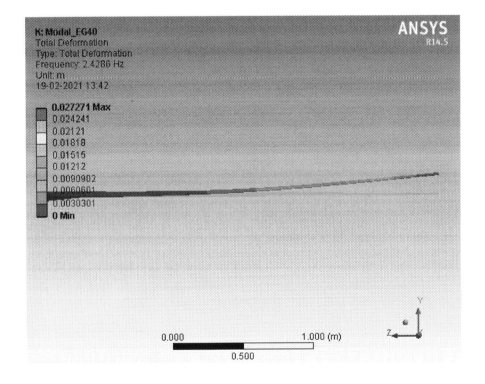

FIGURE 3.5 Natural Frequency and Tip Deflection for the EG_40 Composite for Mode 1

Modal Analysis of a Wind Turbine Blade

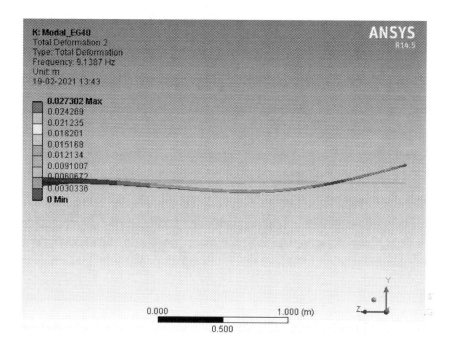

FIGURE 3.6 Natural Frequency and Tip Deflection for the EG_40 composite for Mode 2

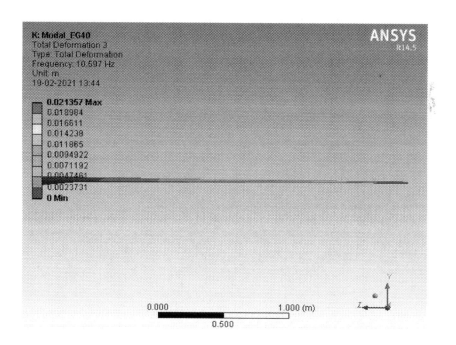

FIGURE 3.7 Natural Frequency and Tip Deflection for EG_40 Composite for Mode 3

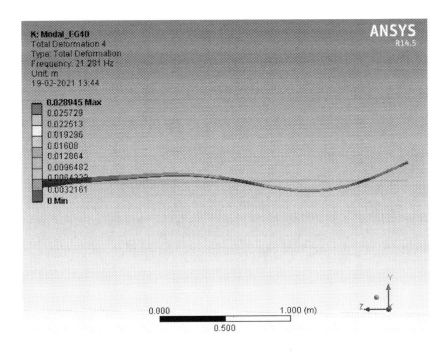

FIGURE 3.8 Natural Frequency and Tip Deflection for the EG_40 Composite for Mode 4

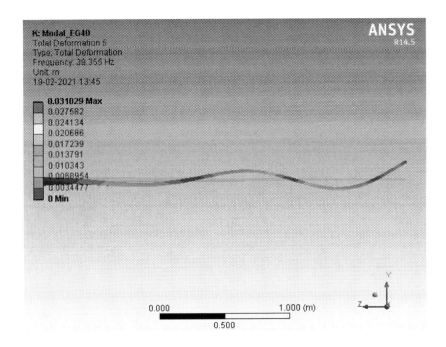

FIGURE 3.9 Natural Frequency and Tip Deflection for the EG_40 Composite for Mode 5

Modal Analysis of a Wind Turbine Blade

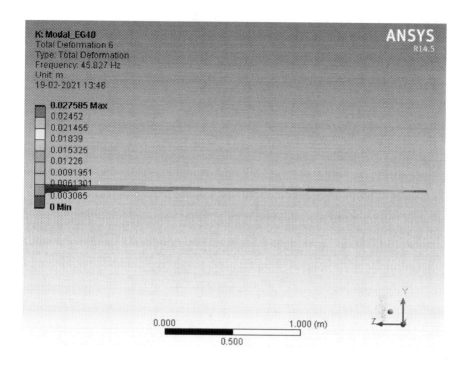

FIGURE 3.10 Natural Frequency and Tip Deflection for EG_40 Composite for Mode 6

TABLE 3.1
Natural Frequency of Blade for Different Modes

Composite	Natural Frequency (Hz)					
	Mode 1	Mode 2	Mode 3	Mode 4	Mode 5	Mode 6
EG_10	1.83	6.9	8	16.06	29.7	34.59
EG_20	2.2	8.29	9.62	19.32	35.73	41.6
EG_30	2.34	8.8	10.21	20.5	37.91	44.14
EG_40	2.43	9.14	10.6	21.28	39.35	45.83
EG_50	2.37	8.92	10.34	20.76	38.39	44.71
EGG_8	2.39	9	10.44	20.97	38.78	45.16
EGG_16	2.26	8.51	9.87	19.81	36.64	42.67
EGG_24	2.17	8.17	9.47	19.03	35.18	40.97

The trends in the variation of the natural frequency and the tip deflection with respect to the mode of the vibration for the series of material considered are identical. Table 3.1 indicates the fundamental and higher natural frequency for a series of material for the different modes of vibration. It is evident that for all materials, the fundamental natural frequency is the lowest natural frequency obtained for any mode shape. For modes 2 through 6, the value of the natural frequency gradually increased. The lowest rise in the natural frequency is observed from mode 2 to mode 3, followed by mode 5 to mode 6. Similar to the natural frequency, the variation of tip the deflection for different modes of vibration for the different materials under consideration is identical. The blade tip deflection for all the materials with respect to the mode of vibrations is indicated in Table 3.2. It is evident that for all the materials, the blade tip deflections for mode 3 and mode 6 decrease considerably compared to the corresponding previous mode. The lowest amplitude of vibrations is observed for mode 3 and mode 6.

Figures 3.11 and 3.12 show the natural frequency and tip deflection for different materials with respect to the mode of vibration. For all the materials, the maximum

TABLE 3.2
Tip Deflection of the Blade for Different Modes

Composite	Tip Deflection (mm)					
	Mode 1	Mode 2	Mode 3	Mode 4	Mode 5	Mode 6
EG_10	30.74	30.78	24.08	32.63	34.98	31.1
EG_20	29.61	29.65	23.19	31.43	33.7	29.96
EG_30	28.32	28.35	22.18	30.06	32.23	28.62
EG_40	27.27	27.3	21.35	28.95	31.03	27.59
EG_50	26.91	26.95	21.08	28.57	30.63	27.23
EGG_8	26.87	26.9	21.04	28.52	30.57	27.18
EGG_16	26.13	26.16	20.46	27.73	29.73	26.43
EGG_24	25.36	25.39	19.86	26.92	28.86	25.65

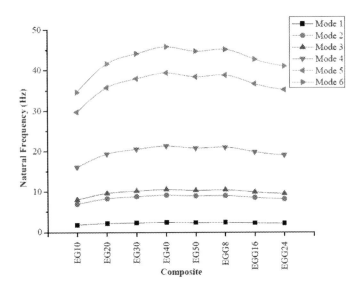

FIGURE 3.11 Variation of the Natural Frequency of the Blade for Different Materials

Modal Analysis of a Wind Turbine Blade

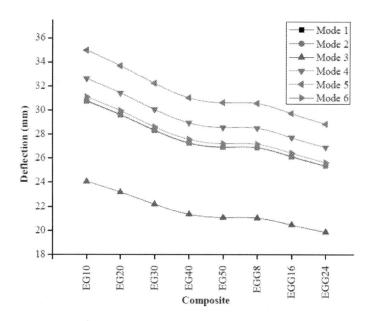

FIGURE 3.12 Variation of the Tip Deflection of the Blade for Different Materials

rise in the natural frequency is observed from mode 4 to mode 5. However, a very marginal rise in the natural frequency is observed from mode 2 to mode 3. For modes 1, 2 and 3, the natural frequency for all the materials is fairly constant. From mode 4 onward, the natural frequency for the different materials varies considerably. The natural frequency gradually increased for materials EG_10 to EG_40. Then, for EG_40, EG_50 and EGG_8, the natural frequency remains fairly constant. Furthermore, for EGG_16 and EGG_24, a gradual decrease in the natural frequency is observed. Irrespective of the mode of vibration, the maximum blade tip deflection is recorded for material EG_10. Furthermore, the gradual decrease in the tip deflection is evident for all the modes of vibration. However, a flattening of the curve for the tip deflection is also observed for materials EG_40, EG_50 and EGG_8 for all the mode shapes.

REFERENCES

1. A. D. Peacock, D. Jenkins, M. Ahadzi, A. Berry, S. Turan, Micro wind turbines in the UK domestic sector, Energy and Buildings 40 (2008) 1324–1333.
2. C. Jadranka, E.T. Horst, K. Syngellakis, M. Niel, P. Clement, R. Heppener, E.P. Pierano, Urban Wind Turbines: Guidelines for Small Wind Turbines in the Built Environment, Intelligent Energy Europe, www.urbanwind.net/pdf/SMALL_WIND_TURBINES_GUIDE_final.pdf.
3. J.S. Rajadurai, T. Christopher, G. Thanigaiyarasu, B. Nageswara Rao, Finite element analysis with an improved failure criterion for composite wind turbine blades, Forschungim Ingenieurwesen 72 (2008) 193–207.

4. W. H. Wu, W. B. Young, Structural analysis and design of the composite wind turbine blade, Applied Composite Materials 19 (2012) 247–257.
5. E. Neu, F. Janser, A. A. Khatibi, C. Braun, A. C. Orifici, Operational modal analysis of a wing excited by transonic flow, Aerospace Science and Technology 49 (2016) 73–79.
6. H. Boudounit, M. Tarfaoui, D. Saifaoui, Modal analysis for optimal design of offshore wind turbine blades, Materials Today: Proccedings 30 (2020) 998–1004.
7. P. Dhatrak, P. Choudhary, Comparative study on vibration characteristics of aircraft wings using finite element method, Materials Today: Proceedings (in press), https://doi.org/10.1016/j.matpr.2020.07.229.
8. A. Kumar, A. Dwivedi, V. Paliwal, P. P. Patil, Free vibration analysis of Al 2024 wind turbine blade designed for Uttarakhand region based on FEA, Procedia Technology 14 (2014) 336–347.
9. P. J. Schubel, R. J. Crossley, Wind turbine blade design, Energies 5 (2012) 3425–3449.
10. M. J. Pawar, A. Patnaik, R. Nagar, Experimental investigation and numerical simulation of granite powder filled polymer composites for wind turbine blade: A comparative analysis, Polymer Composites 38 (2017) 1335–1352.
11. J. Manwell, J. McGowan, A. Rogers, Wind Energy Explained, Theory, Design and Application, Wiley, West Sussex (2009).
12. M. Jureczko, M. Pawlak, A. Mezyk, Optimisation of wind turbine blades, Journal of Material Processing Technology 167 (2005) 463–471.

4 A Comprehensive Review of Ultrasonic Machining
A Tool for Machining Brittle Materials

T.G. Mamatha, Mudit K. Bhatnagar, Vansh Malik, Siddharth Srivastava, and Mohit Vishnoi

CONTENTS

4.1	Introduction	36
4.2	Understanding USM	37
	4.2.1 Working of USM and the Mechanism of Machining	37
	4.2.2 Different Elements of USM	38
	4.2.2.1 Acoustic Head	38
	4.2.2.2 Ultrasonic Transducers and Generators	38
	4.2.2.3 Ultrasonic Tool Assembly	38
	4.2.2.4 Abrasives	38
	4.2.3 Modified Variants of USM	39
	4.2.4 Various Process Parameters Associated with USM	39
	4.2.5 Various Performance Parameters	39
	4.2.5.1 Material Removal Rate	40
	4.2.5.2 Tool Wear Rate	41
	4.2.5.3 Surface Roughness	41
	4.2.5.4 Dimensional Accuracy	41
4.3	Chronological Literature Study	41
	4.3.1 USM Study of Glass	42
	4.3.2 USM Study of Silicon and Its Related Materials	46
	4.3.3 USM Study of Ceramics	48
	4.3.4 USM Study of Composites	52
	4.3.5 USM Study of Varied Materials	55
	4.3.6 USM Study of Crystals	58
	4.3.7 Comparative Work Piece USM Studies	59
4.4	Discussion	63
	4.4.1 Quantitative Analysis of Literature Study	63
	4.4.2 Effect of Process Parameters on Performance Parameters	63

DOI: 10.1201/9781003093213-4

		4.4.2.1	Effect of Static Pressure	63
		4.4.2.2	Effect of Vibrational Amplitude	64
		4.4.2.3	Effect of Vibrational Frequency	64
		4.4.2.4	Effect of Tool Rotation	64
		4.4.2.5	Effect of Abrasives	65
		4.4.2.6	Effect of Coolants	65
4.5	Comparison between Conventional and Nonconventional Techniques for the Machining of Brittle Materials			65
	4.5.1	Conventional Processes		65
		4.5.1.1	Ductile Regime Grinding	65
		4.5.1.2	Lapping	65
	4.5.2	Nonconventional Processes		65
		4.5.2.1	Machining of Metallic Brittle Materials	65
		4.5.2.2	Machining of Nonmetallic Brittle Materials	66
	4.5.3	Comparison between Conventional and Nonconventional Processes		66
4.6	Conclusion and Future Aspects			66
References				67

4.1 INTRODUCTION

A nonthermal nonconventional material removal machining process that has emerged as a center of keen interest for researchers worldwide is ultrasonic machining (USM). Both nonmetallic and electrically conductive materials with a hardness over 40 HRC and low ductility, for example, nickel alloys, silicon nitride, titanium alloys, and inorganic glasses, among others, are preferred materials for machining [1]. Holes of the order of 76 μm can be made, but the ratio for the depth to the diameter of the machined hole remains limited to approximately 3:1. USM is also known by many other terms, such as ultrasonic abrasive machining, ultrasonic dimensional machining, ultrasonic cutting, and ultrasonic drilling [2–12].

The machining of glass, semiconductors and other brittle materials in industries is performed with USM. This is done by providing them a complex geometry, adequate accuracy and a fine surface finish. The machining operation is time-consuming and requires high precision. Tight tolerances and dimensions, along with good quality and appreciable surface parameters, are easily achieved at such low input [13].

USM machines have increased material removal as a result of their higher cutting force and heat generation. USM's ability to machine intricate profiles on glass has many applications. USM is also used in the manufacturing of carbide dies. Electrical discharge machining (EDM) electrodes have been manufactured and redressed with the aid of USM [1]. Applications of USM further include its use in the production of finely detailed and accurate graphite electrodes without cracking and chipping by virtue of its three-dimensional machining action and reduced machining time. USM is also employed in the machining of ceramics, silicon crustal and fused quartz [14, 15]. Applications of USM extend to various disciplines such as medical applications [16–18] and electronic applications [19–22], as well as automobile and aerospace [23–26]. The following sections elucidate various aspects of USM in depth in addition to articulated discussions on research in the fields.

4.2 UNDERSTANDING USM

4.2.1 Working of USM and the Mechanism of Machining

In USM, the machining mechanism entails the interactivity of hard abrasive particles with the workpiece. The working of USM involves the transformation of high-frequency electrical energy into vibrational energy via transducers, which then is propagated to the workpiece via a tool assembly [27–30]. A static load (controlled) is enforced on the workpiece by the tool assembly and an environment of abrasive slurry is introduced for the tool to transmit the vibration energy contained by it into a cutting action. The vibration of the tool leads to momentum transfer to the abrasive particles contained in the slurry, which then strike the surface and cause material removal by microchipping [11]. The motion of the abrasive particles is governed by the vibration of the tool, which is at an ultrasonic frequency and small amplitude. A continuous and constant flow of the slurry gives fresh abrasive grains for the machining process to continue and flushes away the workpiece residue during the machining process. Material is removed primarily by two mechanisms:

a. Forced impact by the tool on the minute particles in contact with the tool and workpiece material
b. Accelerated grain impact caused by tool vibration [31]

USM is done by a high-frequency vibrating tool (about 20 kHz). A slurry medium is introduced between the longitudinally vibrating tool and the workpiece. The material removal takes place under the effect of the consecutive reciprocating force applied by the tool and the abrasive particles in the suspended medium. The removal is in the form of minute particles. A reversed profile cavity is formed in the workpiece during material removal. The impact actions of abrasives along with the high-speed reciprocation of the tool are the primary mechanism behind USM material removal. The alternating current at 18–24 kHz is supplied to the transducer by the power supply; a periodic change in length of the transducer produces resonance [1]. The vibrational amplitude of the transducer is observed to be from 0.005–0.01 mm.

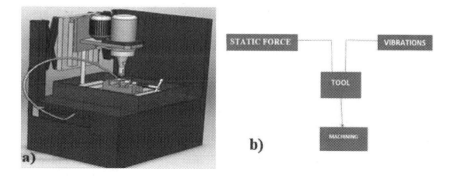

FIGURE 4.1 Representations of (a) the Ultrasonic Machining Model and (b) the Schematic Block Diagram

The vibrations are perpendicular to the machining surface and axial to the tool. Its amplitude is varied using a concentrator. The high-frequency vibration and amplitude of 0.03 mm, in addition to abrasive particles such as boron carbide, aluminum oxide and the like, machine the otherwise brittle material. Material removal is in the form of tiny particles caused by the repetitive action of the tool and the slurry medium present between the workpiece and the tool [14].

4.2.2 Different Elements of USM

4.2.2.1 Acoustic Head

It is a significant part of the machine that works on the magnetostrictive standard due to its high effectiveness and durability in the 15–30 kHz range, low supply voltage and simpler cooling system. Its main task is to vibrate the tool. To meet the demand of a high-frequency electric current, a generator is provided; a transducer converts this into oscillations to obtain high-frequency vibrations at the tool end, a holder holds the head. These vibrations are further amplified before transmitting to the tool by a concentrator [32].

4.2.2.2 Ultrasonic Transducers and Generators

With conventional generator and transducer system, the researcher needs to manually tune the frequency to achieve resonance. In recent advancements, resonances following generators have become available, which can adjust the output frequency automatically [3]. Transducers conventionally vibrate in two modes, in either a longitudinal or a compressive mode [33]; they are generally magnetostrictive [9, 34–36], or a piezoelectric transducer is used. Piezoelectric devices [37] are used in industrial applications in which a magnetostrictive transducer allows greater flexibility and it accommodates tool wear. Their main drawback is in the form of electrical losses, and they have low energy efficiencies. Piezoelectric transducers triumph over others due to their high energy efficiency and the ability to function without further cooling [15].

4.2.2.3 Ultrasonic Tool Assembly

In a properly designed tool, the vibrational amplitude at the free end is maximum for a particular frequency [38]. The desirable properties of a tool are good fatigue strength, good elastic properties and high wear resistance. In addition, properties such as toughness and hardness are preferred depending on the application [39]. A significant reduction in side friction was observed when an abrasive was fed through the center of the tool and horn during a deep-drilling operation [40].

4.2.2.4 Abrasives

The slurry is conventionally pumped by either suction or jet flow across the tool face. It plays multiple roles; first, it plays the role of the abrasive material required in addition to its role as a coolant for the workpiece and tool assembly. Moreover, fresh abrasive pumped across the machining zone removes the debris along with it. The slurry plays a vital role in efficient energy transfers [41–44]. To reduce tool wear, it should be manufactured from considerably ductile materials such as brass, mild steel, stainless steel and others [1]. A static load exerted by the feed system (spring, pneumatic, counter-static weight) holds the tool against the workpiece [45].

A Comprehensive Review of USM

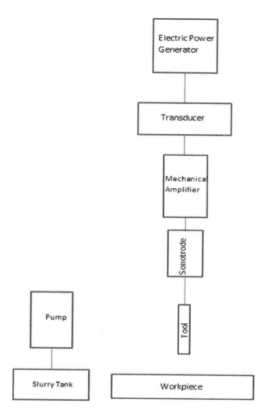

FIGURE 4.2 Block Diagram Representation of USM

4.2.3 Modified Variants of USM

For the improvement of efficiency of the conventional USM, multiple modifications were made. These include the incorporation of other techniques, such as EDM, along with USM for added efficiency. A detailed survey is depicted in Figure 4.3.

4.2.4 Various Process Parameters Associated with USM

Process parameters are defined as the input parameters, varying which affects the performance parameters or output parameters. Figure 4.4 shows the various process parameters and performance parameters associated with USM.

4.2.5 Various Performance Parameters

The efficiency of a machining process is measured against its performance parameters and their closeness to desired values. Broadly, the performance parameters, namely, rates of tool wear, machining rates (MRR), surface roughness (SR), and dimensional accuracy, are most frequently examined.

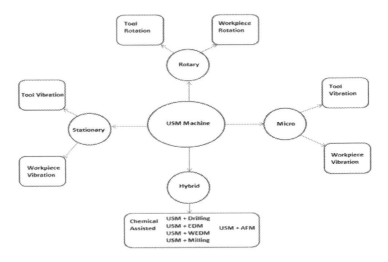

FIGURE 4.3 Different Variants of USM

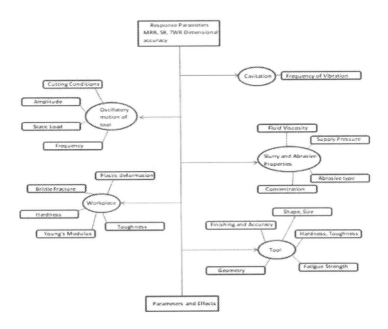

FIGURE 4.4 Various Process Parameters and Performance Parameters

4.2.5.1 Material Removal Rate

The indentation of hard abrasive powder over brittle work material causes material removal. The indentation of the abrasive powder occurs because of the transmission of vibrational energy via the tool. Conventionally, the MRR grows with the increase in tool vibration amplitude; there exists a critical value for the amplitude at which the MRR decreases.

Some investigators have suggested that there is a constant and linear relationship at a constant amplitude that is maintained up to a frequency of 400 Hz; that is, MRR αf^2. A linear relationship between material removal and frequency was observed at high frequencies up to 5 kHz. A rapid decrease in material removal was observed above a threshold value. A higher MRR was obtained when abrasive particles having a hardness greater than the workpiece material with a larger grit size were used. The increase of the grit size of the abrasive or the concentration of particles in the slurry gives an ideal MRR. Increasing any of the parameters even further results in larger particles reaching the cutting zone and a consequent reduction in MRR [34, 41, 46].

4.2.5.2 Tool Wear Rate

Tool wear is an important variable affecting the performance of the process (MRR and accuracy) and, in combination with other factors, such as tool material, abrasive type and slurry feeding methods, interferes with the actual machining operation. A higher MRR is observed in tools whose hardness has been augmented by work hardening, as the penetration of abrasive grains into the tool decrease. In addition, the machining zone periphery will experience greater material removal as a result of the formation of a convex surface in the workpiece. It will result in a dish formation caused by plastic deformation at the central face of the tool. It was observed post-experiment that the degree of hardening was lowest at the center of the tool material and highest at the periphery. This further illuminated that materials such as brass and copper were unsuitable as a tool material as their soft material properties developed burrs at higher amplitude [47–49].

4.2.5.3 Surface Roughness

Altering input parameters of rotary ultrasonic machine (RUM) affects the SR of a machined workpiece and drilled holes. Numerous examinations on RUM have been performed to comprehend the SR affected by various parameters [50].

4.2.5.4 Dimensional Accuracy

Dimensional accuracy coupled with surface quality holds significance in the machining of micro-channels as well as micro-pockets for various applications. The measurement of dimensional accuracy is not limited to measuring depth error, width error and edge chipping [51–53]. Dimensional accuracy also plays a significant role in micromachining of glass [54].

4.3 CHRONOLOGICAL LITERATURE STUDY

An abundance of literature is available on USM since its inception. This section dives in depth into the various experimentations performed on USM. To articulate the evolution of USM, this section seeks to divide the study into subsections based on the machined workpiece.

4.3.1 USM Study of Glass

Table 4.1 elucidates the work done on glass using USM. A shift from conventional USM to RUM was done with time.

TABLE 4.1
Chronological Study of USM of Glass

References	Tool Material	Substrate Material	Type of USM	Input Parameters	Output Parameters	Findings/Conclusion
Zhou et al. 2019 [55]	Tool-diamond abrasive	Confirmatory test on quartz glass was performed	Giant Magnetostrictive-rotary ultrasonic machining	1. Ultrasonic Frequency 2. Ultrasonic power 3. Spindle speed 4.	1. Material Removal 2. Load effect 3. Critical cutting ability 4.	1. Load played a significant role in actual resonance frequency. 2. A decrease in cutting force was observed with successions in the amplitude of the ultrasonic vibrations. 3. Reduction of penetration depth per rotation was observed with the increase of spindle speed, which further reduced the cutting force.
Kumar et al. 2017 [56]	Tungsten carbide	Borosilicate glass	STMUSM and RTMUSM	1. Work feed rate 2. Spindle speed (rpm) 3. Particle concentration in slurry medium (%) 4. Power rating (%) 5. Abrasive size (mesh number)	1. MRR 2. DOC 3. Tool wear	1. Increase in work feed, power rating, and concentration has appositive effect up to an extent, which depicts the ideal machining point for the apparatus being machined by stationary tool micro ultrasonic machining (STMUSM) and rotary tool micro ultrasonic machining (RTMUSM). 2. In micro-USM, rotating tool not only reduces tool wear, also improves machining output parameters.

A Comprehensive Review of USM

References	Tool Material	Substrate Material	Type of USM	Input Parameters	Output Parameters	Findings/Conclusion
Wang et al. 2018 [57]	Diamond core tool	Quartz glass	R-USM	1. Spindle speed 2. Feed rate	1. Reducing edge chipping	1. 60–80% reduction in edge chipping and higher stability can be achieved by using compound tool. 2. A reduction in the tool's taper face critical characteristic angle, hole exit quality, and compound tool effectiveness can be improved by using a tool having a step face.
Wang et al. 2017 [58]	Diamond core drill Bonding material: iron	Quartz glass sapphire	R-USM	Tool RPM Feed rate (mm/min) Vibrational frequency (kHz)	1. Ultrasonic power 2. Ultrasonic vibration	1. Heat production due to vibrations, it was found that it results in the decrease of resonant frequency. 2. Increased feed rates on harder materials led to even higher production of heat.
Wang et al. 2017 [59]	Diamond core tool	Quartz glass	Longitudinal-torsional coupled rotary ultrasonic machining (LTC-RUM)	1. Spindle speed 2. Feed rate 3. Ultrasonic amplitude	1. Surface roughness 2. Edge chipping size 3. Cutting force	1. The LTC-RUM reduced the cutting force by 55% on an average compared to the Con-RUM. Both cutting forces of LTC-RUM and Con-RUM increased as the feed rate increased. 2. RUM compared to Con-RUM decreased as the spindle speed increased.
Wang et al. 2016 [60]	Diamond abrasive coated diamond core drill	Quartz glass Sapphire	Rotary ultrasonic machining	1. Spindle speed (rpm) 2. Feed rate (mm/min)	Critical feed rate	1. Cutting force and feed rates are directly proportional. 2. Higher vibrational amplitude was observed at resonant frequency.
Wang et al. 2016 [61]	Diamond core tool	Quartz glass	Rotary ultrasonic machining	1. Spindle speed 2. Feed rate 3. Frequency 4. Ultrasonic amplitude	Edge chipping	1. The edge-chipping size is dependent on the shape of the tool. 2. Increased feed rates result in instantaneous decrease in ultrasonic vibration and increased edge-chipping size.
Zhang et al. 2014 [62]	Diamond tool	Optical K9 glass	Rotary ultrasonic machining	1. Spindle speed 2. Feed rate 3. Ultrasonic power	1. Cutting force 2. Chipping size 3. Surface roughness	1. RUM has a lower cutting force, when compared to diamond drilling. 2. The chipping size decreased as the spindle speed increased and increased as the feed rate increased.

(*Continued*)

TABLE 4.1
Chronological Study of USM of Glass (Continued)

References	Tool Material	Substrate Material	Type of USM	Input Parameters	Output Parameters	Findings/Conclusion
Schorderet et al. 2013 [63]	1. Tungsten carbide 2. HSS	Glass	Rotary ultrasonic machine	1. Tool shape 2. Grit size 3. Abrasive (B_4C) 4. Slurry concentration 5. Frequency 6. Tool rotation 7. Depth of drill	1. Drill speed 2. Surface finish	1. Twisted gouges drill presented improved results when compared with a plain drill. 2. Twist drills provide a much more consistent output along the hole depth.
Lv et al 2013 [64]	Diamond	BK7 glass	Rotary ultrasonic machine	1. Grain size 2. Amplitude 3. Vibration frequency 4. Rotation speed 5. Feed speed 6. Cutting (15 μm)	1. Subsurface damage 2. Surface roughness	1. The subsurface damage depth was inversely proportional to the Poisson's ratio. 2. The hardness of the brittle material determines the Young's modulus and fracture toughness.
Lv et al. 2013 [65]	Hollow tool with metal-bonded diamond abrasives	K9 optical glass	RUM	1. Rotation speed 2. Feed rate 3. Speed 4. Cutting depth 5. Frequency 6. Amplitude	1. Surface morphology 2. Material removal mechanisms	1. Pulverization was a result of the impact of the abrasive particles at the vertex of the sinusoidal trajectory. 2. The surface finish degrades with the increase in feed rates.
Gong et al. 2009 [66]	Hollow diamond grinding tools	K9 optical glass	RUM	1. Vibration amplitude 2. Static force 3. Rotational speed	1. Tool wear 2. Material removal rate	RUM had less tool wear than grinding.
Choi et al. 2007 [67]	Tungsten carbide	Glass	Chemical-assisted ultrasonic machine	1. Abrasive (SiC) 2. Static load 3. HF concentration 4. Frequency 5. Amplitude	1. Machining time 2. Machining depth 3. MRR 4. Surface roughness	1. HF acid improved the MRR by up to double the initial values. 2. CUSM gave 40% better surface finish results.

A Comprehensive Review of USM

References	Tool Material	Substrate Material	Type of USM	Input Parameters	Output Parameters	Findings/Conclusion
Kuo et al. 2007 [68]	Diamond core drill	Glass	Rotary ultrasonic machining	1. Vibration frequency 2. Vibration amplitude 3. Feed rate 4. Depth of mill 5. Spindle speed	Surface roughness	1. An increase the feed rate resulted in increased MRR 2. An increased milling depth and feed rate result in faster tool wear (TW). 3. RUM reduces the frictional resistance, making faster milling speed possible.
Yan et al. 2002 [69]	Slurry medium consisted of silicon carbide grains mixed in kerosene	Borosilicate glass (Pyrex Corning 7740)	EDM along with micro-USM	1. Vibrational amplitude 2. Rotational speed 3. Feed Rate, 4. Slurry Concentration	1. Micro holes 2. Diameter variation between the entrance and exit	Dimensional variance between entrance and exit was influenced most by Ultrasonic vibration amplitude, slurry concentration and rotation of tool speed.
Komaraiah et al. 1991 [70]	SS (Dia 3 mm)	Glass	Rotary ultrasonic machine	1. Static load 2. Grain size 3. Slurry concentration 4. Amplitude	Machining time	In rotary USM, the speed of rotation was found proportional to the penetration rate.
Kainth et al. 1979 [71]	Mild steel	Glass	Conventional ultrasonic machine	1. Vibration amplitude 2. Slurry concentration 3. Frequency 4. Static load 5. Mesh size	MRR	The effect of static load and amplitude of vibration was observed.
Kaczmarek et al. 1966 [72]	1. NW2 tool steel 2. NW2 die steel	Glass	Conventional ultrasonic machine	1. Slurry concentration 2. Vibration amplitude 3. Frequency 4. Static load 5. Grain size	1. Wear rate 2. Abrasive charge 3. Production cost	Optimum abrasive charge for minimum cost per unit volume machined can be determined.

4.3.2 USM Study of Silicon and Its Related Materials

Table 4.2 elucidates the work done on silicon and its compounds.

TABLE 4.2
Chronological Study of USM of Silicon and Its Materials

References	Tool Material	Substrate Material	Type of USM	Input Parameters	Output Parameters	Findings/Conclusion
Sreehari et al. 2018 [73]	Solid cylindrical tungsten carbide rod	Silicon	RUM	1. Slurry medium 2. Slurry concentration 3. Feed rate	1. Overcut 2. Straycut	1. Improved surface finish is attained with increased feed rates. 2. Output parameters are greatly affected by slurry related factors such as medium viscosity.
Cong et al. 2012 [74]	Metal-bonded diamond abrasives	Silicon	RUM	1. Tool rotation speed (rpm) 2. Feed rate (mm/s) 3. Ultrasonic power (%)	1. Edge chipping 2. Cutting force	1. A major effect of the input variables was observed on the edge chipping phenomena. 2. higher cutting force always led to larger edge chipping.
Zarepour et al. 2012 [75]	Tungsten	Single crystal silicon Fused quartz	Micro ultrasonic machining	Amplitude frequency	Material removal rate	1. Three modes of material removal were observed: pure ductile, partial ductile, and pure brittle. 2. Pure brittle material was removed in silicon at 3–4 mm amplitude.
Churi et al. 2007 [76]	Glass	Silicon carbide	Rotary ultrasonic machine	1. Spindle speed 2. Feed rate 3. Ultrasonic power 4. Grit size	1. Cutting force 2. Surface roughness 3. Chipping size (9–22 mm)	Increase in cutting force by reduction of workpiece feed and tool rotation speed.

A Comprehensive Review of USM

References	Tool Material	Substrate Material	Type of USM	Input Parameters	Output Parameters	Findings/Conclusion
Yu et al. 2006 [77]	Tungsten (Dia 95 μm)	Silicon	Micro-ultrasonic machining	1. Static load without tool rotation 2. Static load with tool rotation	Machining speed with respect to tool rotation	The machining speed decreases after a certain value of static load.
Zeng et al. 2005 [78]	Metal-bonded diamond core drill	Silicon carbide	Rotary ultrasonic machining	1. Spindle speed 2. Feed rate 3. Vibration frequency 4. Coolant pressure 5. Vibration power supply 6. Grit size	1. TW 2. Cutting force	1. Bond fracture and attrition wear was observed in RUM of SiC. 2. Grain fracture in ceramic materials was not observed. 3. TW on the end face is much more severe than that on the lateral face.
Yu et al. 2004) [79]	Tungsten	Silicon	Micro ultrasonic machining	1. Vibration frequency 2. Vibration amplitude 3. Static load 4. Spindle speed 5. Grain size 6. Tool feed	1. SR 2. MRR 3. Machining speed	1. The tool shape remains constant, and the tool wear has been reimbursed. 2. Increasing static load results in improved MRR. 3. The larger static load meant larger pressure on the abrasive grains which resulted in an increased MRR.
Komaraiah et al. 1993 [80]	Nimonic 80A, thoriated tungsten	SiC (220 mesh)	Ultrasonic machine	1. Abrasive type 2. Abrasive size 3. Amplitude 4. Frequency	Tool wear	Tool hardness is directly proportional to the MRR.
Khairy et al. 1990 [81]	Stainless steel (18% Cr, 8% Ni, 0.1% C)	Silicon carbide (23 μm)	Ultrasonic machine	1. Penetration depth 2. Ultrasonic power 3. Frequency	1. MRR 2. Longitudinal tool wear	A special design was proposed for the abrasive free-striking mechanism.

4.3.3 USM Study of Ceramics

Ceramics form a significant part of our daily lives. Table 4.3 focuses on various works done on Ceramics of different kinds.

TABLE 4.3
Chronological Study of USM of Ceramics

References	Tool Material	Substrate Material	Type of USM	Input Parameters	Output Parameters	Findings/Conclusion
Abdo et al. 2020 [82]	Nickel-bonded diamond	Zirconia (ZrO2) Ceramics	(RUM)	1. Cutting speed 2. RPM 3. Feed rate 4. Depth of cut 5. Amplitude 6. Frequency	1. Surface roughness 2. Edge chipping 3. Depth error 4. Width error	1. The dimensional accuracy with respect to the depth error and width error was observed to be less than 10% and 13%, respectively. 2. Surface finish is better with increased cutting speed and worse with high feed and depth of cut.
AlKawaz et al. 2020 [83]	—	Fully sintered zirconia ceramic	Rotary Ultrasonic Machining	1. Spindle speed 2. Ultrasonic frequency 3. Depth of cut 4. Feed	1. Surface quality 2. Cutting force 3. Edge chipping size	1. Ultrasonic vibrations aid in improving the surface and subsurface quality of completely sintered zirconia dental materials. 2. The cutting force decreased at higher tool RPM.
Abdo et al. 2020 [84]	Diamond/nickel base	Alumina ceramic	Rotary ultrasonic machining	1. Spindle speed 2. Feed rate 3. Depth of cut 4. Frequency 5. Amplitude 6. Coolant pressure	1. Surface roughness 2. MRR 3. Tool wear	1. The Ra factor varies in conjugation with the tool overlapping percentage. 2. Scanning electron microscopy (SEM) images showed the smooth surface morphology with higher overlapping percentages.

References	Tool Material	Substrate Material	Type of USM	Input Parameters	Output Parameters	Findings/Conclusion
Sandá et al. 2020 [85]	Metal-bonded diamond abrasives	ZrO2-NbC and ZrO2-WC ceramic	Rotary ultrasonic machining	1. Rotational speed 2. Feed rate 3. Axial depth of cut (mm) 4. Radial depth of cut (mm)	1. MRR 2. Cutting force 3. Surface roughness 4. Surface topography	In ZrO2-NbC the presence of the ductile fracture mode was more evident than in ZrO2-WC, where the brittle fracture mode seemed to be dominant.
Abdo et al. 2019 [86]	Nickel-bonded diamond RUM tools	Biolox forte ceramic	Rotary ultrasonic machining	1. Spindle speed 2. Feed rate 3. Depth of cut 4. Vibration amplitude 5. Vibration frequency	1. Surface roughness 2. Surface morphology 3. Edge chipping analysis 4. Channels accuracy 5. Tool wear analysis	1. Good surface finish of Ra = 0.21 μm and Rt = 2.3 μm was achieved by employing RUM process to the biolox forte material. 2. Smoother surface morphology was observed which was attributed to shallower pits and lesser amount of brittle cracking. In addition, higher levels of tool rotation along with low levels of cut depth, vibrational amplitude, and feed rate contributed to lower exit-edge chipping.
Singh et al. 2017 [87]	Tool-Metal-bonded diamond-impregnated core drill. Coolant-BC 20 SW water	Alumina ceramic	Series 10 knee mill rotary USM	1. Power, 2. Coolant pressure 3. Feed rate 4. Tool rotation	1. Chipping size 2. Material removal rate	1. Feed rate was a commanding parameter for MRR. 2. Ultrasonic power and coolant pressure is the most important parameter for CS.

(*Continued*)

TABLE 4.3
Chronological Study of USM of Ceramics (Continued)

References	Tool Material	Substrate Material	Type of USM	Input Parameters	Output Parameters	Findings/Conclusion
Singh et al. 2016 [88]	Metal-bonded (bond type: B) diamond-impregnated core drill	Alumina ceramic	Rotary ultrasonic machining	1. Spindle speed 2. Feed rate 3. Coolant pressure 4. Ultrasonic power	1. Chipping size 2. Material removal rate	1. Considered parameters were impactful for the MRR but had no significant effect on CS. 2. Increased spindle speed caused into improved machining rates. Furthermore, higher coolant pressure reported into enhanced machining. 3. Improved spindle speed caused into smaller CS.
Singh et al. 2016 [89]	B Bond-type metal-bonded hollow diamond drill	Quartz ceramic (SiO_2)	Rotary ultrasonic machining	1. Feed rate 2. Tool rotation 3. Coolant pressure 4. Power	1. MRR 2. Size of chip	Examined process parameters had significant effect on MRR but their effect on CS was trivial.
Liu et al. 2014 [90]	CVD diamond-coated drill	Alumina oxide ceramic	rotary ultrasonic machining	1. Feed rate 2. Ultrasonic power 3. Spindle speed	1. Tool wear 2. Exit crack	1. Feed rates, axial vibrations and cutting speed were optimized. 2. From the experimental results, the wear of the tool and exit cracks were found to have been decreased.
Liu et al. 2011 [91]	Coredrills with metal-bond diamond-abrasive particles	Alumina	Rotary ultrasonic machining	1. Spindle speed 2. Feed rate 3. Ultrasonic 4. Vibration amplitude 5. Abrasive size 6. Abrasive concentration	Cutting force	Cutting force decreased as abrasive size, vibration amplitude, and spindle speed increased.

References	Tool Material	Substrate Material	Type of USM	Input Parameters	Output Parameters	Findings/Conclusion
Cong et al. 2011 [92]	Metal-bonded diamond abrasives	Ceramics	Rotary ultrasonic machining	1. Tool rotation speed 2. Feed rate 3. Workpiece material 4. Ultrasonic power 5. Cutting tool	Vibration amplitude	The vibration amplitude without RUM machining had previously observed effects of vibrational amplitude on rotary USM.
Churi et al. (2009) [93]	Diamond core drills	Macor	Rotary ultrasonic machining	1. Spindle speed 2. Mesh size 3. Feed rate 4. Slurry concentration 5. Ultrasonic vibration power	1. Cutting force 2. SR 3. Chipping size	1. Spindle speed increased; cutting force and chipping size decreased while SR decreased initially and then increased. 2. Feed rate increased, cutting force and chipping size increased while SR increased initially and then decreased.
Zeng et al. (2006) [94]	Diamond core drill	Advanced ceramic	Conventional ultrasonic machine and rotary ultrasonic machine	1. Vibration frequency (20 kHz) 2. Amplitude 3. Spindle speed 4. Grit size	1. MRR 2. TWR	Low material removal rate (MRR) and low accuracy.

4.3.4 USM STUDY OF COMPOSITES

Work done on composites, from metal to polymer-based, were studied and are represented in Table 4.4.

TABLE 4.4
Chronological Study of USM of Composites

References	Tool Material	Substrate Material	Type of USM	Input Parameters	Output Parameters	Findings/Conclusion
Wang et al. 2020/ [95]	Bond: metal Abrasive material: diamond	Carbon-fiber and epoxy-resin	RUM	1. Depth of cut (mm) 2. Feed rate (mm/s) 3. Spindle rotation	1. Cutting force 2. Surface roughness	1. In RUM surface machining with elliptical ultrasonic vibration, reduction in Fx, Fz, and Sa was achieved by keeping smaller depth of cut federate values and higher tool rpm. 2. Reduction in cutting forces, cutting contact area per unit time, machined defects and increasing effective cutting contact time helps in machined surface quality.
Wang et al. 2020 [96]	—	CFRP	Rotary ultrasonic machining	1. Ultrasonic amplitude 2. Abrasive-grain concentration and size 3. Depth of cut 4. Feed rate 5. Tool rotation speed	1. Cutting forces 2. Critical indentation depth	Both Fx and Fy decreased by increase in ultrasonic amplitude, tool rotation speed, particle size and reduction in feed rate, depth of cut, abrasive concentration, or increasing shows reduction in Fx and Fy both.
Ning et al. 2019 [97]	Single-grain diamond tool	CFRP composites	Rotary USM	1. Ultrasonic vibration amplitude 2. Ultrasonic power 3. Feed rate	1. Cutting force 2. Surface roughness	1. Brittle material removal was observed. 2. Increased scratching depth resulted in jagged cross-section profiles.

References	Tool Material	Substrate Material	Type of USM	Input Parameters	Output Parameters	Findings/Conclusion
Wang et al. 2019 [98]	Abrasive material: diamond Bond: metal	Carbon fiber–reinforced plastic composites	RUM	1. Horizontal ultrasonic amplitude 2. Horizontal ultrasonic frequency 3. Spindle speed 4. Feed rate	1. Material removal 2. Surface formation	Increased ultrasonic frequencies leads to decrease in Fx and Fz.
Fernando et al. 2017 [99]	Diamond-abrasive-coated core drill	Carbon fiber–reinforced plastic	Rotary ultrasonic machine	Ultrasonic power, tool rotation speed, feed rate, tool angle	Torque, Cutting force, Delamination of hole	Tool angle has higher significance on the delamination as compared to the cutting force.
Zha et al. 2017 [100]	Electroplated diamond drills	Silicon carbide–reinforced aluminum matrix composites (high fraction)	Rotary ultrasonic machining	1. Spindle speed 2. Ultrasonic amplitude 3. Penetration angle 4. Frequency of ultrasonic vibration	Material removal mechanism	The Al matrix was enhanced by the ultrasonic vibration, and the SiC reinforcements removed more easily because of many micro cracks.
Ning et al. 2017 [101]	Diamond-abrasive tool	CFRP composite	Rotary ultrasonic machining	Ultrasonic power, feed rate, tool rotation speed.	Ultrasonic vibrational amplitude	Ultrasonic amplitude measurement was performed which was one of the most significant variable process parameter.
Kataria et al. 2016 [102]	1. Stainless steel 2. Silver steel 3. Nimonic 80A	WC-Co composite	Sonic mill USM	1. Tool material 2. Power rating 3. Tool profile 4. Grain size	Material removal, TWR, surface morphology	1. Higher fracture toughness led to low material removal rate in high WC-Co concentration composite. 2. Nimonic80A was observed to have best performance. 3. The hollow tool performed better.
Ning et al. [103]	Tool-Metal-bonded diamond-core drill	Carbon fiber–reinforced plastic with epoxy resin	Sonic mill rotary USM	1. Feed rate 2. Tool rotation	1. Cutting force 2. Torque 3. Surface roughness 4. Hole diameter 5. Material removal rate	Comparative study between RUM and grinding in which grinding showed higher cutting force and torque than RUM.

(Continued)

TABLE 4.4
Chronological Study of USM of Composites (Continued)

References	Tool Material	Substrate Material	Type of USM	Input Parameters	Output Parameters	Findings/Conclusion
Ding et al. 2014 [104]	Sintered diamond core drill	Carbon fiber–reinforced silicon carbide matrix (C/SiC) composites	Rotary USM	1. Coolant pressure 2. Frequency 3. Federate 4. Spindle rotation speed 5. Vibration amplitude	1. Drilling force 2. Torque 3. SR	1. Reduction in drilling force and torque was observed. 2. RUSM gave better exit holes. 3. SR is lower in holes drilled by RUSM.
Liu et al. 2014 [105]	CVD diamond-coated drill	Glycol phthalate thermoplastic	Rotary ultrasonic machine	1. Vibrations of tool rotation 2. Spindle speed (3000 rpm) 3. Amplitude (8 μm) 4. Ultrasonic frequency (40 kHz)	Reduction of microchipping	Exit chipping and severe tool wear in hard material machining were investigated and optimized.
Ding et al. 2014 [106]	Sintered diamond core drill	Carbon fiber–reinforced silicon carbide matrix (Si/C) composites	Rotary USM	1. Spindle rotation speed 2. Feed rate 3. Vibration frequency 4. Amplitude	1. Drilling force 2. Surface roughness	1. Drilling force and torque have a combined effect on the exit quality of the holes. 2. There was no stable variation in the Ra factor the holes drilled.
Yuan et al. 2014 [107]	Diamond abrasive (metal bond)	C/SiC composite	Rotary USM	1. Spindle speed 2. Cutting depth (mm) 3. Feed rate (mm/s)	1. Cutting force 2. Ductile-mode percentage	1. Ductile-mode percentage was reduced with the increase in the cutting depth or feed rate. 2. Maximum penetration depth dominantly affected the ductile-mode percentage.
Feng et al. 2012 [108]	Diamond	CFRP	RUM	1. Vibration amplitude 2. Rotation speed 3. Feed rate	1. Chips 2. Edge quality 3. Thrust force 4. Machined surface 5. Material removal rate 6. Tool wear	The diameter variation coefficient was below 1% for all produced holes. The thickness of chipping and the chip size were about 200 μm and 600 μm, respectively.
Cong et al. 2012 [109]	Metal-bonded diamond abrasives core drill	Carbon fiber reinforced plastic composites	RUM	1. Ultrasonic power 20–60% 2. Tool rotation speed 1000–4000 rpm 3. Feed rate 0.1–0.5 mm/s	1. Cutting 2. Temperature thermocouple and fiber optic sensor methods 3. Laser scan using fiber optic sensor	A reduction in heat produced was observed with the decrease in the ultrasonic power and feed rate; hence, the working temperatures reduced.

4.3.5 USM Study of Varied Materials

TABLE 4.5
Chronological Study of USM of Miscellaneous Materials

References	Tool Material	Substrate Material	Type of USM	Input Parameters	Output Parameters	Findings/Conclusion
Marichamy et al. 2020 [110]	Boron carbide and titanium carbide mixed slurry	Duplex brass metal matrix	SONIC type USM	1. Power, 2. Slurry concentration 3. Grit size	1. MRR 2. SR 3. Fretting wear	Speed of rotation was the most influential parameter followed by temperature for fret wear.
Fernando et al. 2020 [111]	Diamond core drill	1. Igneous rock type 2. Basalt rock	RUM	1. USM power 2. RPM 3. Feed rate (mm/s)	Cutting force	1. Cutting force increased with feed rate, abrasive size, slurry concentration, and size of tool. 2. Cutting force reduced with increase in RPM and amplitude.
James et al. 2017 [112]	1. Polycrystalline diamond (PCD) 2. Tungsten carbide	CFRP/Titanium stacks	Micro ultrasonic machining	1. Tool diameter 2. Workpiece thickness 3. Slurry concentration 4. Grain size 5. Frequency 6. Amplitude 7. Feed rate	1. MRR 2. Tool wear 3. Hole size variation	The parameter study revealed that larger grain size resulted in higher MRR, while the copper tool showed higher tool wear compared to tungsten carbide tools.
Kuruc et al. 2015 [113]	Diamond	Poly-crystalline cubic boron nitride (PCBN)	RUM	1. Amplitude 2. Frequency	Surface roughness	1. The machined surface showed the absence of ledges as the depth of the cut was not higher than the amplitude of the tool vibration.
Cong et al. 2014 [114]	CFRP/Ti stacks	Ti—6Al—4V	Rotary ultrasonic machine	1. Cutting force 2. Hole size variation 3. Vibrational frequency 4. Tool rotation	Surface roughness	In comparison to other methods of CFRP/Ti stacks drilling, an elongated tool life with better finished surface, reduced groove depth, and reduced variation in hole size due to reduced cutting force was observed in rotational USM.

(*Continued*)

TABLE 4.5
Chronological Study of USM of Miscellaneous Materials (Continued)

References	Tool Material	Substrate Material	Type of USM	Input Parameters	Output Parameters	Findings/Conclusion
Cong et al. 2013 [115]	CFRP (Carbon fiber–reinforced plastic)	Ti–6Al–4V	Rotary ultrasonic machine	1. Feed rate 2. Frequency 3. Ultrasonic power 4. Tool rotation speed	MRR	Cycle time is reduced by using variable feed rate.
Cong et al. 2012 [116]	CFRP (Carbon fiber–reinforced plastic)	Titanium	Rotary ultrasonic machine	1. Tool rotation 2. Feed rate 3. US power	Power consumption	As tool rotation increased, power consumption percentage decreased.
Kumar et al. 2009 [117]	Cemented carbide	Pure titanium (ASTM grade-I)	Ultrasonic machine	1. Abrasive type 2. Slurry concentration 3. Power rating 4. Grit size	MRR	Tool material of a higher hardness helps in achieving ideal MRR in USM of titanium.
Kumar et al. 2009 [118]	1. HCS 2. HSS 3. Cemented carbide 4. Titanium 5. Titanium alloy	Grade 1 ASTM titanium	USM	1. TWR 2. Particle type 3. Grit size 4. Power rating	MRR	1. Power rating factor was the most crucial factor followed by abrasive type and slurry grit size. 2. Tool material factor was least contributing factor for MRR.
Tawakoli et al. 2009 [119]	Diamond disc dresser	42CrMo4	Ultrasonic-assisted dry grinding	1. Feed speed 2. Cutting speed 3. Depth of cut 4. No coolant 5. Frequency 6. Amplitude	1. MRR 2. Surface roughness 3. Specific grinding force	Workpieces machined with superimposed ultrasonic vibration shows 60% reduction in normal grinding.

A Comprehensive Review of USM

References	Tool Material	Substrate Material	Type of USM	Input Parameters	Output Parameters	Findings/Conclusion
Churi et al. 2007 [120]	Diamond core drills	Titanium alloy (Ti-6Al-4V)	Rotary USM	1. Grit size 2. Diamond concentration 3. Frequency 4. Spindle speed 5. Feed rate	1. TWR 2. Cutting force 3. Roughness	1. Tool with lower grit size of mesh #60/80 gave higher cutting force and SR but lower TW. 2. Lower diamond grit concentration tool gave lower SR and TW but a higher cutting force.
Singh et al. 2007 [121]	HSS	Ti15	Ultrasonic machine	1. Grit size (320) 2. Ultrasonic power	1. MRR 2. TWR	As grit size of alumina slurry is increased, MRR and TWR reduced.
Azarhoushang et al. 2006 [122]	TiAlN-coated carbide drills (Dormer-R522) and TiN-coated carbide drills (Dormer-R550)	Inconel 738-LC	Rotary ultrasonic machining	1. Vibration amplitude 3–10 μm 2. Spindle speed 250–500 rpm 3. Feed rate 0.5–1 mm/s	Drilled hole surface roughness Drill entrance circularity Drilled hole cylindricity	1. Improved surface roughness and circularity of machined parts were observed.
Ghahramani et al. 2001 [123]	PSM1	Aluminum oxide	Ultrasonic machine	1. Variation of tensile stress 2. Compressive stress with distance	Dynamic abrasives impacts (high-frequency vibrations) using single and double impact test	Lower impact velocities led to machining below plastic zone.
Thoe et al. 1999 [124]	Mild Steel	ceramic coated nickel alloy	Ultrasonic machine	Slurry concentration	Hole quality	Ultrasonic vibration during EDM increased workpiece MRR.
Neppiras et al. 1964 [125]	HSS	Titanium alloys	Ultrasonic machine	Rate of feed	Surface quality	Cutting rates increased very rapidly with increasing oscillatory motion of the tool

4.3.6 USM Study of Crystals

Table 4.6 focuses on work done on crystals and machining using USM.

TABLE 4.6
Chronological Study of USM of Crystals

References	Tool Material	Substrate Material	Type of USM	Input Parameters	Output Parameters	Findings/Conclusion
Ning et al. 2017 [126]	CFRP	Zirconia	Rotary ultrasonic machine	1. Frequency 2. Feed rate 3. Inner and outer diameter of tool	1. Cutting force 2. MRR	1. Ultrasonic power increases vibrational amplitude, tool rotation speed decreased, and feed rate increased.
Kuruc et al. 2015 [127]	Diamond core drill	Poly-crystalline cubic boron nitride	Rotary USM	1. RPM 2. Amplitude 3. Depth cut 4. Vibration frequency 5. Feed rate	Surface roughness	Low surface roughness was achieved post RUM machining.
Wang et al. 2009 [128]	Diamond core drill	Potassium dihydrogen phosphate (KDP) crystal	Rotary ultrasonic machine	1. Spindle speed 2. Feed rate	1. Surface roughness	Surfaces became rougher as the feed rate increased.
Hocheng et al. 1999 [129]	Tubular stainless steel (SUS 304)	Zirconia 3 mol% Y,O, stabilized ZrO	Ultrasonic drilling	1. Abrasive particle size 2. Amplitude 3. Static load	1. MRR 2. Hole clearance 3. Surface roughness	1. Machining rate augmented vibrational amplitude. 2. With an applied load, a decrease in hole clearance was observed. 3. With static load, TWR decreased.
Pei et al. 1998 [130]	—	Magnesia-stabilized zirconia	RUM	1. Rotational speed 2. Vibration frequency 3. Vibration amplitude	MRR	1. MRR varies with different process parameters. It increased as the tool rotation increased. 2. MRR increased as the tool abrasive particle decreased.

4.3.7 Comparative Work Piece USM Studies

Table 4.7 focuses on various comparative studies. These studies focus on comparisons between tools or between workpieces of different kinds.

TABLE 4.7
Chronological Comparative Study of USM

References	Tool Material	Substrate Material	Type of USM	Input Parameters	Output Parameters	Findings/Conclusion
Haashiret al. 2020 [131]	B$_4$C particles added in the slurry Tool-Diamond cutter	Boron carbide material (B$_4$C), brass, glass	AP 1000 Sonic Mill USM	1. Feed rate 2. Grit size 3. Slurry concentration	1. MRR 2. Tool wear rate, 3. Over cut 4. Circularity 5. Taper angle	1. The least MRR of B$_4$C was found and glass. 2. The overcut was least for brass and highest for glass.
Wang et al. 2020 [132]	304 stainless steel	Float glass, alumina, and silicon carbide	Micro ultrasonic machining (micro-USM)	1. Vibration frequency 2. Vibration amplitude 3. Feed depth 4. Machining force 5. Tool feed rate 6. Slurry flow rate	1. MRR 2. Machining force 3. Roughness 4. Surface crack and damage 5. TWR	1. The machining rate in increased in order of SiC, Al$_2$O$_3$, and glass. 2. The tool wear rate decreased substantially in case of glass. 3. Better form accuracy on the glass workpiece.
Haashir et al. 2019 [133]	Stainless steel	Boron carbide (B4C) Glass Brass	RUM	1. Feed rate 2. Grit size 3. Slurry concentration	1. MRR 2. Machining rate 3. Tool wear rate 4. Surface quality	1. MRR of glass was found to be highest followed by brass and Boron carbide. 2. CE was found to be least for brass.
Singh et. al 2019 [134]	1. HCHCrS 2. HSS 3. HCS	BS476 heat-resistance glass	RUM	1. Tool 2. Abrasive size 3. Concentration 4. HF concentration	1. Surface morphology	Abrasive slurry, HF acid percentage and grit size played a consequential role for performance parameter optimization.

(Continued)

TABLE 4.7
Chronological Comparative Study of USM (Continued)

References	Tool Material	Substrate Material	Type of USM	Input Parameters	Output Parameters	Findings/Conclusion
Wang et al. 2017 [135]	Tool-Diamond drill	1. Quartz glass 2. Sapphire	Ultrasonic 50 DGM RUM	1. Spindle rotation speed 2. Material feed rate 3. Vibration amplitude	Critical feed rate	1. Feed rate, ultrasonic amplitude and rpm were observed, and a relationship was established.
Wang et al. 2017 [136]	1. 304 stainless steel 2. 1045 carbon steel 3. Tungsten carbide	1. Float glass 2. SiC 3. SS304 4. STEEL1045 5. WC	USM	1. Tool feed rate 2. Slurry flow rate 3. Vibration frequency 4. Vibration amplitude 5. Tool feed depth 6. Machining force	1. MRR 2. TWR	1. The wear of abrasive particles is fast when using a hard tungsten carbide (WC) tool. 2. Steels with a high flexibility such as 304 stainless steel are good choices as tool materials in micro-USM.
Wang et al. 2016 [137]	1. Diamond core drill 2. Step diamond core drills	1. Quartz glass 2. Sapphire	Rotary ultrasonic machining	1. Spindle speed 2. Feed rate	1. Edge chipping	1. The step drill reduced the edge-chipping size up to 60%. 2. Higher step thickness and higher critical step height were found to be in conjugation with one another.
Nath et al. 2012 [138]	Stainless steel	1. SiC 2. ZrO_2 3. Al_2O_3	USM	1. Feed Rate; 2. Cutting length 3. Spindle Speed 4. Depth of cut	1. Chipping, hole roughness 2. Surface morphology	The cracks were 2 to 4 times larger than the average radius of the used abrasive.
Kuo Kei-lin et al. 2012 [139]	Electroplated diamond	1. Glass 2. Quartz 3. Ceramics 4. Diamond hard alloy 5. Hardened steel 6. Titanium alloy lead 7. Mild steel 8. Copper	RUM	1. Vibration 2. Amplitude 3. Rotation speed 4. Depth cut 5. Feed rate	1. Surface roughness 2. Tool wear	1. Cut depth and feed rate lead to increased surface roughness. 2. Increasing the depth and feed rate resulted in an enhanced milling resistance and tool wear.

A Comprehensive Review of USM

References	Tool Material	Substrate Material	Type of USM	Input Parameters	Output Parameters	Findings/Conclusion
Nath et al. 2012 [140]	Stainless steel	1. SiC 2. Zirconia 3. Alumina	Ultrasonic machine	1. Grit size 2. Amplitude 3. Vibration frequency	1. Wall roughness 2. Subsurface damage	Hole integrity was optimized by integrating smaller grit size.
Singh et al. 2007 [141]	1. SS, 2. HCS 3. HSS 4. Carbide 5. Titanium	1. TITAN15 2. TITAN31	Ultrasonic Machine	1. Amplitude 2. Frequency 3. Tool 4. Abrasive concentration 5. Slurry type 6. Slurry temperature 7. Power rating	MRR	The optimized results were obtained by at 90% power rating and SS tool.
Guzzo et al. 2003 [142]	Stainless steel	1. Alumina 2. Zirconia 3. LiF 4. quartz 5. Soda-lime glass	RUM	Structural and mechanical properties of workpiece materials	1. Cutting rate 2. Surface roughness 3. MRR	1. MRR decreased with machining depth for workpiece materials in which hardness was the same order of magnitude than the hardness of abrasive grits. 2. Brittle microcracking was the dominant mechanism for machining.
Pei et al. 1995 [143]		1. Hot-pressed silicon nitride 2. Magnesia-stabilized Zirconia 3. Reaction-bonded silicon nitride	Rotary ultrasonic machining	1. Vibration amplitude 2. Rotational speed 3. Feed rate 4. Depth of cut 5. Machining pressure	MRR	The machining performance of RUM improved by a factor of 10 in comparison with conventional USM.
Spur et al. 1995 [144]	Steel (St 37)	1. Silicon carbide 2. Glass 3. Alumina 4. Titanium diboride 5. Hot-pressed silicon nitride	CNC-controlled rotary ultrasonic machining	1. Tool rotation 2. Grit size (280 B$_4$C) 3. Frequency 4. Vibration amplitude	1. Production rate 2. Tool wear rate 3. Precision 4. SR 5. Deviation from nominal diameter	1. Tough material had greater TW and lesser machining rates and precision of the holes. 2. A lower MR tends to give a lesser SR.

(*Continued*)

TABLE 4.7
Chronological Comparative Study of USM (Continued)

References	Tool Material	Substrate Material	Type of USM	Input Parameters	Output Parameters	Findings/Conclusion
		6. Tetragonal zirconium oxide stabilized with cerium 7. Tetragonal zirconium oxide stabilized with yttrium			6. Material removal mechanism	3. Tough materials' material removal was based on plasticity.
Narasimha et al. 1993 [145]	1. Mild steel 2. Titanium 3. SS 4. Maraging steel 5. Silver steel 6. Thoriated tungsten 7. Nimonic	1. Porcelain (soft) 2. Glass 3. Ferrite 4. Alumina 5. Tungsten carbide	Conventional ultrasonic machine and rotary ultrasonic machine	1. Tool material 2. Tool motion 3. Mesh size 4. Frequency 5. Slurry concentration 6. Static load 7. Amplitude	1. MRR 2. Tool wear 3. Surface roughness	Higher MRR was accompanied by high TWR.
Venkatesh et al. 1983 [146]	1. SS 2. MS 3. Tungsten carbide	1. Glass 2. Tungsten carbide	Ultrasonic machine	1. Vibration frequency 2. Slurry concentration 3. Static load 4. Grit size	1. MRR 2. SR 3. TW	1. Abrasive-jet machining (AJM) and USM were influential machining processes. 2. Boron carbide is the best abrasive. 3. Stainless steel is found to be the best tool material, as it offers strong resistance to wear.
Kremer et al. 1981 [147]	1. Steel 2. Tool steel XC 38	1. Glass 2. Hard steel 3. Graphite ELLOR 9	Conventional ultrasonic machine	1. Vibration amplitude 2. Slurry concentration 3. Frequency 4. Static pressure 5. Static force 6. Machining time	1. MRR 2. Depth of cut 3. Surface quality	Variation of MRR with varying amplitude, static pressure, and frequency was observed.

4.4 DISCUSSION

4.4.1 Quantitative Analysis of Literature Study

In the trend captured from the quantitative analysis of the literature reviewed, we can narrow down the fields of research and scope of further exploration. Analysis of the literature present informed us on the behavior of USM with different materials and its effectiveness. New improvements to USM are still being explored. Figure 4.5 represents various percentage distributions of the studies conducted on USM. The highest number of studies (20%) was conducted on glass, followed by 18% (comparative analysis), 17% (miscellaneous), 16% (composites), 15% (ceramics), 9% (silicon), and the least (5%) on crystals.

4.4.2 Effect of Process Parameters on Performance Parameters

4.4.2.1 Effect of Static Pressure

RUM drilling performance is strongly affected by the static pressure, although an increase in static pressure will lead to an increase in MRR to its maximum value. In comparison, a decrease, on the other hand, produced poor surface finish and higher tool wear were encountered while machining of advanced ceramics like Al_2O_3, SiC, Si_3N_4, ZrO_2, and B_4C, among others [33, 39, 45, 80, 148]. Similar problems were observed while machining technical glasses like glass plate, porcelain and borosilicate glass at high static pressure [149]. Increased MRR, reduced tool

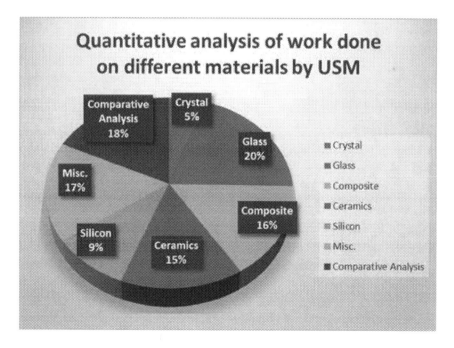

FIGURE 4.5 Quantitative Analysis of Literature

wear and hole clearance were found with C/SiC composite was post the increase in static pressure [150].

4.4.2.2 Effect of Vibrational Amplitude

MRR first increases up to its maximum then drops as registered by the increasing vibrational amplitude. The decrease in MRR encountered was attributed to an exponential increase in alternating the load on the diamond abrasives and a decrease in bond strength. The change in vibration amplitude had no significant effect on hole clearance (SR). The highest material removal was obtained at optimized values of vibration amplitude. Higher hole clearance was observed with augmented values of vibrational amplitude in the case of hard-to-machine composites [151].

4.4.2.3 Effect of Vibrational Frequency

A broad frequency range was observed in our literature review [40]. The load (static) in conjunction with resonance frequency had a crucial role in machining [55]. A decrease in the resonance frequency was observed as a result of adding a coolant. In addition to this, a significant change in resonance frequency was attributed to properties such as the hardness of the material and tool feed rate [58]. The frequency had a crucial and consequential effect on MRR in conjunction with amplitude and static pressure [147]. Figure 4.6 represents the working of USM in close.

4.4.2.4 Effect of Tool Rotation

The effect of tool rpm on the RUM drilling of some advanced ceramics were studied for MRR, which shows direct relation ergo, with increase in tool rpm MRR will also increase [141].

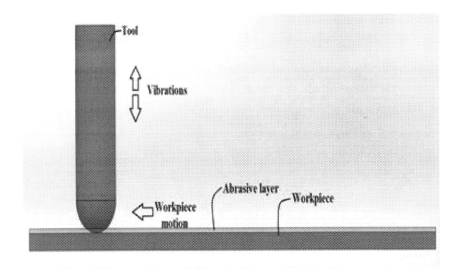

FIGURE 4.6 Working of USM

4.4.2.5 Effect of Abrasives

In the case of advanced ceramics, with an increase in abrasive concentration, the MRR first increases to its maximum and then continuously falls has been observed [146]. This was attributed to the significant lowering in mechanical strength of a diamond-impregnated layer [152], which, on further increases, reduces tool wear and increases tool wear. SR also follows the same trend [40]; it elevates to its maximum and then decreases as the particle size increases. With an increase in bond strength, MRR is reduced, and tool wear is also reduced particularly; stronger diamond particles need stronger binders. For hard-to-machine materials, high-strength synthetic diamonds are preferred as this provides improved outputs in comparison with weaker synthetic diamonds. In the case of natural diamonds, both tool wear (TW) and MRR are lower. A natural diamond provides a higher surface finish than a high-strength diamond [153].

4.4.2.6 Effect of Coolants

Through various experiments and investigations conducted on different coolant types and coolant pressures to optimize the performance of RUM, it was observed that coolant pressure had a trivial to no effect on MRR, although by applying optimal pressure, the lowest surface roughness was achieved [15]. A mixture of synthetic coolant and tap water showed better results than water-based coolants in RUM drilling operations. The introduction of an air-operated double-diaphragm pump into a cooling system reduced machined SR [45, 154].

4.5 COMPARISON BETWEEN CONVENTIONAL AND NONCONVENTIONAL TECHNIQUES FOR THE MACHINING OF BRITTLE MATERIALS

4.5.1 CONVENTIONAL PROCESSES

4.5.1.1 Ductile Regime Grinding

To deform the workpiece, traditional processes with hard abrasive particles are used to get the desired shape of the workpiece. Abrasives act as a medium for cutting material from the workpiece. A coolant is necessarily required for this process to safeguard the workpiece against thermally damaging it.

4.5.1.2 Lapping

Lapping and polishing may be among the oldest manufacturing processes, but they seem to be the ones least well supplied with predictive process models. The mechanisms of lapping and polishing involve four components: the workpiece, fluid, granule, and lap.

4.5.2 NONCONVENTIONAL PROCESSES

4.5.2.1 Machining of Metallic Brittle Materials

The machining of metallic brittle materials, which have good electrical conductivity are machined by putting their electrical conductivity to use. Nonconventional machining processes, namely, EDM, are useful for machining brittle materials of any hardness with high accuracy. Machining of brittle materials with complex shapes and thin workpieces can be done without any distortion.

4.5.2.2 Machining of Nonmetallic Brittle Materials

For the machining of materials that have low electrical conductivity, processes such as USM can be used. Intricate shapes and profiles can be machined using USM with high accuracy. Precision machining of materials such as ceramics, stones, glass, and the like, which are brittle and possess hardness, is possible with USM.

Using USM, we can machine brittle and hard materials independent of their electrical conductivity.

4.5.3 Comparison between Conventional and Nonconventional Processes

TABLE 4.8
Comparison between Conventional and Nonconventional Processes

Conventional Processes	Nonconventional Processes
Materials with higher factor of hardness have lower machinability using conventional processes.	Materials of higher factor of hardness can be machined.
Higher material removal rates with faster machining process.	Lower material removal rates along with the process being slower.
Limitations of size and shape are present and machining on a micro level is not feasible.	Intricate profiles can be machined.
There are not many variations possible for conventional machining processes.	Additive machining by addition of powder in EDM and addition of abrasives in USM is also possible.
Machined parts usually need finishing after being machined.	With lower MRR, the surface finish is usually better.
Lower power consumption for similar machining magnitude.	Higher power consumption.
Skilled operators are required.	High degree of automation.

4.6 CONCLUSION AND FUTURE ASPECTS

The USM process possesses a wide array of possibilities in the field of machining brittle materials. The review showcased a chronological evolution in the machining of brittle materials through USM and investigated different modifications and variants of USM. The different process parameters and performance parameters were discussed, and their relationships were examined. The evolution in the study of USM was formulated in tabular form.

Summarizing the conclusion
- Feed rate and tool rotation were among the significant process parameters in determining the surface quality and MRR.

- Conventional USM can be modified for improved efficiency or can be used in accordance with other nonconventional techniques for improved efficiency.
- During the process of machining titanium, increased spindle rotation had the effect that the SR and cutting force decreased considerably. The subsequent decrease and later increase in the cutting force values were observed with the increase in ultrasonic vibration power. A steady decrease in the SR values was clear with an increase in ultrasonic vibration power.
- Machining of silicon carbide showed that the most profound effects on chip size, SR, and cutting force were that of feed rate and rotation.
- The effect of ultrasonic vibrations and grit size was consequential on chipping and SR.
- For various applications, the optimum values for tool toughness and hardness, in addition with high tool wear resistance, proper elastic, and fatigue strength, were deemed necessary.
- Abrasive medium should have good thermal conductivity, high wettability, low viscosity, and high specific heat for an optimized cooling. In addition, the medium should have a density comparable with that of the abrasive particles.

Future Scope
- Many brittle materials like silicon wafers require more experimental work, and a thorough study of process parameters and performance parameters is needed for evaluation.
- The addition of powdered additives and different types of abrasives were discussed. Further study on their effects on performance parameters is required.
- Experimental determination of the relations between process parameters, process conditions and ratio of brittle fracture portion to plastic flow portion is needed.
- A better elucidation of the effects and influences of process parameters on the modes of material removal will be possible through more experimentation.
- The power consumption for USM is high, and optimizations for the same are required.

Limitations
- The production of such frequencies for machining is not very efficient given to the limitations of present transducer technology.
- The power consumed to the material removal rate ratio is very poor.
- At higher frequencies, the thermal gradient increases, which results in excessive heating of the tool.

REFERENCES
1. Abdullin, I., Bagautinov, A., & Ibragimov, G. (1988). Improving surface finish for titanium alloy medical instruments. *Biomedical Engineering*, 22(2), 48–50.

2. Thoe, T. B., Aspinwall, D. K., & Wise, M. L. H. (1998). Review on ultrasonic machining. *International Journal of Machine Tools and Manufacture, 38*(4), 239–255.
3. Gilmore, R. (1990). *Ultrasonic machining of ceramics.* SME Paper, MS 90-346, 12 p.
4. Rosenberg, L. D., Kazantsev, V. F., Makarov, L. O., & Yakhimovich, D. F. (1964). *Ultrasonic cutting* (Translated from Russian by JES Bradley). New York: Consultants Bureau Enterprises Inc.
5. Kohls, J. B. (1984). Ultrasonic manufacturing process: Ultrasonic machining (USM) and ultrasonic impact grinding (USIG). *Carbide and Tool Journal, 16*(5), 12–15.
6. Haslehurst, M. (1981). *Manufacturing technology,* Cambridge University Press (3rd ed., pp. 270–271).
7. Soundararajan, V., & Radhakrishnan, V. (1986). An experimental investigation on the basic mechanisms involved in ultrasonic machining. *International Journal of MTDR, 26*(3), 307–321.
8. Satyanarayana, A. & Krishna Reddy, B. G. (1984). Design of velocity transformers for ultrasonic machining. *Electrical India, 24*(14), 11–20.
9. Drozda, T. J. & Wick, C. (1983). Non-traditional machining (chapter 29). In *Tool and manufacturing engineers handbook* (Desk ed., pp. 1–23). Dearborn, MI: Soc. Manuf Engrs. ISBN No. 0872633519.
10. Perkins, J. (1927). *An outline of power ultrasonics.* Technical Report by Kerry Ultrasonics, 7 p.
11. Moreland M. A., Ultrasonic impact grinding: What it is: What it will do, *1984 Proc. – 22nd Abrasive Engg. Soc. Conf.: Abrasives and Hi-Technology, A 2-way Street,* 1984, pp. 111–117.
12. Kremer, D. (1981). The state of the art of ultrasonic machining. *Annals of the CIRP, 30*(1), 107–110.
13. www.engineeringenotes.com/manufacturing-science/ultrasonic-machining/ultrasonic-machining usm-mechanics-process-parameters-elements-tools-characteristics/52038
14. Ahmed, Y., Cong, W. L., Stanco, M. R., Xu, Z. G., Pei, Z. J., Treadwell, C., & Li, Z. C. Rotary ultrasonic machining of alumina dental ceramics: a preliminary experimental study on surface and subsurface damages. *Journal of Manufacturing Science and Engineering, Transactions of the ASME, 134*(6), 064501.
15. Zhang, Q., Zhang, J., Jia, Z., & Ai, X. (1998). Fracture at the exit of the hole during the ultrasonic drilling of engineering ceramics. *Journal of Materials Processing Technology, 84*(1–3), 20–24.
16. Heredia-Rivera, U., Ferrer, I., & Vázquez, E. (2019). Ultrasonic molding technology: Recent advances and potential applications in the medical industry. *Polymers, 11*(4), 667.
17. Yaqoob, A. A., Ahmad, H., Parveen, T., Ahmad, A., Oves, M., Ismail, I. M., . . . & Mohamad Ibrahim, M. N. (2020). Recent advances in metal decorated nanomaterials and their various biological applications: A review. *Frontiers in Chemistry, 8,* 341.
18. James, S., & Panchal, S. (2019). Finite element analysis and simulation study on micromachining of hybrid composite stacks using micro ultrasonic machining process. *Journal of Manufacturing Processes, 48,* 283–296.
19. Zhang, J., Long, Z., Wang, C., Ren, F., & Li, Y. (2019). Novel optimization approach in ultrasonic machining: Unilateral compensation for resonant vibration in primary side. *IEEE Access, 7,* 34131–34140.
20. Zhang, J. G., Long, Z. L., Ma, W. J., Hu, G. H., & Li, Y. M. (2019). Electromechanical dynamics model of ultrasonic transducer in ultrasonic machining based on equivalent circuit approach. *Sensors, 19*(6), 1405.

21. Kumar, S., Doloi, B., & Bhattacharyya, B. (2019, November). Experimental investigation into micro ultrasonic machining of quartz. In *IOP Conference Series: Materials Science and Engineering* (Vol. 653, No. 1, p. 012027), International Conference on Advances in Materials and Manufacturing Engineering 15–19 March 2019, Kalinga Institute of Industrial Technology (Deemed to be University), Bhubaneswar, India.
22. Sharma, A., Jain, V., & Gupta, D. (2019). Comparative analysis of chipping mechanics of float glass during rotary ultrasonic drilling and conventional drilling: For multi-shaped tools. *Machining Science and Technology, 23*(4), 547–568.
23. James, S., & Panchal, S. (2019). Parametric study of micro ultrasonic machining process of hybrid composite stacks using finite element analysis. *Procedia Manufacturing, 34*, 408–417.
24. Zhang, D., Wang, H., Hu, Y., Chen, X., Cong, W., & Burks, A. R. (2019, June). Rotary ultrasonic machining of CFRP composites: Effects of carbon fiber reinforcement structure. In *ASME 2019 14th International Manufacturing Science and Engineering Conference*. American Society of Mechanical Engineers Digital Collection, June 10–14, 2019, Erie, Pennsylvania, USA.
25. Zhang, J., Long, Z., Wang, C., Zhao, H., & Li, Y. (2020). Compensation modeling and optimization on contactless rotary transformer in rotary ultrasonic machining. *Journal of Manufacturing Science and Engineering, 142*(10).
26. Han, X., Zhang, D., & Song, G. (2020). Review on current situation and development trend for ultrasonic vibration cutting technology. *Materials Today: Proceedings, 22*, 444–455.
27. Moreland, M. A. (1988). Versatile performance of ultrasonic machining. *Ceramic Bulletin, 67*(6), 1045–1047.
28. Farago, F. T. (1980). *Abrasive methods engineering* (Vol. 2, pp. 480–481). Industrial Press, New York.
29. Scab, K. H. W. et al. (1990). *Parametric studies of ultrasonic machining*. SME Tech. Paper, MR 90-294, 11 p.
30. Balamuth, L. (1964). Ultrasonic vibrations assist cutting tools. *Metalworking Products, 108*(24), 75–77.
31. Kainth, G. S., Nandy, A., & Singh, K. (1979). On the mechanics of material removal in ultrasonic machining. *International Journal of Machine Tool Design and Research, 19*(1), 33–41.
32. Cong, W. L., Pei, Z. J., Deines, T., Wang, Q. G., & Treadwell, C. (2010). Rotary ultrasonic machining of stainless steels: Empirical study of machining variables. *International Journal of Manufacturing Research, 5*(3), 370–386.
33. Debnath, K., Singh, I., & Dvivedi, A. (2015). Rotary mode ultrasonic drilling of glass fiber-reinforced epoxy laminates. *Journal of Composite Materials, 49*(8), 949–963.
34. Nishimura, G. (1954). Ultrasonic machining—Part I. *Journal of Faculty of Engineering Tokyo University, 24*(3), 65–100.
35. McGeough, J. A., 1988, *Advance Methods of Machining*, Chapman & Hall, New York, NY, USA.
36. Jay, F. (1977). *IEEE standard dictionary of electrical and electronics terms*. Jay, Frank Institute of Electrical and Electronics Engineers.
37. Moore D., Ultrasonic impact grinding, *Proc. Non-traditional Machining Conf.*, Cincinnati, 1985, pp. 137–139.
38. Legge, P. (1966). Machining without abrasive slurry. *Ultrasonics, 4*, 157–162.
39. Cong, W., Pei, Z. J., Van Vleet, E. G., & Wang, Q. (2009). Surface roughness in rotary ultrasonic machining of stainless steels, IIE Annual Conference. Proceedings; Norcross (2009): 1477–1482.

40. Lee, T. C., Zhang, J. H., & Lau, W. S. (1998). Machining of engineering ceramics by ultrasonic vibration assisted EDM method. *Material and Manufacturing Process, 13*(1), 133–146.
41. Gilmore, R. (1989). *Ultrasonic machining and orbital abrasion techniques.* SME Technical Paper (Series) AIR, NM89-419, pp. 1–20
42. Moreland, M. A. (1991). Ultrasonic machining. In *Engineering material handbook: Ceramics and glasses* (Vol. 4, pp. 359–362). ASM International. ISBN No. 0871702827.
43. Neppiras, E. A. (1956). Report on ultrasonic machining. *Metalworking Production, 100*, pp. 1283–1288, 1333–1336, 1377–1382, 1420–1424, 1464–1468, 1554–1560, 1599–1604.
44. Weller, E. J. (1984). *Nontraditional machining processes.* Society of Manufacturing Engineers.
45. Gong, H., Fang, F. Z., & Hu, X. T. (2010). Kinematic view of tool life in rotary ultrasonic side milling of hard and brittle materials. *International Journal of Machine Tools & Manufacture, 50*, 303–307.
46. Scab, K. H. W. *et al.* (1990). *Parametric studies of ultrasonic machining.* SME Tech. Paper, MR90-294, 11 p.
47. Yang, X., & Liu, R. (1999). Machining titanium and its alloys. *Machining Science and Technology, 3*(1), 107–139.
48. Yishuang, H. (1990). The effect of ultrasonic vibration to EDM. *Metals and Machines Overseas, 4*, 1–5.
49. Zang, H., Su, Q., Mo, Y., Cheng, B. W., & Jun, S. (2010). Ionic liquid [EMIM] OAc under ultrasonic irradiation towards the first synthesis of trisubstituted imidazoles. *Ultrasonics Sonochemistry, 17*(5), 749–751.
50. Pei, Z. J., Ferreira, P. M., Kapoor, S. G., & Haselkorn, M. B. A. C. (1995). Rotary ultrasonic machining for face milling of ceramics. *International Journal of Machine Tools and Manufacture, 35*(7), 1033–1046.
51. Basem M. A. Abdo et al., 2020 IOP Conf. Ser.: Mater. Sci. Eng. 727 012009.
52. Abdo, B., Mian, S. H., El-Tamimi, A., Alkhalefah, H., & Moiduddin, K. (2020). Micromachining of biolox forte ceramic utilizing combined laser/ultrasonic processes. *Materials, 13*(16), 3505.
53. Singh, A. M., Majhi, R., & Patowari, P. K. (2021). Machinability study for slot cutting on glass using ultrasonic machining process. In *Recent advances in mechanical engineering* (pp. 771–778). Singapore: Springer.
54. Singh, V., Saraswat, P., & Joshi, D. (2020). Experimental investigation of edge chipping defects in rotary ultrasonic machining of float glass. *Materials Today: Proceedings, 44*(6), 4462–4466.
55. Zhou, H., Zhang, J., Feng, P., Yu, D., & Cai, W. (2019). An output amplitude model of a giant magnetostrictive rotary ultrasonic machining system considering load effect. *Precision Engineering, 60*, 340–347.
56. Kumar, S., & Dvivedi, A. (2017). Experimental investigation on drilling of borosilicate glass using micro-USM with and without tool rotation: A comparative study. *International Journal of Additive and Subtractive Materials Manufacturing, 1*(3–4), 213–222.
57. Wang, J., Feng, P., & Zhang, J. (2018). Reducing edge chipping defect in rotary ultrasonic machining of optical glass by compound step-taper tool. *Journal of Manufacturing Processes, 32*, 213–221.
58. Sindhu, D., Chandna, P., & Thakur, L. (2018). Rotary Ultrasonic Machining of Hard to Machine Materials—A Review. *International Journal of Advanced Engineering Research and Applications, 4*(2).

59. Feng, P., Wang, J., Zhang, J., & Zheng, J. (2017). Drilling induced tearing defects in rotary ultrasonic machining of C/SiC composites. *Ceramics International, 43*(1), 791–799.
60. Wang, J., Feng, P., Zhang, J., Zhang, C., & Pei, Z. (2016). Modeling the dependency of edge chipping size on the material properties and cutting force for rotary ultrasonic drilling of brittle materials. *International Journal of Machine Tools and Manufacture, 101*, 18–27.
61. Wang, J., Zha, H., Feng, P., & Zhang, J. (2016). On the mechanism of edge chipping reduction in rotary ultrasonic drilling: A novel experimental method. *Precision Engineering, 44*, 231–235.
62. Zhang, C., Cong, W., Feng, P., & Pei, Z. (2014). Rotary ultrasonic machining of optical K9 glass using compressed air as coolant: A feasibility study. *Proceedings of the Institution of Mechanical Engineers, Part B: Journal of Engineering Manufacture, 228*(4), 504–514.
63. Schorderet, A., Deghilage, E., & Agbeviade, K. (2013). Tool type and hole diameter influence in deep ultrasonic drilling of micro-holes in glass. *Procedia CIRP, 6*, 565–570.
64. Lv, Dong Xi, Yong Jian Tang, Yan Hua Huang, and Hong Xiang Wang. "Effects of High Frequency Vibration on the Surface Quality in Rotary Ultrasonic Machining of Glass BK7." Applied Mechanics and Materials 427–429 (September 2013): 187–90. https://doi.org/10.4028/www.scientific.net/amm.427-429.187
65. Huang, Yan Hua, Dong Xi Lv, Yong Jian Tang, Hong Xiang Wang, and Hai Jun Zhang. "Experimental Investigation on Surface/Subsurface Damage in Rotary Ultrasonic Machining of Glass BK7." Applied Mechanics and Materials 325–326 (June 2013): 1357–61. https://doi.org/10.4028/www.scientific.net/amm.325-326.1357
66. Gong, H., Fang, F. Z., & Hu, X. T. (2010). Kinematic view of tool life in rotary ultrasonic side milling of hard and brittle materials. *International Journal of Machine Tools & Manufacture, 50*, 303–307.
67. Choi, J. P., Jeon, B. H., & Kim, B. H. (2007, March 6). Chemical-assisted ultrasonic machining of glass. *Journal of Materials Processing Technology.* Advances in Materials and Processing Technologies, July 30th–August 3rd 2006, Las Vegas, Nevada, *191*(1–3), 153–156.
68. Kuo, Kei Lin. "A Study of Glass Milling Using Rotary Ultrasonic Machining." Key Engineering Materials 364–366 (December 2007): 624–28. https://doi.org/10.4028/www.scientific.net/kem.364-366.624
69. Yan, B. H., Wang, A. C., Huang, C. Y., & Huang, F. Y. (2002). Study of precision micro-holes in borosilicate glass using micro EDM combined with micro ultrasonic vibration machining. *International Journal of Machine Tools and Manufacture, 42*(10), 1105–1112.
70. Komaraiah, M., & Narasimha Reddy, P. (1991). Rotary ultrasonic machining—A new cutting process and its performance. *International Journal of Production Research, 29*(11), 2177–2187.
71. Kainth, G. S., Nandy, A., & Singh, K. (1979). On the mechanics of material removal in ultrasonic machining. *International Journal of Machine Tool Design and Research, 19*(1), 33–41.
72. Kaczmarek, J., Kops, L., & Shaw, M. C. (1966). Ultrasonic grinding economics. *ASME Journal of Engineering for Industry, 88*, 449–454.
73. Sreehari, D., & Sharma, A. K. (2018). On form accuracy and surface roughness in micro-ultrasonic machining of silicon microchannels. *Precision Engineering, 53*, 300–309.
74. Cong, W. L., Feng, Q., Pei, Z. J., Deines, T. W., & Treadwell, C. (2012). Edge chipping in rotary ultrasonic machining of silicon. *International Journal of Manufacturing Research, 7*(3), 311–329.

75. Zarepour, H., & Yeo, S. H. (2012). Predictive modeling of material removal modes in micro ultrasonic machining. *International Journal of Machine Tools and Manufacture*, 62, 13–23.
76. Churi, N. J., Pei, Z. J., Shorter, D. C., & Treadwell, C. (2007). Rotary ultrasonic machining of silicon carbide: Designed experiments. *International Journal of Manufacturing Technology and Management*, 12(1–3), 284–298.
77. Yu, Z., Ma, C., An, C., Li, J., & Guo, D. (2012). Prediction of tool wear in micro USM. *CIRP Annals*, 61(1), 227–230.
78. Zeng, W. M., Li, Z. C., Pei, Z. J., & Treadwell, C. (2005). Experimental observation of tool wear in rotary ultrasonic machining of advanced ceramics. *International Journal of Machine Tools and Manufacture*, 45(12–13), 1468–1473.
79. Yu, Z. Y., Rajurkar, K. P., & Tandon, A. (2004). Study of 3D micro-ultrasonic machining. *Journal of Manufacturing Science and Engineering*, 126(4), 727–732.
80. Komaraiah, M., & Reddy, P. N. (1993). Relative performance of tool materials in ultrasonic machining. *Wear*, 161(1–2), 1–10.
81. Khairy, A. B. E. (1990). Assessment of some dynamic parameters for the ultrasonic machining process. *Wear*, 137(2), 187–198.
82. Abdo, B. M., El-Tamimi, A., & Alkhalefah, H. (2020). Parametric analysis and optimization of rotary ultrasonic machining of zirconia (ZrO2) ceramics. Volume 727, International Conference on Materials Science and Manufacturing Engineering (MSME 2019) 7–9 November 2019, Guangzhou, China.
83. AlKawaz, M. H., Kasim, M. S., & Izamshah, R. (2020). Effect of spindle speed on performance measures during rotary ultrasonic machining of fully sintered zirconia ceramic. *Journal of Mechanical Engineering Research and Developments*, 43(4), 381–387.
84. Abdo, B., El-Tamimi, A., & Nasr, E. A. (2020). Rotary ultrasonic machining of alumina ceramic: An experimental investigation of tool path and tool overlapping. *Applied Sciences*, 10(5), 1667.
85. Sandá, A., & Sanz, C. (2020). Rotary ultrasonic machining of ZrO2-NbC and ZrO2-WC ceramics. *International Journal of Machining and Machinability of Materials*, 22(2), 165–179.
86. Abdo, B. M., Anwar, S., & El-Tamimi, A. (2019). Machinability study of biolox forte ceramic by milling microchannels using rotary ultrasonic machining. *Journal of Manufacturing Processes*, 43, 175–191.
87. Singh, R. P., & Singhal, S. (2017). Investigation of machining characteristics in rotary ultrasonic machining of alumina ceramic. *Materials and Manufacturing Processes*, 32(3), 309–326.
88. Singh, R. P., & Singhal, S. (2016). Rotary ultrasonic machining: A review. *Materials and Manufacturing Processes*, 31(14), 1795–1824.
89. Singh, R. P., & Singhal, S. (2018). Experimental investigation of machining characteristics in rotary ultrasonic machining of quartz ceramic. *Proceedings of the Institution of Mechanical Engineers, Part L: Journal of Materials: Design and Applications*, 232(10), 870–889.
90. Liu, J. W., Baek, D. K., & Ko, T. J. (2014). Chipping minimization in drilling ceramic materials with rotary ultrasonic machining. *International Journal of Advanced Manufacturing Technology*, 72(9–12), 1527–1535.
91. Liu, D., Cong, W. L., Pei, Z. J., & Tang, Y. (2012). A cutting force model for rotary ultrasonic machining of brittle materials. *International Journal of Machine Tools and Manufacture*, 52(1), 77–84.
92. Cong, W. L., Pei, Z. J., Mohanty, N., Van Vleet, E., & Treadwell, C. (2011). Vibration amplitude in rotary ultrasonic machining: A novel measurement method and effects of process variables. *Journal of Manufacturing Science and Engineering*, 133(3).

93. Churi, N. J., Pei, Z. J., Shorter, D. C., & Treadwell, C. (2009). Rotary ultrasonic machining of dental ceramics. *International Journal of Machining and Machinability of Materials*, 6(3–4), 270–284.
94. Zeng, Wei Min, Xi Peng Xu, and Zhi Jian Pei. "Rotary Ultrasonic Machining of Advanced Ceramics." Materials Science Forum 532–533 (December 2006): 361–64. https://doi.org/10.4028/www.scientific.net/msf.532-533.361
95. Wang, H., Hu, Y., Cong, W., Hu, Z., & Wang, Y. (2020). A novel investigation on horizontal and 3D elliptical ultrasonic vibrations in rotary ultrasonic surface machining of carbon fiber reinforced plastic composites. *Journal of Manufacturing Processes*, 52, 12–25.
96. Wang, H., Pei, Z. J., & Cong, W. (2020). A mechanistic cutting force model based on ductile and brittle fracture material removal modes for edge surface grinding of CFRP composites using rotary ultrasonic machining. *International Journal of Mechanical Sciences*, 176, 105551.
97. Ning, F., Wang, H., & Cong, W. (2019). Rotary ultrasonic machining of carbon fiber reinforced plastic composites: A study on fiber material removal mechanism through single-grain scratching. *International Journal of Advanced Manufacturing Technology*, 103(1), 1095–1104.
98. Wang, H., Hu, Y., Cong, W., & Hu, Z. (2019). A mechanistic model on feeding-directional cutting force in surface grinding of CFRP composites using rotary ultrasonic machining with horizontal ultrasonic vibration. *International Journal of Mechanical Sciences*, 155, 450–460.
99. Fernando, PKSC, Zhang, M., Pei, Z., & Cong, W. "Rotary Ultrasonic Machining: Effects of Tool End Angle on Delamination of CFRP Drilling." *Proceedings of the ASME 2017 12th International Manufacturing Science and Engineering Conference* collocated with the JSME/ASME 2017 6th International Conference on Materials and Processing. Volume 1: Processes. Los Angeles, California, USA. June 4–8, 2017. V001T02A015. ASME. https://doi.org/10.1115/MSEC2017-2863
100. H. Zha, J. Zhang and P. Feng, "An experimental study on rotary ultrasonic drilling small-diameter holes of high-volume fraction silicon carbide-reinforced aluminum matrix composites (SiCp/Al)," 2017 14th International Bhurban Conference on Applied Sciences and Technology (IBCAST), 2017, pp. 34-39, doi:10.1109/IBCAST.2017.7868032
101. Ning, F., Wang, H., Cong, W., & Fernando, P. K. S. C. (2017). A mechanistic ultrasonic vibration amplitude model during rotary ultrasonic machining of CFRP composites. *Ultrasonics*, 76, 44–51.
102. Kataria, R., Kumar, J., & Pabla, B. S. (2016). Experimental investigation and optimization of machining characteristics in ultrasonic machining of WC—Co composite using GRA method. *Materials and Manufacturing Processes*, 31(5), 685–693.
103. Ning, F. D., Cong, W. L., Pei, Z. J., & Treadwell, C. (2016). Rotary ultrasonic machining of CFRP: A comparison with grinding. *Ultrasonics*, 66, 125–132.
104. Ding, K., Fu, Y., Su, H., Chen, Y., Yu, X., & Ding, G. (2014). Experimental studies on drilling tool load and machining quality of C/SiC composites in rotary ultrasonic machining. *Journal of Materials Processing Technology*, 214(12), 2900–2907.
105. Liu, J. W., Baek, D. K., & Ko, T. J. (2014). Chipping minimization in drilling ceramic materials with rotary ultrasonic machining. *International Journal of Advanced Manufacturing Technology*, 72(9–12), 1527–1535.
106. Ding, K., Fu, Y., Su, H., Chen, Y., Yu, X., & Ding, G. (2014). Experimental studies on drilling tool load and machining quality of C/SiC composites in rotary ultrasonic machining. *Journal of Materials Processing Technology*, 214(12), 2900–2907.

107. Yuan, S., Fan, H., Amin, M., Zhang, C., & Guo, M. (2016). A cutting force prediction dynamic model for side milling of ceramic matrix composites C/SiC based on rotary ultrasonic machining. *International Journal of Advanced Manufacturing Technology, 86*(1), 37–48.
108. Feng, Q., Cong, W. L., Pei, Z. J., & Ren, C. Z. (2012). Rotary ultrasonic machining of carbon fiber-reinforced polymer: Feasibility study. *Machining Science and Technology, 16*(3), 380–398.
109. Cong, W. L., Pei, Z. J., Feng, Q., Deines, T. W., & Treadwell, C. (2012). Rotary ultrasonic machining of CFRP: A comparison with twist drilling. *Journal of Reinforced Plastics and Composites, 31*(5), 313–321.
110. Marichamy, S., Babu, K. V., Madan, D., & Ganesan, P. (2020). Ultrasonic machining and fretting wear of synthesized duplex brass metal matrix. *Materials Today: Proceedings, 21*, 734–737.
111. Fernando, P. C., Pei, Z. J., & Zhang, M. (2020). Mechanistic cutting force model for rotary ultrasonic machining of rocks. *The International Journal of Advanced Manufacturing Technology, 109*(1), 109–128.
112. James, S., & Sonate, A. (2018). Experimental study on micromachining of CFRP/Ti stacks using micro ultrasonic machining process. *The International Journal of Advanced Manufacturing Technology, 95*(1), 1539–1547.
113. Kuruc, M., Vopát, T., & Peterka, J. (2015). Surface roughness of poly-crystalline cubic boron nitride after rotary ultrasonic machining. *Procedia Engineering, 100*, 877–884.
114. Cong, W. L., Pei, Z. J., & Treadwell, C. (2014). Preliminary study on rotary ultrasonic machining of CFRP/Ti stacks. *Ultrasonics, 54*(6), 1594–1602.
115. Cong, W. L., Pei, Z. J., Deines, T. W., Liu, D. F., & Treadwell, C. (2013). Rotary ultrasonic machining of CFRP/Ti stacks using variable feedrate. *Composites Part B: Engineering, 52*, 303–310.
116. Cong, W. L., Zou, X., Deines, T. W., Wu, N., Wang, X., & Pei, Z. J. (2012). Rotary ultrasonic machining of carbon fiber reinforced plastic composites: An experimental study on cutting temperature. *Journal of Reinforced Plastics and Composites, 31*(22), 1516–1525.
117. Kumar, J., Khamba, J. S., & Mohapatra, S. K. (2009). Investigating and modeling tool-wear rate in the ultrasonic machining of titanium. *International Journal of Advanced Manufacturing Technology, 41*(11), 1107–1117.
118. Kumar, J., & Khamba, J. S. (2009). An investigation into the effect of work material properties, tool geometry and abrasive properties on performance indices of ultrasonic machining. *International Journal of Machining and Machinability of Materials, 5*(2–3), 347–366.
119. Tawakoli, T., Azarhoushang, B., & Rabiey, M. (2009). Ultrasonic assisted dry grinding of 42CrMo4. *International Journal of Advanced Manufacturing Technology, 42*(9), 883–891.
120. Churi, N. J., Pei, Z. J., & Treadwell, C. (2007). Rotary ultrasonic machining of titanium alloy (Ti-6Al-4V): Effects of tool variables. *International Journal of Precision Technology, 1*(1), 85–96.
121. Singh, R., & Khamba, J. S. (2007). Investigation for ultrasonic machining of titanium and its alloys. *Journal of Materials Processing Technology, 183*(2–3), 363–367.
122. Azarhoushang, B., & Akbari, J. (2007). Ultrasonic-assisted drilling of Inconel 738-LC. *International Journal of Machine Tools and Manufacture, 47*(7–8), 1027–1033.
123. Ghahramani, B., & Wang, Z. Y. (2001). Precision ultrasonic machining process: A case study of stress analysis of ceramic (Al2O3). *International Journal of Machine Tools and Manufacture, 41*(8), 1189–1208.

124. Thoe, T. B., Aspinwall, D. K., & Killey, N. (1999). Combined ultrasonic and electrical discharge machining of ceramic coated nickel alloy. *Journal of Materials Processing Technology, 92,* 323–328.
125. Neppiras, E. A., & Foskett, R. D. (1965). Ultrasonic machining. *Ultrasonics, 3*(2), 99–99.
126. Ning, F., Wang, H., Cong, W., & Fernando, P. K. S. C. (2017). A mechanistic ultrasonic vibration amplitude model during rotary ultrasonic machining of CFRP composites. *Ultrasonics, 76,* 44–51.
127. Kuruc, M., Vopát, T., & Peterka, J. (2015). Surface roughness of poly-crystalline cubic boron nitride after rotary ultrasonic machining. *Procedia Engineering, 100,* 877–884.
128. Wang, Q., Cong, W., Pei, Z. J., Gao, H., & Kang, R. (2009). Rotary ultrasonic machining of potassium dihydrogen phosphate (KDP) crystal: An experimental investigation on surface roughness. *Journal of Manufacturing Processes, 11*(2), 66–73.
129. Hocheng, H., Kuo, K. L., & Lin, J. T. (1999). Machinability of zirconia ceramics in ultrasonic drilling. *Materials and Manufacturing Processes, 14*(5), 713–724.
130. Pei, Z. J., & Ferreira, P. M. (1998). Modeling of ductile-mode material removal in rotary ultrasonic machining. *International Journal of Machine Tools and Manufacture, 38*(10–11), 1399–1418.
131. Haashir, A., Debnath, T., & Patowari, P. K. (2020). A comparative assessment of micro drilling in boron carbide using ultrasonic machining. *Materials and Manufacturing Processes, 35*(1), 86–94.
132. Wang, J., Fu, J., Wang, J., Du, F., Liew, P. J., & Shimada, K. (2020). Processing capabilities of micro ultrasonic machining for hard and brittle materials: SPH analysis and experimental verification. *Precision Engineering, 63,* 159–169.
133. T. Debnath, A. Haashir, and P.K. Patowari, Parametric study of micro-hole drilling in glass using ultrasonic machining, in International Conference on Precision, Meso, Micro and Nano Engineering (2017).
134. Singh, K. J., & Kapoor, J. (2019). Chemical assisted USM of acrylic heat resistant glass. In *Manufacturing engineering* (pp. 167–183). Singapore: Springer.
135. Wang, J., Feng, P., Zhang, J., Cai, W., & Shen, H. (2017). Investigations on the critical feed rate guaranteeing the effectiveness of rotary ultrasonic machining. *Ultrasonics, 74,* 81–88.
136. Wang, J., Shimada, K., Mizutani, M., & Kuriyagawa, T. (2018). Tool wear mechanism and its relation to material removal in ultrasonic machining. *Wear, 394,* 96–108.
137. Wang, J., Feng, P., & Zhang, J. (2016). Reduction of edge chipping in rotary ultrasonic machining by using step drill: A feasibility study. *International Journal of Advanced Manufacturing Technology, 87*(9), 2809–2819.
138. Zhang, X., Kumar, A. S., Rahman, M., Nath, C., & Liu, K. (2011). Experimental study on ultrasonic elliptical vibration cutting of hardened steel using PCD tools. *Journal of Materials Processing Technology, 211*(11), 1701–1709.
139. Kuo, K. L., & Tsao, C. C. (2012). Rotary ultrasonic-assisted milling of brittle materials. *Transactions of Nonferrous Metals Society of China, 22,* s793–s800.
140. Nath, C., Lim, G. C., & Zheng, H. Y. (2012). Influence of the material removal mechanisms on hole integrity in ultrasonic machining of structural ceramics. *Ultrasonics, 52*(5), 605–613.
141. Singh, R., & Khamba, J. S. (2007). Investigation for ultrasonic machining of titanium and its alloys. *Journal of Materials Processing Technology, 183*(2–3), 363–367.
142. Guzzo, P. L., Raslan, A. A., & De Mello, J. D. B. (2003). Ultrasonic abrasion of quartz crystals. *Wear, 255*(1–6), 67–77.

143. Pei, Z. J., Prabhakar, D., Ferreira, P. M., & Haselkorn, M. (1995). Rotary ultrasonic drilling and milling of ceramics. *Ceramic Transactions, 49,* 185–185.
144. Spur, G., & Holl, S. E. (1996). Ultrasonic assisted grinding of ceramics. *Journal of Materials Processing Technology, 62*(4), 287–293.
145. Komaraiah, M., & Reddy, P. N. (1993). A study on the influence of workpiece properties in ultrasonic machining. *International Journal of Machine Tools and Manufacture, 33*(3), 495–505.
146. Venkatesh, V. C. (1983). Machining of glass by impact processes. *Journal of Mechanical Working Technology, 8*(2–3), 247–260.
147. Kremer, D., Saleh, S. M., Ghabrial, S. R., & Moisan, A, The state of: the art of USM. In *31st CIRP Ass., Canada (1981).*
148. F. Benedict Gary, Book on Non Traditional Manufacturing Processes Marcel Dekker, Inc, New York (1987), pp. 67–86.
149. Adithan, M., & Laroiya, S. C. (1997). Investigations on the performance of ultrasonic drilling process with special reference to precision machining of advanced ceramics. *Jurnal Teknologi, 26*(1), 1–12.
150. Patwardhan, A. V. (2012). *Experimental investigation of hard and brittle materials machining using micro rotary ultrasonic machining* (Thesis). The Graduate College at the University of Nebraska.
151. Das, S., Mazumder, P., Doloi, B., & Bhattacharyya, B. (2018). Fabricating tool for ultrasonic machining using reverse engineering. *International Journal of Computer Integrated Manufacturing, 31*(3), 296–305.
152. Kennedy, D. C., & Grieve, R. J. (1975). Ultrasonic machining-a review. *Production Engineer, 54*(9), 481–486.
153. Ya, G., Qin, H. W., Yang, S. C., & Xu, Y. W. (2002). Analysis of the rotary ultrasonic machining mechanism. *Journal of Materials Processing Technology, 129*(1–3), 182–185.
154. Hegade, T., Jnaneshwar, R. K., Nagarjuna, L. R., Babu, P. P., Srikanth, A. K., & Nanjundeswaraswamy, T. S. (2019). Process characteristics in ultrasonic machining. *International Journal of Engineering Research & Technology (IJERT), 8*(11).

5 Analysis of Mechanical and Sliding Wear Performance of AA6061–SiC/Gr/MD Hybrid Alloy Composite Using a PSI Approach for Rotor Applications

Ashiwani Kumar, Amar Patnaik, Mukesh Kumar, Vikas Kukshal, M.J. Pawar, and Vikash Gautam

CONTENTS

5.1	Introduction	78
5.2	Materials and Methods	79
	5.2.1 Design Concept and Developed Materials Procedure	79
	5.2.2 Physical and Mechanical Characterizations	80
	5.2.3 Sliding Wear Analysis	80
	5.2.4 Taguchi Design of Experimental Approach	81
	5.2.5 Methodology of Preference Index Method	82
5.3	Results and Discussions	84
	5.3.1 Physical and Mechanical Analysis	84
	5.3.2 Wear Studies of Alloy Composite	86
	5.3.2.1 Influence of Specific Wear Rate/Coefficient of Friction on Sliding Distance of Composite	86
	5.3.3 Taguchi and ANOVA Analysis of Alloy Composite	87
	5.3.4 Ranking Analysis of AS Alloy Composite Using PSI Approach	90
5.4	Conclusion	92
	Acknowledgments	92
	References	92

DOI: 10.1201/9781003093213-5

5.1 INTRODUCTION

AA6061 alloy is widely utilized in various technological areas like marine, aircraft, automotive, and the like because of its excellent strength, stiffness, and corrosion properties. AA6061 alloy shows excellent mechanical properties (hardness, tensile strength, flexural strength, and impact strength) and wear analysis of composites. AA6061 materials indicate the minimum resistance to wear performance under dry conditions. Base materials indicate the various element combinations and that shows the superior mechanical performance of the material. Research scholars are used for sliding wear applications. AA6061 alloy composite performance can be improved to enhance in ceramic reinforcements like silicon carbide/graphite/marble dust (SiC/Gr/MD) and improve the wear and friction characteristics of composites. Many authors have found that enhanced in ceramic particulates with enhanced wear and mechanical properties [1–2]. Aluminum metal–based composites may be a fantastic option when compared to other monolithic alloys [3]. Sharma et al. [4] examined the mechanical and tribological properties of AA6061 alloy/rare earth particulate/SiC/aluminum oxide (Al_2O_3) based composite and obtained the properties of mechanical like tensile strength, hardness, flexural strength, and impact strength of composite were enhanced with the addition the particulates. Aruri et al. [5] investigated the wear and mechanical properties SiC/Gr/Al_2O_3 ceramic particulates filled with AA6061 alloy composites and found the tensile strength, hardness, and wear performance increased with an increase in the addition of filler content. Kiran et al. [6] studied the effect of SiC/Al_2O_3 on sliding wear behavior of a composite and reported the property of wear resistance improves with an increase in ceramic particulates. Similar outcomes are reported by Kumar et al. [7, 8], and mechanical and wear properties improve with an increase in cobalt (Co)/nickel (Ni) content. Chen et al. [9] reported the influence of SiC/titanium carbide (TiC) particulate–based aluminum alloy composite and found the mechanical characteristics of the composites were enhanced with an increase in filler content. Prabu et al. [10] studied the effect of SiC on mechanical property of AA7075 composite and obtained the different property (like hardness, compressive) grows with the addition in filler content. Similar finding is reported by Kumar and Kumar [11] found that mechanical and tribological properties improve with incremental increase in boron carbide (B_4C)/rice rush content. Kumar et al. [12] examined the assessment of wear and mechanical performances of Ni particulate–based AA7075 alloy composites. They found the tensile, compressive, hardness and wear properties also improved. Many researchers' results are reported the mechanical and tribological performance improved with the addition of particulate content [13, 14]. Kumar et al. [15] reported the optimal ranking of SiC particulate–based AA2024 alloy composite evaluated via multi criteria decision making (MCDM) based criteria (like preference selection index [PSI], analytic hierarchy process [AHP], and technique for order of preference by similarity to ideal solution [TOPSIS]) and found the ranking of particular composite. Mukesh and Ashiwani [16] evaluated the optimization and ranking order of SiO_2/SiC particulate based AA2024 alloy composite via the preference index method and the design of the Taguchi method. Based on previous research outcomes, the various multi-decision criteria methods are applied to evaluate the optimization problems of composite [17, 18]. The most popular optimization method (like PSI) is effectively implemented in various sectors like design and operation research [19–22]. Therefore, the present study focuses on

1. The manufacture of SiC/MD/Gr particulate–filled AA6061 alloy composite;
2. The analysis of wear, mechanical, and physical characterizations of an SiC/MD/Gr-reinforced AA6061 alloy composite; and
3. The ranking order of fabricated composite evaluated via preference index method.

5.2 MATERIALS AND METHODS

5.2.1 Design Concept and Developed Materials Procedure

The composite design concept and fabricated material characterization are found from the previous literature study. AA6061 alloy was in rod form, as supplied by Bharat Aerospace Metals, Mumbai, India, having copper (~0.28 wt.%), magnesium (Mg ~1.0 wt.%), Si (~0.6 wt.%), chromium (Cr ~0.2 wt.%), and Al remaining wt.%. A graphite particulate size of 98.5 µm and an MD/SiC particle size of 10 µm was bought from S. S. Pvt. Ltd. Jaipur, India [6]. The five composite details are AS-0 (0 wt.% of SiC, constant 3 wt.% of GR, 3 wt.% of MD and the rest AA6061 alloy); AS-1 (1 wt.% of SiC, a constant 3 wt.% of Gr, 3 wt.% of MD, and the rest AA6061 alloy); AS-3 (3 wt.% of SiC, constant 3 wt.% of GR, 3 wt.% of MD, and the rest AA6061 alloy); AS-5 (5 wt.% of SiC, constant 3 wt.% of GR, 3 wt.% of MD, and the rest AA6061 alloy); AS-7 (7 wt.% of SiC; constant 3 wt.% of GR, 3 wt.% of MD, and the rest AA6061). The fabrication of the AA6061–SiC/Gr/MD hybrid alloy composites was done using a high vacuum furnace apparatus and developed composite procedure discussed later [23]. The pure aluminum alloy rods were cleaned and cut; the small parts were heated in the graphite vessel via using electric furnace equipment as per industry standard. The AA6061 alloy material was heated in induction at 790 °C for 30 minutes; thereafter, the alloy material temperature was slowly lowered to 650 °C. Impurities present in developed material were removed with the incorporation of flux through the slag. The different ceramic-based particulate reinforcement was initially heated individually in an induction oven such as muffle apparatus at 870 °C for 3 hours per industrial standard. To improve the wettability of the reinforcement phase with mixed of 2 wt.% Mg powder in the molten alloy material, and the uniform mix with Mg powder during the reinforcing phase of the molten alloy were obtained via steel stirrer at a speed of about 300 rpm for 20 minutes. The developed material was poured into a rigid cast iron vessel with size of 150 × 90 × 10 mm^3, and it was permitted it to solidify to environmental conditions in the air for 45 minutes. Testing specimens were cut using a WEDM apparatus and with selected dimensions of specimen as per ASTM standards; thereafter, a polishing process was performed on testing samples, and they were evaluated for physical, mechanical, and wear characterizations also.

5.2.2 Physical and Mechanical Characterizations

Testing specimens for experiment density (ρ_e) were computed using immersion techniques following the approach of Archimedes as per 792-ASTMD [24], while the specimens for theoretical density (ρ_t) were measured via using the approach of the rule of mixture (Agarwal and Broutman) by using Equation 5.1. The test specimens of void fraction (V) or porosity was achieved through Equation 5.2 [25].

$$\rho_t = \frac{1}{\frac{W_{Al7075}}{\rho_{Al7075}} + \frac{W_{SiC}}{\rho_{SiC}} + \frac{W_{Gr}}{\rho_{Gr}} + \frac{W_{Cu}}{\rho_{Cu}}}, \tag{5.1}$$

where W and ρ depicted as the weight of the specimen and testing specimens of density composite, respectively.

$$V = \frac{\rho_t - \rho_e}{\rho_t} \tag{5.2}$$

The testing specimens of tensile strength (sample size of 150 × 10 × 10 mm³; span length of specimen = 65 mm as per D3039–76 ASTM) were computed via the Universal Testing approach (UTM) of alloy composites. UTM was used to calculate the flexural strength of the alloy composite (specimen size = 127 × 12.5 × 4 mm³, specimen of span length = 70 mm, D2344–84-ASTM). The toughness of the specimens (size of 55 × 10 × 10 mm³, 45° of the notch, depth of 2.5 mm, as per D256-ASTM) was computed using the impact tester apparatus (Charpy V-notch approach). The Vickers hardness of the specimens (size of 10 * 10 * 75 mm³) as per ASTM-E18 was computed through using the apparatus of hardness tester and the strength of the compressive samples (size of 10 * 10 * 10 mm³, as per ASTM E 9–09) was computed by using UTM equipment [26].

5.2.3 Sliding Wear Analysis

A wear (pin-on-disk) apparatus (Ducom; Model TR-20; G-99 ASTM; Bangalore, India) was used for the sliding wear characterization. The contributing factors and their levels are represented in Table 5.2 and Figure 5.3. The machine parameters (pin-on-disk apparatus) are taken as (specimen size = 25 × 10 × 10 mm³; EN-31 of sliding disk material, disk dimension as (8 mm of thickness, diameter of 165 mm, hardness of 62 HRC, surface roughness of 0.6 µRa). The diameter of the disk was taken as 55 mm in the research. The disc surface area was cleaned using acetone and abrasive paper before any experiment run. The electronic balance (precision ± 0.001 mg) is utilized to compute the reduction of mass of specimen earlier and later the experiment run. The software was used for data acquisition [11, 25, 26]. The specific wear rate (mm³/N-m) of testing samples might be measured using Equation 5.3:

$$W_s = \frac{\Delta m}{\rho \times V_s \times t \times F}, \tag{5.3}$$

where Δm depicts the mass loss of samples, ρ depicts the density testing sample (g/mm³), t shows the time of the experiment(s), V_s refers to the sliding velocity of the sample (m/s), and F denotes load of samples [8].

5.2.4 Taguchi Design of Experimental Approach

The optimization techniques of a Taguchi design is an important statistical tool frequently used by authors in the experimental work [26, 7]. This approach develops a clear knowing of the experimental studies. It helps identify the contribution control variables particular to the test, such as the dry sliding wear test simulation; these parameters are sliding distance, sliding velocity, normal load, and environmental temperature, among others, and their suitable levels along with uncontrolled sound parameters like humidity and environmental temperature. These may affect the experimental response and uncontrolled factors. It aids in the estimation of the response with minimal test trials. It ranks the involvement of the contribution factors in observing the output response. It might be validated via using the analysis of variance (ANOVA) techniques. It estimates the optimal combination finding optimal response and compares the model through running confirmation experiment test runs. The concept of an orthogonal array is widely used aspect of experimental permutations and combination for the outcomes of robust design. Table 5.1 shows the input control parameters and their levels as finalized for this experimental work.

Furthermore, the L25 orthogonal array was utilized for the design of experiment (DOE) in place of $5^4 = 25 \times 25$ runs using a full factorial design. The investigated specific wear rate and the evaluated coefficient of friction (COF) value are transformed into an S/N ratio using Equation 5.4. Here, the S/N ratio for the lowest specific wear rate selected by using the Smaller to Better approach. The SB characteristic equation is

$$\frac{S}{N} = -10\log\frac{1}{F}\sum D^2, \quad (5.4)$$

where F depicts the number of observations and D denotes the observed data.

TABLE 5.1
Working Range of Selected Parameters

Control Factor	I	II	III	IV	V	Unit
A: Normal load	10	20	20	30	40	N
B: Filler content	AS-0	AS-1	AS-3	AS-5	AS-7	wt.%
C: Sliding velocity	0.6	0.8	0.1	1.2	1.4	m/s
D: Sliding distance	500	750	1000	1250	1500	m

5.2.5 Methodology of Preference Index Method

Preference index method is a theoretical systematic approach for designers to evaluate the most suitable material for particular application. The performance selection index (PSI) method is described in the various steps [21, 22]:

Step 1: Problem Structure: It requires collection and organisation of data input to a PSI algorithm like possible alternatives, evaluating criteria, their implications on goal as depicted in Figure 5.1 [27–30].

Step 2: Formulate decision matrix: Any problem expressed the m-alternatives denotes multi-alternative and n-criteria indicates multi-criteria. $D_{m \times n}$ order matrix shows product of decision matrix.

$$D_{m \times n} = \begin{matrix} & \begin{matrix} C_1 & C_2 & \cdots & C_n \end{matrix} \\ \begin{matrix} A_1 \\ A_2 \\ \vdots \\ A_m \end{matrix} & \begin{bmatrix} p_{11} & p_{12} & \cdots & p_{1n} \\ p_{21} & p_{22} & \cdots & p_{2n} \\ \vdots & \vdots & \ddots & \vdots \\ p_{m1} & p_{m2} & \cdots & p_{mn} \end{bmatrix} \end{matrix}$$

where $C_1, C_2, \ldots, C_n$ are the n-criteria and $A_1, A_2, \ldots, A_m$ are the m-alternatives

The element p_{ij} is performance value of ith alternative and jth attribute (C_j), where $i = 1, 2, \ldots, m$ and $j = 1, 2, \ldots, n$.

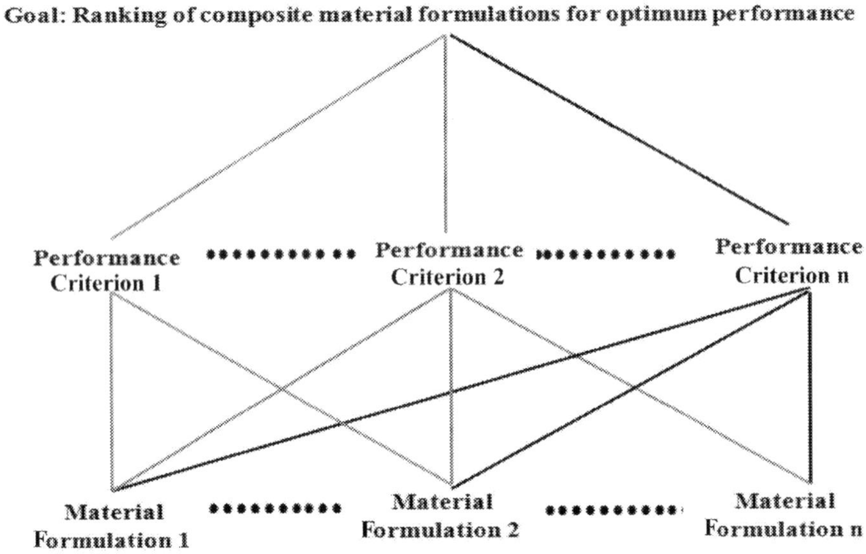

FIGURE 5.1 Hierarchy Structure of the Problem

Step 3: Normalization: Normalization is a method of transforming the decisive data range of 0 to 1. Hence, find normalized matrix (R_{ij}): If the expectancy is larger the better and *smaller the better*, then the performance of the original attribute can be normalized using Equations 5.5 and 5.6:

$$Rij = \frac{x_{ij}}{x_j^{max}} \tag{5.5}$$

$$Rij = \frac{x_j^{min}}{x_{ij}}, \tag{5.6}$$

where x_{ij} are the attribute measures ($i = 1, 2, 3, \ldots, m$ and $j = 1, 2, 3, \ldots, n$).

Step 4: Evaluate PV_j value: To find out R_{ij}, used to evaluate PV_j (preference variation value) for every criterion with the idea of the sample variance analogy via Equation 5.7:

$$PVj = \sum_{i=1}^{N}\left(R_{ij} - \bar{R}_j\right)^2, \tag{5.7}$$

where $\bar{R}_j$ denotes mean criterion j; that is, $\bar{R}_j = \frac{1}{n}\sum_{i=1}^{N} R_{ij}$.

Step 5: Evaluate (ψ_j): To compute ψ_j (overall preference value) for each criterion using Equation 5.8:

$$\Psi j = \frac{\Phi_j}{\sum_{j=1}^{n}\Phi_j}, \tag{5.8}$$

where $\Phi_j = 1 - \Sigma PV_j$ represents the preference value of deviation for each criteria. For consistency, $\Sigma \psi_j = 1$.

Step 6: Evaluate of (I_i): The PSI (I_i) for each alternative is computed using Equation 5.9:

$$I_i = \sum_{j=1}^{n}\left(R_{ij} \times \Psi_j\right). \tag{5.9}$$

Step 7: The computed values of preference selection index (I_i): The computed values of preference selection index (I_i) are used to rank the alternatives in descending order; that is, the highest to lowest I_i values are used to rank the alternatives from first to last and then making related interpretations or recommendation.

5.3 RESULTS AND DISCUSSIONS

5.3.1 Physical and Mechanical Analysis

The influence of the physical characterization of AA6061/SiC/Gr/MD alloy composites is depicted as Figure 5.2 It was observed that the experimental density decreased with a rise in the SiC content as compared to the theoretical density of the composite. It may be attributed to the reduction in experimental density and an increase in SiC owing to the casting defect of developed composite. Similar findings are obtained by different authors/researchers [7, 8, 24]. Therefore, the theoretical density is greater than the actual/experimental density of the composite. The experiment/actual density trend is AS-7 > AS-5 > AS-3 > AS-1 > AS-0. The voids/porosity content of AA6061/SiC/Gr/MD alloy composites is indicated in Figure 5.2. It was noticed that the voids/porosity content reduces with enhance in SiC content. Thus, the value of void content reduces in the range of 0.74% to 0.73% for AS-0 to AS-3. The reduction in voids content because of the suitable particle dispersion and distribution into matrix while the voids/porosity content for AS-5 and AS-7 alloy composite was enhanced to 1.44% and 1.79%, respectively. Generally, the void content value is accepted if lower than 10% for industrial standards. It may be that the greatest porosity leads to poor interfacial bonding strength between SiC and alloy matrix. Improvement in hardness of composite due to decrease in the value of voids content /porosity. The voids content order is AS-7 > AS-3 > AS-0 > AS-1 > AS-5 [7, 24].

The influences of SiC on the hardness of AS based composite are shown the Figure 5.2. The hardness of composite rises with an increase in SiC. The hardness of AS-7 alloy composite reached a higher value of 79 HV while it was 65 HV for

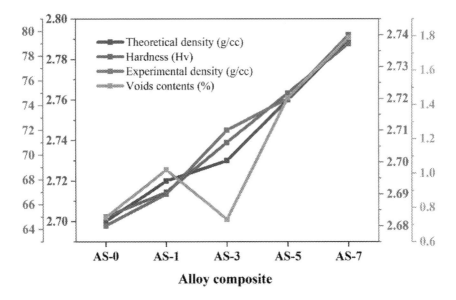

FIGURE 5.2 Physical Characterization of the Alloy Composites

AS-0. It was observed that the hardness of the AS alloy composite is 50 HV for AS-0 (Table 5.3). On the addition of 1 wt.% SiC-reinforced with alloy composites increases in hardness (67 HV). Furthermore, increments of 3, 5, and 7 wt.% SiC/AA6061 composite shows continuously growth in the hardness of the composite. Same observations are reported by many authors [11]. Thus, the trend of hardness is AS-0 < AS-1< AS-3 < AS-5 < AS-7. The rise in hardness may be attributed to the modification in plastic deformation and ceramic particles present in the alloy composite. It may be another reason for the increase in hardness of composite due to the important phenomenon of porosity is absent in current fabrication process. Similar trends were observed by different researchers [7, 23] that the adding different fillers (like $SiC/B_4C/AlB_2$) increases the hardness. Hence, hardness improves because of the mechanism influence existence in reinforcement into base matrix. The influence of mechanical characterization as (tensile, compressive, flexural and impact strength) of SiC particulate–filled aluminum alloy composites are illustrated in Figure 5.3. It was observed the tensile/compressive/impact strength of composites increase with an increase in SiC content while flexural strength diminishes with an increase in SiC. It was noticed the tensile strength shows the higher for 7 wt.% SiC (440 MPa) and it was 292 MPa for AS-0. On addition of 1 wt.%, 3 wt.% and 5 wt.% AS of tensile strength is 335 MPa, 398 MPa and 435 MPa. An increase in tensile strength of composites might be the massy interfacial bonding between aluminum alloy matrix and SiC. Nonappearance of pores in designed composites may also a reason for improved in tensile strength of alloy composites. For the reason of enhance in tensile strength; addition of various particulate like SiC, B_4C, and Gr, among others, are reported by various materialists/researchers. It is concluded the high dislocation density, thermal

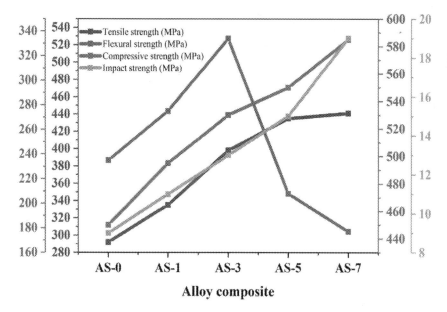

FIGURE 5.3 Mechanical Characterization of the Alloy Composites

expansion, voids and cracks increase with an increase in tensile strength of alloy composite [23, 24, 11].

The influence of compressive strength of SiC particulate based alloy composites is depicted in Figure 5.3. The compressive strength reduces with rise in SiC. Thus, it observed the compressive strength trend as AS-7 > AS-5 > AS-3 > AS-1 > AS-0. Similar results are reported the strength increase due to mechanism of composite [11, 7]. The influence of SiCon the impact strength of composites is displayed in Figure 5.3. It was noticed that the impact strength rises with a growth in SiC content. The similar results are described by Ravi Kumar et al. [23] for SiC-filled Al 7075 composites. Thus, impact strength trend as AS-0 < AS-1 < AS-3 < AS-5 < AS-7. The effect of flexural strength on SiC/Gr/MD-filled AA6061 composites. Flexural strength was increased (235 MPa to 334 MPa) with an increase in SiC while it was abruptly decreased (334 MPa to 177 MPa) for composite. The same trends are reported by different scientists [23, 11].

The flexural strength trend is AS-7 < AS-5 < AS-0 < AS-1 < AS-2. Therefore, because of poor attachment and improper distribution in the casting process, the effect of flexural strength can decrease. The flexural strength reduction may be attributed to an increase cracks, propagation, and pores in base [8].

5.3.2 Wear Studies of Alloy Composite

5.3.2.1 Influence of Specific Wear Rate/Coefficient of Friction on Sliding Distance of Composite

The effect of the wear rate on various sliding distances (500–1500 m) of SiC/Gr/MD-filled 6061 alloy composite is presented in Figure 5.4(a). The wear rate enhances with increase in sliding distance for all respective composites. The wear rate order is AS-0 > AS-3 > AS-5 > AS-7. The wear rate of SiC-reinforced-based alloy composite is lower for AS-7, and the wear rate is higher for AS-0. It is found the wear rate illustrates the lesser value of the lesser sliding distance (500 m), but beyond the increased 1250- to 1500-m sliding distance, specific wear was increased. It might be that the wear rate improved with an enhancement in the sliding distance due to the lower section of sliding disk being stable; then the particles were fractured, and the worn debris particles are generated through the counter surface. The rise in sliding distance is generally owing to the increase in area of sliding disk surface and raised in the specific wear rate [24]. The impact of sliding distance (500–1500 m) on the COF of SiC reinforced with AA6061 composites are presented in Figure 5.4b. The COF improves with addition of SiC(0–7 wt.%) content. The coefficient of friction of reinforced based composites shows higher for AS-0 and lower value of COF is 0.16 for AS-12. However, a reduction in the wear rate of an AS alloy composite is generally due to micrograph structure, specific size and distribution of the grain shape. It is a major support to improve the tribological properties of various applications. The presence of SiC reinforced an increase with an increase in COF of the composite. Hence, it may be attributed that the COF increases due to the grain growing mechanism and hardness increase in wear resistance of the composite [7, 26].

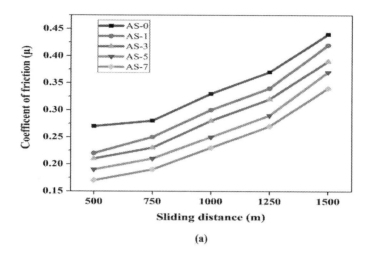

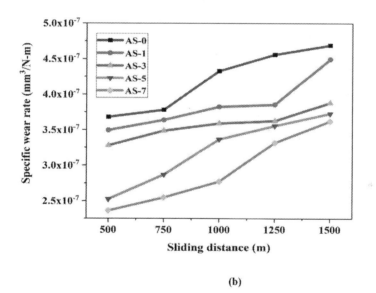

FIGURE 5.4 Variation of Specific Wear Rate/Coefficient of Friction of Alloy Composites with Sliding Distance

5.3.3 Taguchi and ANOVA Analysis of Alloy Composite

The Taguchi method is utilized an optimal tool for evaluating robustness and S/N ratio shows valuable factor in parametric design. This methodology introduces the Signal, which depicts the need target, that is, wear rate, and Noise which depicts the undesirable value. S/N proportion test outcomes are depicted in Tables 5.2 and 5.3. The whole average means the S/N proportion test outcomes

TABLE 5.2
Experimental Results (L25 Orthogonal Array) of SWR and S/N ratio of AS Alloy Composite

Level	Normal Load (N)	Filler wt.%	Sliding Distance (m)	Sliding Velocity (m/s)	SWR (mm³/Nm)	S/N Ratio (dB)
1	10	0	250	0.6	0.000293688	70.642
2	10	1	500	0.8	0.000295714	70.583
3	10	3	750	0.1	0.000150351	76.458
4	10	5	1000	1.2	0.000219059	73.189
5	10	7	1250	1.4	0.000122676	78.225
6	20	0	500	0.1	8.45627E-05	81.456
7	20	1	750	1.2	0.000200476	73.959
8	20	3	1000	1.4	8.35673E-05	81.559
9	20	5	1250	0.6	6.57491E-05	83.642
10	20	7	250	0.8	0.000677277	63.385
11	30	0	750	1.4	0.000175485	75.115
12	30	1	1000	0.6	9.92857E-05	80.062
13	30	3	1250	0.8	1.50351E-05	96.458
14	30	5	250	0.1	7.96864E-05	81.972
15	30	7	500	1.2	4.52113E-05	86.895
16	40	0	1000	0.8	1.51844E-05	96.372
17	40	1	1250	0.1	2.48571E-05	92.091
18	40	3	250	1.2	0.000132281	77.570
19	40	5	500	1.4	0.000205122	73.760
20	40	7	750	0.6	0.000132676	77.544
21	50	0	1250	1.2	8.45627E-06	101.456
22	50	1	250	1.4	5.76191E-05	84.789
23	50	3	500	0.6	8.35672E-05	81.559
24	50	5	750	0.8	0.000226802	72.887
25	50	7	1000	0.1	0.000137371	77.2421

TABLE 5.3
Response Data for S/N (Smaller Is Better)

Level	NL (N)	FC (wt.%)	SD (m)	SV (m/s)
1	73.82	85.01	75.67	81.84
2	76.80	80.30	78.85	78.69
3	84.10	82.72	75.19	79.94
4	83.47	77.09	81.68	82.61
5	83.59	76.66	90.37	78.69
Delta	10.28	8.35	15.18	3.92
Rank	2	3	1	4

of composite is 80.35 dB, respectively. The Taguchi-based test outcomes are analyzed via MINITAB 17 software. First, any test outcomes are utilized in this mathematical model as estimates for the examination of the model's performance; the proper interaction between the level and the factors should be considered. Thus, the interaction effect is computed via using a model of factorial parametric design. Impact of control factors on wear rate are depicted in Figure 5.5. The significant trend of control factors order is sliding distance (SD) > normal load (NL) > reinforcement (R) > sliding velocity (V) [8, 31]. The experimental results (L25 orthogonal array) of specific wear rate (SWR) and S/N ratio of AS alloy composite are listed in Table 5.2.

The response data for S/N (smaller is better) are listed in Table 5.3.

The ANOVA experimental results are performed for Taguchi L25 DOE on SiC/MD/Gr-filled Al6061 alloy composites and are illustrated in Table 5.4. The ANOVA test outcomes help measure the percentage of contribution for each control factor,

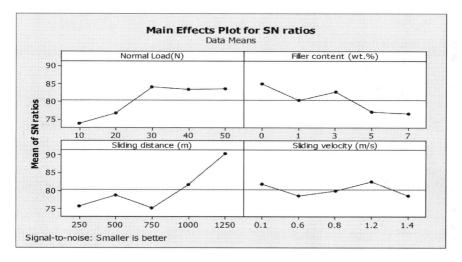

FIGURE 5.5 Variation of S/N Ratios of Input Control Factors

TABLE 5.4
ANOVA Experimental Results of Alloy Composites

Source	DF	Adj SS	Adj MS	F-Value	P-Value	P %
Normal load (N)	4	496.63	125.41	2.47	0.137	28.92
Filler wt.%	4	194.16	49.04	0.96	0.497	9
Sliding distance (m)	4	742.83	188.21	3.73	0.071	37.39
Sliding velocity (m/s)	4	68.78	18.45	0.39	0.859	4.64
Error	8	420.46	55.43			20.05
Total	24	1922.86				100

that is, specific wear rate. P-test illustrates a percentage contribution in production for a single factor. The P data–based performance of parameter order is sliding distance (P = 0.071), normal load (P = 0.137), filler content (P = 0.497), and sliding velocity (P = 0.859), and the optimized orders are evaluated by using the ANOVA techniques. The ANOVA experimental results of the alloy composites are listed in Table 5.4.

5.3.4 Ranking Analysis of AS Alloy Composite Using PSI Approach

The outcome of various performance defining attributes shows that Si/Gr/MD reinforcement in the AA6061 alloy causes more complex performance characterization variations (Table 5.5).

In PSI methodology, the alternatives the performance criteria's (PCs) are organized in the form of decision matrix via Equation 5.3. The stepwise computations follow:

Step 1: The evaluated performance data are organized in decision matrix and are represented in Table 5.6.
Steps 2–5 Table 5.7 shows the preference variation value (PVj) of alloy composites.
Steps 6–7: Table 5.8 depicts the overall preference index (Ij) of alloy composites. The overall preference index (Ij) of the alloy composites is listed in Table 5.8.

However, the decision matrix is normalized for profitable and non-profitable PCs using Equation 5.4. Furthermore, the normalization preference variation (Rij) computed via Equation 5.5 and Equation 5.7 are used to compute the deviation in preference variation (PVj). The values from Equation 5.8 (ψ_j) for all PCs are depicted in Table 5.7. Experimental optimization results are shown in Table 5.8. It was noticed that the PSI of the composition/alternative PCs-6 highest (0.1287),

TABLE 5.5
Details of Performance Criteria/Alternatives of Alloy Composites

S. No.	Criteria/ Alternatives Performance	Preference
1.	TS (MPa)	Higher
2.	FS (MPa)	Higher
3.	IM (J)	Higher
4.	Hardness (HV)	Higher
5.	CS (MPa)	Higher
6.	D (g/cc)	Lower
7.	VD (%)	Lower
8.	SWR (mm³/Nm)	Lower
9.	COF	Lower

TABLE 5.6
Decision Matrix of Composites

Criteria	Material Alternatives—>	Composition (wt.%)				
		AS-0	AS-1	AS-3	AS-5	AS-7
PC-1	TS (MPa)	292	355	398	435	440
PC-2	FS (MPa)	235	275	334	208	177
PC-3	IM (J)	9	11	13	15	19
PC-4	HRD (HV)	65	67	71	75	79
PC-5	Compression strength (Mpa)	450	495	530	550	585
PC-6	Experimental density (g/cc)	2.68	2.69	2.71	2.73	2.74
PC-7	Void content (%)	0.74	0.73	1.09	1.44	1.79
PC-8	Specific wear rate (mm^3/Nm)	0.005	0.0015	0.0019	0.0027	0.0030
PC-9	Coefficient of friction	0.015	0.023	0.027	0.028	0.037

TABLE 5.7
The Preference Variation Value (PVj) of the Alloy Composites

	Material Alternatives—>	0%	1%	3%	5%	7%	SUM(PVj)	Φj	Ψj
PCs	Property								
PC-1	TS (MPa)	0.05	0.09	0.00	0.01	0.03	0.0846	0.9264	0.1177
PC-2	FS (MPa)	0.00	0.01	0.07	0.01	0.05	0.1253	0.8877	0.1147
PC-3	IS (J)	0.04	0.02	0.00	0.00	0.08	0.1398	0.8802	0.1137
PC-4	HRD (HV)	0.01	0.01	0.00	0.01	0.02	0.0532	0.9588	0.1239
PC-5	Compression strength (Mpa)	0.00	0.00	0.00	0.00	0.00	0.0143	0.9787	0.1270
PC-6	Experimental density (g/cc)	0.00	0.00	0.00	0.00	0.00	0.0005	0.9896	0.1285
PC-7	Void content (%)	0.01	0.06	0.00	0.31	0.07	0.4755	0.5445	0.0682
PC-8	Specific wear rate (mm^3/Nm)	0.00	0.10	0.00	0.02	0.06	0.1648	0.8452	0.1072
PC-9	Coefficient of friction	0.18	0.00	0.01	0.02	0.06	0.2594	0.7606	0.0962

TABLE 5.8
The Overall Preference Index (Ij) of the Alloy Composites

	Material Alternatives—>	0%	1%	3%	5%	7%
PCs	Property					
PC-1	TS (MPa)	0.0788	0.0921	0.1068	0.1175	0.118
PC-2	FS (MPa)	0.0828	0.0931	0.1138	0.0725	0.062
PC-3	IS (J)	0.0602	0.0677	0.0752	0.0827	0.112
PC-4	Hardness (HV)	0.0905	0.0942	0.1031	0.1157	0.122
PC-5	CS (MPa)	0.1157	0.1108	0.1196	0.1245	0.128
PC-6	Experimental density (g/cc)	0.1287	0.1298	0.1277	0.1267	0.127
PC-7	Void content (%)	0.0226	0.0114	0.0337	0.0696	0.013
PC-8	SWR (mm^3/Nm)	0.0707	0.1085	0.0782	0.0617	0.050
PC-9	Coefficient of friction	0.0975	0.0604	0.0487	0.0437	0.035
Overall Preference Index (Ij)		**0.85**	**0.79**	**0.82**	**0.84**	**0.87**
	Preference Ranking	5	4	3	2	1

which is followed by PCs-5 (0.1282). Alternatively, AS composite PCs-7 represents the smallest preferred performance with a PSI value of 0.0114. Thus, it observed that the ranking of fabricate AS is AS-7 > AS-5 > AS-3 > AS-1 > AS-0. The ψ_j values were measured via Equations 5.8 and 5.9, and these values are improved using the PSI method. At last, the entire preference selection index (I_i) of alternative is determined by using Equation 5.9. The AS-7 alternative values is higher compared to other alternatives. Thus, this alternative performance is selected by using the PSI approach [32, 33].

5.4 CONCLUSION

The significant finding from the discussed the influence of SiC/Gr/MD-reinforced-based 6061 aluminum alloy composites are as follows:

1. The high vacuum casting method is employed to manufacture the alloy composite with a fairly equally distribution of SiC.
2. The experimental density of alloy composite samples is set between 2.68 to 2.74 g/cc, while void content is set between 0.74 to 1.79%.
3. Increase in tensile strength, compression strength, hardness and impact strength with increase in particulate (SiC/Gr/MD) content while a reduction in flexural strength and void content were lower for the AS-1 alloy composite.
4. The PSI approach and Taguchi DOE might be used to forecast the specific wear rate of the alloy composite under examination.
5. The increase in sliding distance improvement the wear performance order is AS-7 > AS-5 > AS-3 > AS-1 > AS-0 at steady state experimental condition (sliding velocity = 0.6 m/s; load = 10 N).
6. The study reveals that the PSI approach should be aid in the optimal rotor material formulation selection without conducting a long and expensive experiment. The rank order of the composite is obtained by using the PSI approach.

ACKNOWLEDGMENTS

The authors thank to MNIT Jaipur, Advanced Research Lab for Tribology and Material Research Centre for giving the characterization facilities.

REFERENCES

1. I. Topcu, H.O. Gulsoy, N. Kadiogluand, and A. Gulluoglu, Processing and mechanical properties of B_4C reinforced Al matrix composites, J. Alloys Compd., 482, 516 (2009).
2. A.V. Smith, and D.D.L. Chung, Titanium diboride particle-reinforced aluminium with high wear resistance, J. Mater. Sci., 31, 59–61 (1996).
3. R. Jojith, and N. Radhika, Mechanical and tribological properties of $LM13/TiO_2/MoS_2$ hybrid metal matrix composite synthesized by stir casting, Part. Sci. Technol., 37(5), 570-582, (2019).

4. V.K. Sharma, V. Kumar, and R.S. Joshi, Investigation of rare earth particulate on Tribology and Mechanical properties of Al-6061 alloy composite for aerospace application, J. Mater. Res. Technol., 8(4), 3504 (2019).
5. D. Aruri, K. Adepu, K. Adepu, and K. Bazavada, Wear and mechanical properties of 6061–T6 aluminium alloy surface hybrid composite (SiC+Gr) and SiC+Al$_2$O$_3$) fabricated by friction stir processing, J. Mater. Res. Technol., 2(4) 362 (2013).
6. T.S. Kiran, M.P. Kumar, S. Basavarajappa, and B.M. Viswanatha, Dry sliding wear behavior of heat treated hybrid metal matrix composite using Taguchi techniques, Mater. Des, 63, 294 (2014).
7. A. Kumar, A. Patnaik, and I.K. Bhat, Investigation of nickel metal powder on tribological and mechanical properties of Al7075 alloy composites for gear material, J. Powder Metall., 60, 371 (2017).
8. A. Kumar, A. Patnaik, and I.K. Bhat, Tribology analysis of Cobalt particulate filled Al7075 alloy for Gear Materials: a comparative study, Silicon, 11, 1295 (2019).
9. W. Chen, Y. Liu, C. Yang, D. Zhu, and Y. Li, (SiCp+ Ti)/7075Al hybrid composites with high strength and large plasticity fabricated by squeeze casting, Mater. Sci. Eng. A, 609, 250 (2014).
10. S.B. Prabu, and L. Karunamoorthy, Microstructure-based finite element analysis of failure prediction in particle-reinforced metal–matrix composite, J.Mater.Proc. Tech., 207, 53 (2008).
11. A. Kumar, and M. Kumar, Mechanical and dry sliding wear behaviour of B$_4$C and rice husk ash reinforced Al 7075 alloy hybrid composite for armors application by using Taguchi techniques, J. Mater Today: Proceed., 27, 2617 (2020).
12. A. Kumar, V. Kukshal, and V. Rajashekhara Kiragi, Assessment of mechanical and sliding wear performance of Ni particulate filled 7075 aluminium alloy composite, J. Mater. Today: Proc., (2020) https://doi.org/10.1016/j.matpr.2020.10.556.
13. A. Kumar, M. Kumar, and B. Pandey, Investigations on mechanical and sliding wear performance of Aa7075 – Sic/marble dust/graphite hybrid alloy composites using hybrid ENTROPY-VIKOR method, Silicon, (2021) https://doi.org/10.1007/s12633-021-00996-7.
14. A. Kumar, V. Kumar, A. Kumar, B. Nahak, and R. Singh, Investigation of mechanical and tribological performance of marble dust 7075 aluminium alloy composites, J. Mater. Today: Proc., (2020) https://doi.org/10.1016/j.matpr.2020.10.812.
15. S. Bhaskar, M. Kumar, and A. Patnaik, Application of hybrid AHP-TOPSIS technique in analyzing material performance of silicon carbide ceramic particulate reinforced AA2024 alloy composite, Silicon, 12, 1075 (2019).
16. M. Kumar, and A. Kumar, Application of preference selection index method in performance based ranking of ceramic particulate (SiO$_2$/SiC) reinforced AA2024 composite materials, J. Mater. Today: Proc., 27, 2667 (2020).
17. B.K. Satapathy, A. Majumdar, and B.S. Tomar, Optimal design of fly ash filled composite friction materials using combined analytical hierarchy process and technique for order preference by similarity to ideal solutions approach, Mater. Des., 31, 1937 (2010).
18. N. Kumar, T. Singh, J.S. Grewal, A. Patnaik, and G. Fekete, A novel hybrid AHP-SAW approach for optimal selection of natural fiber reinforced non-asbestos organic brake friction composites, Mater. Res. Express., 6, 065 (2019).
19. V. Mahale, J. Bijwe, and S. Sinha, Application and comparative study of new optimization method for performance ranking of friction materials, Proc. IMEJ. J. Eng. Tribol., 232, 143 (2018).
20. B.K. Satapathy, and J. Bijwe, Performance of friction materials based on variation in nature of organic fibers Part II: optimization by balancing and ranking using multiple criteria decision model (MCDM), Wear, 257, 585 (2004).

21. K. Maniya, and M.G. Bhatt, A selection of material using a novel type decision–making method: preference selection index method, Mater. Des., 31, 1785–1789 (2010).
22. R. Chauhan, T. Singh, N.S. Thakur, and A. Patnaik, Optimization of parameters in solar thermal collector provided with impinging air jets based upon preference selection index method, Renew. Energy, 99, 118 (2016).
23. K.R. Kumar, K. Kiran, and V.S. Balaji, Micro structural characteristics and mechanical behaviour of aluminium matrix composites reinforced with titanium carbide, J. Alloys Compd., 723, 795 (2017).
24. S. Bhaskar, M. Kumar, and A. Patnaik, Silicon. Silicon carbide ceramic particulate reinforced AA2024 alloy composite – part I: evaluation of mechanical and sliding tribology performance, Silicon, 12, 843–865 (2019).
25. D792-13, Standard Test Methods for Density and Specific Gravity (Relative Density) of Plastics by Displacement, ASTM International, 100 Barr Harbor Drive, PO Box C700, West Conshohocken, PA 19428-2959. United States (2013).
26. M. Kumar, and A. Kumar, Sliding wear performance of graphite reinforced AA6061 alloy composites for rotor drum/disk application, J. Mater. Today: Proc., (2019) https://doi.org/10.1016/j.matpr.2019.09. 042.
27. M.P. Borujerri, and H. Gitinavard, Evaluating the sustainable mining contractor selection problems an imprecise last aggregation preference selection problem, J. Sustain. Mining, 16, 207 (2017).
28. V. Behnamv, M.S. Meysam, and S. Ebrahimnejad, Soft computing–based preference selection index method for human resource management, J. Intell. Fuzzy Syst., 26, 393 (2014).
29. A. Jahan, L. Kevin, and M.B. Edwards, Multi Criteria Decision Analysis for Supporting the Selection of Engineering Materials in Product Design, Butterworth-Heinemann, UK (2016) https://doi.org/10.1016/C2012-0-02834-7.
30. D.Petkovic, M. Madic, M. Radovanovic, and V. Gecevska, Application of the preference selection index method for solving marching MCDM problems, Facta. Univ. Ser. Mech. Eng., 15, 97 (2017).
31. C.S. Ramesh, R. Keshavamurthy, and B.H. Channabasappa, Friction and wear behavior of Ni–P coated Si_3N_4 reinforced Al6061 composites, Tribol. Int., 43, 623 (2010).
32. J. Khorshidi, and R. Hassani, A comparative analysis between TOPSIS and PSI method of material selection to achieve a desirable combination of strength and workability in Al/SiC composite, Mater. Des., 52, 999 (2013).
33. M. Panahi, and H. Gitinavard, Evaluating the sustainable mining contractor selection problems: an imprecise last aggregation preference selection method, J. Sustain. Min., 16, 207 (2018).

6 Fabrication and Characterization of Metallic Biomaterials in Medical Applications
A Review

Ganesh Kumar Sharma, Vikas Kukshal,
Deepika Shekhawat, and Amar Patnaik

CONTENTS

6.1 Introduction..95
6.2 Classification of Biomaterials..97
    6.2.1 Metallic-Based Biomaterial 97
    6.2.2 Polymeric-Based Biomaterial..98
    6.2.3 Ceramic-Based Biomaterial...98
    6.2.4 Composite-Based Biomaterial 98
6.3 Fabrication of Biomaterials 99
    6.3.1 Cellulose 99
    6.3.2 Chitin 99
    6.3.3 Chitosan 99
    6.3.4 Alginate..100
6.4 Properties of Biomaterials 100
6.5 Corrosion Resistance 101
6.6 Tribological Characteristics of Biomaterials 101
Conclusion 102
References..103

6.1 INTRODUCTION

In 1960, the first generation of biomaterials was designed and fabricated of regular materials used in biomedical implant. Biomaterial is specifically designed using the classes of purely synthetic materials as metallic biodegradable polymers, nonbiodegradable polymers, and composite-, ceramic- and protein-based natural biomaterials for medical applications [1]. Biomaterial has specifically been described by the Clemson University Advisory Board for biomaterials as "a[n] inequitable and pharmaceutically inert substrate originally developed for transplantation inside or implementation into biological organism" [2]. It is also known as "a non-viable substance used to comprise with biological processes in a medical product" [3]. Metallic biomaterials have the excellent ability to be able to withstand tensile stresses, which can be exceptionally high and even complex in the situation of steel alloys. However, metallic biomaterial is often used for unloaded, exclusive devices such as heart valves, pacemakers, coils, wires and joints. Hip- and knee-end bionic implants, pipes, screws, nails, dental implants, and the like are common examples of such heavily loaded prosthetics. The hip joint allows a wide range of motions that includes a flexion range of 0° to 130°, an extension range of 0° to 30°, an abduction range of 0° to 45°, an abduction range of 0° to 30°, an internal rotation of 30°, and an external rotation of 40°. After implant placement, the durability of the implants is determined by their fracturing and degradation after an essential stage of infection. Furthermore, to use the implants effectively, the mechanical behavior of structural biomaterials in a living body environment, such as fatigue, strength, corrosion, and wear resistance, need to be tested and then enhanced. A metallurgical principle guides the understanding the structure–property relationships and indicates judgments about the implant design. Along with mechanical properties, surface texture properties, such as how the biomolecules will attach to metallic implant surfaces, promoting certain interfacial activities and cellular reactions, also impact the biological performance of the implants. Metallic biomaterial employed with stainless steel, titanium and magnesium alloys has good biodegradability and aesthetic properties. The key criteria that metallic biomaterials must comply with are biocompatibility, biofunctionality, corrosion resistance, ease of application, and sustainability [4]. In different types, metallic biomaterials are used to compensate for weakened structural elements or to recover missing functionality inside the body [5].

Metals are distinguished by greater electroconductivity compared to polymer and ceramic biomaterials and, as such, have been used to situate electrodes in synthetic electrical organs [6]. A first metallic alloy specially developed for biological applications, such as plates and screws, was vanadium steel. This chapter focuses on a fabrication technique that includes casting, forging and machining metallic biomaterials and its characterization processes such as microstructural characteristics, surface characteristics, mechanical behavior and biological properties, corrosion resistance, and tribological characterization. The methodology adopted in the present work is shown in Figure 6.1.

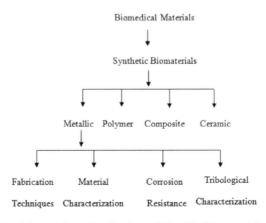

FIGURE 6.1 Methodology Adopted as Review of Metallic Biomaterials Characterization

6.2 CLASSIFICATION OF BIOMATERIALS

Biomaterials are classified into two groups: synthetic biomaterial or natural biomaterial and biological biomaterial. Synthetic biomaterial can be classified as metallic, polymer, ceramic, and composite. Using a wide range of different processing methods, synthetic biomaterial can be prepared. There are many fabrication techniques and processing procedures for the development of synthetic biomaterials. It is possible to describe the output of materials in several aspects. Biomaterial is used in a healthy, secure, economical, and physiologically appropriate manner to make devices to replicate a portion or a feature of the body [7]. Table 6.1 explains the need for biomaterials in various areas of the medical industry, and Table 6.2 defines the need for biomaterials in various areas of the body and the removal of multiple portions with living organisms.

6.2.1 METALLIC-BASED BIOMATERIAL

The application of metal-based materials for medical implants was started in 19th century during the Industrial Revolution. Metallic biomaterials have been identified a wide variety of applications as medical devices, with stainless steel, chromium alloys, titanium alloys, and aluminum alloys due to their attractive mechanical properties, particularly fracture and fatigue strength.

TABLE 6.1
Biomaterial Needs in Various Areas of the Medical Industry [8, 9]

Problem State	Examples
Repair of diseased or defective components	Joint replacements, cancer treatment unit
Aid to repair	Dental implants, stitches and staples
Enhance purpose	Pacemaker cells activity, intracranial lens
Right operational irregularity	Cardiac tissue
Right aesthetic defect	Mammoplasty augmentation
Assist in diagnostic	Samples and cannulas
Assist in rehabilitation	Stents, sinks

TABLE 6.2
Biomaterials in Various Areas of the Body [10, 11]

Organ Systems	Examples
Heart	Cardiovascular pacemaker, artificial valve
Lungs	Oxygenator device
Eye	Contact lens, intraoral lens
Ear	Synthetic staple, occipital cortex implant
Bone	Plate of bone, intra-medullar rod
Kidney	Endoscopes, fistula, dialysis machine

6.2.2 POLYMERIC-BASED BIOMATERIAL

Polymers are essential therapeutic materials and are used as instruments to substitute and regenerate different connective tissues in biomedical science. In the composition and structure of contingent macromolecules, the properties of polymers are described. In addition, versatility involves the production of polymers prepared for specific purposes in different mechanisms and for formulations in a variety of applications.

6.2.3 CERAMIC-BASED BIOMATERIAL

Ceramic materials have long been a part of everyday life, and porcelain, refractory products, cement, and glass are traditional ceramics. Ceramics are also used extensively for the application of biomaterials [12, 13]. The use of ceramics was affected by body inertia, the formability of assays in a variety of materials, high porosity, high strength, and high corrosion properties.

6.2.4 COMPOSITE-BASED BIOMATERIAL

Composites are composed of two or more different material components on a microscopic or macroscopic scale with distinct mechanical properties. However, regarding biomaterials, an individual part of the composite is bioactive, and the interfaces between the components are not affected by the environment of the body.

Biomaterials are generally categorized on the basis of the composition of the substances, as shown in Table 6.3.

TABLE 6.3
Classification of Biomaterials with the Advantages, Disadvantages, and Their Applications

Types of Biomaterials	Advantages	Disadvantages	Application	Reference
Metallic and metal alloy (stainless steel, titanium, Cr-Co alloys, Ti alloy)	High strength of substance, fabrication easy and sterilizing	Destructive Loosening aseptic Excessive modulus of elasticity	Implants of orthopedic origin Clips, screws, and plates	[14]
Polymer (PMMA, nylon polyethylene, polypropylene)	Biodegradable Bioactive Simply moldable and easily usable	Body fluid leachable Hard for sterilization	Orthopedic and dental fluids Implantations Prostheses Tissue engineering Scaffolding	[15]
Ceramics (aluminum oxides, calcium phosphates, titanium oxides, carbon)	Strength of material is high Biocompatibility is high Corrosion resistance	Hard to mold Excessive modulus of elastics	Orthopedic bioactive implants Implants from dentistry Hearing aids that are artificial	[16]
Composites (ceramic coatings with metals, carbon coated material)	Outstanding mechanical properties, corrosive resistance	Expensive cost Laborious methods of fabrication	Porous implants for orthopedics Fillings of dentistry Catheters and gloves in rubber	[17]

6.3 FABRICATION OF BIOMATERIALS

Biomaterials have also played a main role in the transport of cells, biological activities, medicines, and cell proliferation. Natural biomaterials are comprehensive requirements of tissue engineering. To resolve the issue of repeatability, details of the study fabrication techniques of biomaterials should be properly explained. Biomaterials are dependent on cellulose, alginate, collagen, chitin, dextran, chitosan and fibrinogen, silk, extracellular decellularized matrix (dECM) and glycosaminoglycan. Natural extracellular matrix (ECM) elements are imitations of dECM. The general studies on biomaterial development techniques will show a close review of the existing literature; furthermore, specific important technical information is often overlooked and not published. ECM plays an essential part in the regeneration of tissues, such as providing cells with physical support and controlling cellular function, involving development, proliferation, relocation, immune function, and morphology, allowing the human body to rebuild specific cell, tissue and organ defects [18, 19]. Biological cross-linking of biomaterials and their biocompatibility with fabrication techniques are shown in Table 6.4.

6.3.1 CELLULOSE

Cellulose biomaterials are systematic tissue engineering applications. The key problem associated with cellulose biomaterials, however, is their reduced rate of adsorption because the cellulose did not fully degrade subcutaneous in rats [20].

6.3.2 CHITIN

Chitin is really the second-most commonly used in biopolymer and tissue engineering. Chitin nanoparticles are being used to promote cellular function in tissue engineering, including cell growth and permeability [21].

6.3.3 CHITOSAN

Chitosan is biodegradable, biocompatible and has an equivalent glycosaminoglycan (GAG) structure. Chitosan was used by blending to improve the mechanical properties and biostability of natural biomaterials, as well as to stimulate antiseptic activity of biomaterials [22].

TABLE 6.4
Biological Cross-Linking of Biomaterials and Their Biocompatibility with Fabrication Techniques

Component of Biomaterials	Treatment and Agents for Cross-Linking	Techniques of Fabrication	Biocompatibility	References
Microfiber/gelatin cellulose	NHS/EDAC	Dryer with freeze	In vitro nontoxic	[24]
PLA/chitosan	Glutaraldehyde medication	Elektro spinning	In vitro nontoxic	[25]
Chitin/chitosan/alginate/fucoidan	Diglycidyl ether ethylene glycol	Hydrogel	In vivo nontoxic	[26]
Alginate	NHS/EDAC	Dryer with freeze	–	[27]
PLGA/Dextran	Foto cross-link	Elektro spinning	In vitro nontoxic	[28]
Fibroin/gelatin of silk	About Genipin	Printing in 3D	In vitro/vivo nontoxic	[29]
Glycosaminoglycan-collagen	At 105 °C DHT	Dryer with freeze	In vitro nontoxic	[30]

6.3.4 ALGINATE

Alginate is naturally bioactive, environmentally friendly, and electrophonic. Porous alginate scaffolds were produced and tested in vitro. Alginate materials possess long-term support for the cell structure of fibroblasts attached and expanding on alginate scaffolds in round anatomy [23].

6.4 PROPERTIES OF BIOMATERIALS

The physical properties of biomaterial are the most significant requirement regarding the living organism in which it would be implanted. Biomaterials and equipment must ideally have acceptable mechanical and performance specifications that are similar to those of implant replacement. The mechanical characteristics of the designated biomaterials are demonstrated by the mechanical efficiency of the biomaterial as a medical device designed to operate in the environment of the living body. Durability implies the minimum amount of time over which a biomaterial successfully conducts its stated function. To fulfill their intended purpose, biomaterials should have unique physical properties. The physical and mechanical characteristics of the biomaterial are classified in three terms as defined as shown in Table 6.5.

TABLE 6.5
Mechanical and Physical Characteristics of Biomaterials

Mechanical Effectiveness

Biomaterials	Mechanical Properties	Reference
Prosthesis	Hard, strong	[31]
Collagen material	Effective and reliable	[32]
Aortic valve bulletin	Versatile and strong	[33]
Articular spinal cord	Sensitive and thermoplastic elastomers	[34]
Chemotherapy membranes	Solid, versatile, and non-elastomeric	[35]

Mechanical Operability

Biomaterial	Mechanical Duration	Reference
Cannula	3 days	[36]
Bone plates	6 months or 1 year	[37]
Heart valve leaflet	60 times in 1 minute	[38]
Hip joint	For more than 10 years, it must work under heavy loads	[39]

Physical Properties

Biomaterial	Physical Characteristic	Reference
Dialysis tubing	Permeation	[40]
Vertebral cups of the joint replacement	High ductility High lubricity	[41]
Intra-ocular lens	Clarification and diffraction	[42]

6.5 CORROSION RESISTANCE

Corrosion is among the main factors influencing the life and operation of metal and alloy orthodontic instruments used in the body as implants. An implant material's corrosion resistance impacts its performance and longevity and is a primary factor controlling biocompatibility. Corrosion is capable of two consequences; the implant will degrade first, and the result will be premature failure. The second effect is the tissue reactions that lead to the implant releasing corrosion products. Existing n vitro corrosion studies include synthetic biological aqueous alkaline solutions that meet body fluids other than the organic substances. In vivo corrosion includes the implantation of the systems of most alloys greatly increases the tissue-adjacent concentration of multiple ions. The condition of the body is rough and poses many corrosion-control problems. The contacts between both material and tissue are of major importance in contrast to the toxic atmosphere and the substantial loads faced by the implant. Such contact causes the embedded system to be corrosive/ionized. Electrochemical polishing can also improve the corrosion efficiency of medical devices made of stainless steel (electro-polishing). During this method, the component to be polished is immersed in an electrolyte, and a small metal layer is extracted while an electric current is implemented. The corrosion resistance increase can be due to the chosen separation of iron (Fe) and nickel (Ni), creating a surface of chromium (Cr)–rich oxide [43]. Modern metallic biomaterials with enhanced mechanical properties, good biocompatibility and good resistance to corrosion have been found recently. Titanium (Ti) alloys, stainless steel, and cobalt (Co)–Cr alloys are examples of new metallic biomaterial. The higher concentration of Cr provides strong resistance to a large variety of corrosive approaches. Potential methodologies for enhancing the efficiency of orthopedic devices can be summarized as surface alloying ion implantation of Ti, stainless steels and their alloys and surface modification of stainless steel with bioceramic coatings for reducing corrosion and to attain good biocompatibility.

6.6 TRIBOLOGICAL CHARACTERISTICS OF BIOMATERIALS

The recognition of synthetic biomaterials marked a significant development in the field of bioengineering. It put forth the scope and idea of benefiting the patients in relieving pain and discomfort. This not only revolutionized the medical science but also increased the quality of living for people. Tribological processes extensively take place in human organisms, particularly in living body, such as heart valves, skin, teeth, and eyes, and it is relevant to expose the tribological mechanism in the human body and fabricate biomaterials artificially to take the place of damaged tissue to get rid of pain in patients [44, 45]. One of the essential elements in tribology is bio-tribological material, which is strongly linked to human health [46]. Friction and wear have a major role in the analysis of efficiency of biomaterial. The friction and wear tests were performed in ambient conditions employing a ball-on-disc and a pin-on-disk tribometer under various conditions and sliding velocities with lubricant and dry conditions. Tribo-mechanical wear dominates the wear mechanism in the contexts of cracking, adhesion, and abrasion. In this chapter, accomplishments in the

TABLE 6.6
Tribological Characteristics of Biomaterials with Applications

Material	Application	Tribological Characteristics	References
Metal and metal alloys (Ti alloys, TAV alloy, CO-Cr alloy	Total replacement of joints	Resistance to wear and oxidation	[47, 48]
Inorganic: diamond-carbon-like	Coating materials that are biocompatible	Reduction of friction and improved resistance to wear	[49, 50]
Polymer (PMMA, Nylon Polyethylene, Polypropylene	Joint socket, implant with interposition, joint bone	Resistance to wear and oxidation, low frication of coefficient, elastic with a lower wear rate	[51, 52]
Ceramics (aluminum oxides, titanium oxides, calcium phosphates, carbon)	Coating of bone joints	Corrosion and wear resistance	[53, 54]
Composite (ceramic coatings with metals, carbon-coated material)	Joint of bone	Fatigue and wear resistance	[55, 56]

field of research in biomedical material tribology have been briefly reviewed, specifically on biomaterials and suggested evaluation techniques in this area. Tribological characteristics of different biomaterial with their applications are shown in Table 6.6.

CONCLUSION

The basic criteria for the successful implementation of metallic implants in bone fractures and substitutions are mechanical behavior and biocompatibility of body tissues. Basically, some materials used in engineering are metal and their alloys, ceramics, polymers, and composite, but metal and ceramics play a vital role in the biomedical field. Researchers have contributed to finding many engineering innovations that have been developed to enhance the tribological efficiency of synthetic nanostructured implants, such as protective coatings, hydrothermal synthesis, and micro implants. This review considered metallic implants for biomedical applications, emphasizing their merits and demerits when implanted inside the human body. As polymer implants fail in representing bioactivity in vivo ambiance and the brittle nature of ceramic materials makes them improper for implant applications, metallic implants have been considered as the most promising materials for biomedical applications owing to their mechanical properties, surface texture properties, and biocompatibility promoting the growth of biomolecules on the surface of implant optimizing molecular and cellular reactions. Another reason for opting for metallic implants is their intrinsic material properties including tensile strength, elastic modulus, and fatigue strength that ultimately affect the implant performance and its long-term success. However, faulty structure properties might also lead to structure failure; other reasons include surgical errors and inadequate mechanical design. On this basis, discussions on biomaterials and their fabrication, properties, corrosion resistance, and tribological characteristics have been carried out in this chapter.

REFERENCES

1. Bao Ha, T. L., Minh, T., Nguyen, D., & Minh, D. (2013). Naturally derived biomaterials: preparation and application. *Regenerative Medicine and Tissue Engineering*. doi:10.5772/55668.
2. J. Black, The education of the biomaterialist: Report of a survey, *Journal of Biomedical Materials Research*, 16(2) (1982), 159–167, http://doi.org/10.1002/jbm.820160208.
3. B.D. Ratner, A.S. Hoffman, F.J. Schoen, J.E. Lemons (eds.), *Biomaterials Science: An Introduction to Materials in Medicine*. San Diego: Academic (2004), ISBN 0-12582463-7.
4. J. Black and G. Hastings (eds.), *Handbook of Biomaterial Properties*. London: Chapman & Hall (1998), ISBN 0-412-60330-6.
5. L.T. Kuhn, Biomaterials. In *Introduction to Biomedical Engineering* (3rd ed.). Boston: Academic Press (2012).
6. C. Hsu, G. Parker and R. Puranik, Implantable devices and magnetic resonance imaging. *Heart, Lung and Circulation*, 21 (2012), 358–363.
7. J.D. Bronzino, *The Biomedical Engineering Handbook* (2nd ed., Vol. 1) (2000), ISBN: 3-540-66351-7.
8. P. Paridaand S.C. Mishra, Biomaterials in medicine. In *UGC Sponsored National Workshop on Innovative Experiments in Physics, 9–10, 2012*. Rourkela: Neelashaila Mahabidyalaya (2012), http://hdl.handle.net/2080/1624.
9. R. Langer and J.P. Vacanti, Tissue engineering. *Science*, 260(5110) (May 14, 1993), 920–926.
10. N. Hassan, P. Habibollah and A. Abolhassan, Properties of the amniotic membrane for potential use in tissue engineering. *European Cells & Materials*, 15 (2008), 88–89.
11. E. Zhang, D. Yin, L. Xu, L. Yang and K. Yang, Microstructure, mechanical and corrosion properties and biocompatibility of Mg-Zn-Mn alloys for biomedical application. *Materials Science and Engineering: C*, 29(3) (2009), 987–993.
12. D. Shekhawat, et al. Bioceramic composites for orthopaedic applications: A comprehensive review of mechanical, biological, and microstructural properties. *Ceramics International*, 47(3) (2021), pp. 3013–3030, ISSN 0272-8842, https://doi.org/10.1016/j.ceramint.2020.09.214.
13. D. Shekhawat, A. Singh, A. Bhardwaj, A, Patniak, In *IOP Conference Series: Materials Science and Engineering, Volume 1017, International Conference on "Advances in Materials Processing & Manufacturing Applications" (iCADMA 2020)*, 5th–6th November 2020, Jaipur, India.
14. O. Addison, A.J. Davenport, R.J. Newport, S. Kalra, M. Monir, et al., Do 'passive' medical titanium surfaces deteriorate in service in the absence of wear?? *Journal of the Royal Society Interface*, 7 (2012), 3161–3164.
15. J.B. Park and R.S. Lakes, *Biomaterials: An Introduction*. New York: Springer (2007).
16. M. Vallet-Regí, Ceramics for medical applications. *Journal of the Chemical Society, Dalton Transactions*, 2 (2001), 97–108.
17. A.D. Theocharis, S.S. Skandalis, C. Gialeli and N.K. Karamanos, Extracellular matrix structure. *Advanced Drug Delivery Reviews*, 97 (2016), 4–27.
18. W. Yang, L. Li, G. Su, Z. Zhang, Y. Cao, X. Li, et al., A collagen telopeptide binding peptide shows potential in aiding collagen bundle formation and fibril orientation, *Biomaterials Science*, 5(9) (2017), 1766–1776.
19. M. Martson, J. Viljanto, T. Hurme, P. Laippala and P. Saukko, Is cellulose sponge degradable or stable as implantation material? An in vivo subcutaneous study in the rat. *Biomaterials*, 20 (1999), 1989–1995.
20. Z. Ge, S. Baguenard, L.Y. Lim, A. We and E. Khor, Hydroxyapatite–chitin materials as potential tissue engineered bone substitutes. *Biomaterials*, 25 (2004), 1049–1058.

21. S. Ullah, I. Zainol, S.R. Chowdhury and M. Fauzi, Development of various composition multicomponent chitosan/fish collagen/glycerin 3D porous scaffolds: Effect on morphology, mechanical strength, biostability and cytocompatibility. *International Journal of Biological Macromolecules*, 111 (2018), 158–168.
22. L. Shapiro and S. Cohen, Novel alginate sponges for cell culture and transplantation. *Biomaterials*, 18 (1997), 583–590.
23. Q. Xing, F. Zhao, S. Chen, J. McNamara, M.A. DeCoster and Y.M. Lvov, Porous biocompatible three-dimensional scaffolds of cellulose microfiber/gelatin composites for cell culture. *Acta Biomaterials*, 6 (2010), 2132–2139.
24. T. Xu, H. Yang, D. Yang and Z.-Z. Yu, Polylactic acid nanofiber scaffold decorated with chitosan island like topography for bone tissue engineering. *ACS Applied Materials & Interfaces*, 9 (2017), 21094–21104.
25. K. Murakami, H. Aoki, S. Nakamura, S.-I. Nakamura, M. Takikawa, M. Hanzawa, et al., Hydrogel blends of chitin/chitosan, fucoidan and alginate as healing-impaired wound dressings. *Biomaterials*, 31 (2010), 83–90.
26. M. Costantini, C. Colosi, J. Jaroszewicz, A. Tosato, W. Swieszkowski, M. Dentini, et al., Microfluidic foaming: A powerful tool for tailoring the morphological and permeability properties of sponge-like biopolymeric scaffolds. *ACS Applied Materials & Interfaces*, 7 (2015), 23660–23671.
27. H. Pan, H. Jiang and W. Chen, Interaction of dermal fibroblasts with electrospun composite polymer scaffolds prepared from dextran and poly lactide-co-glycolide. *Biomaterials*, 27 (2006), 3209–3220.
28. W. Shi, M. Sun, X. Hu, B. Ren, J. Cheng, C. Li, et al., Structurally and functionally optimized silk-fibroin—gelatin scaffold using 3D printing to repair cartilage injury in vitro and in vivo. *Advanced Materials*, 29 (2017), 1701089.
29. C.M. Murphy, M.G. Haugh and F.J. O'Brien, The effect of mean pore size on cell attachment, proliferation and migration in collagen—glycosaminoglycan scaffolds for bone tissue engineering. *Biomaterials*, 31 (2010), 461–466.
30. M. Mehdizadeh and J. Yang, Design strategies and applications of tissue bioadhesives. *Macromolecular Bioscience*, 13(3) (2013), 271–288.
31. M. Niinomi, Recent metallic materials for biomedical applications. *Metallurgical and Materials Transactions A*, 33(3) (2002), 477–86.
32. D.F. Williams, Review: tissue-biomaterial interactions. *Journal of Materials Science*, 22(10) (1987), 3421–3445.
33. K. Anselme, Osteoblast adhesion on biomaterials, review. *Biomaterials* 21 (2000), 667–681.
34. M. Mehdizadeh and J. Yang, Design strategies and applications of tissue bioadhesives. *Macromolecular Bioscience*, 13(3) (2013), 271–288.
35. J.M. Anderson, Biological responses to materials. *Annual Review of Materials Research*, 31 (2001), 81–110.
36. J.M. Anderson, Mechanism of inflammation and infection with implanted devices. *Cardiovascular Pathology*, 2 (1993), 33S–41S.
37. J.M. Anderson, Inflammatory response to implants. *ASAIO Journal*, 11 (1988), 101–107.
38. R.Z. Cotran, V. Kumar and S.L. Robbins, *Pathologic Basis of Disease* (6th ed., pp. 50–112). Philadelphia, PA: Saunders (1999).
39. J.R. Davis (Ed.), Properties and selection: nonferrous alloys and special-purpose materials, *ASM Handbook, vol. 2, ASM International*, Materials Park, OH (1990).
40. J.M. Anderson, Inflammatory response to implants. *ASAIO Journal*, 11 (1988), 101–107.
41. S.V. Dorozhkin, Biocomposites and hybrid biomaterials based on calcium orthophosphates. *Biomatter*, 1(1) (2011), 3–56.

42. Baldev Raj, U Kamachi Mudali, TM Sridhar - Corrosion of bio implants. Sadhana, 2003(June 2003), 601–637, http://doi.org/10.1007/BF02706450.
43. N. Eliaz, Biomaterials and corrosion. In U. Kamachi Mudali and B. Raj (eds.), *Corrosion Science and Technology: Mechanism, Mitigation and Monitoring* (pp. 356–397). New Delhi: Narosa Publishing House (2008). CRC Press (USA) and Alpha Science International (Europe) (2009).
44. Z.R. Zhou, Z.M. Jin, Biotribology: recent progresses and future perspectives, *Biosurface and Biotribology*, 1(1) (2015), 3–24, ISSN 2405-4518, https://doi.org/10.1016/j.bsbt.2015.03.001.
45. D. Shekhawat, A. Singh and A. Patnaik, Tribo-behaviour of biomaterials for hip arthroplasty. *Materials Today: Proceedings* (2021), ISSN 2214–7853, https://doi.org/10.1016/j.matpr.2020.11.420.
46. M. Niinomi, Recent metallic materials for biomedical applications, *Metallurgical and Materials Transactions A* 33 (2002), 477, https://doi.org/10.1007/s11661-002-0109-2.
47. P.A. Lilley, P.S. Walker, G.W. Blunn, Wear of titanium by soft tissue. *Transactions of the 4th Word Biomaterials Congress, Berlin, April 24–28, 1992* (p. 227).
48. M. Fellah, M. Labaïz, O. Assala, L. Dekhil, A. Taleb, H. Rezag, A. Iost, Tribological behavior of Ti-6Al-4V and Ti-6Al-7Nb alloys for total hip prosthesis, *Advances in Tribology*, 2014 (2014), 13, Article ID 451387, https://doi.org/10.1155/2014/451387.
49. E. Confortoa, B.-O. Aronssonb, A. Salitoc, C. Crestou D. Caillard, Rough surfaces of titanium and titanium alloys for implants and prostheses, *Materials Science and Engineering C*, 24 (2004), 611–618.
50. Y. Luo, M. Zhi Rong and M. Qiu Zhang, Tribological behavior of epoxy composites containing reactive SiC nanoparticles, *Journal of Applied Polymer Science*, 104 (2007), 2608.
51. M.Z. Rong, M.Q. Zhang, G. Shi, Q.L. Ji, B. Wetzel and K. Friedrich, Graft polymerization onto inorganic nanoparticles and its effect on tribological performance improvement of polymer composites, *Tribology International*, 36 (2003), 697–707.
52. M.G. Faga, A. Vallée, A. Bellosi, M. Mazzocchi, N.N. Thinh, G. Martra and S. Coluccia, Chemical treatment on alumina–zirconia composites inducing apatite formation with maintained mechanical properties, *Journal of the European Ceramic Society*, 32 (2012), 2113–2120.
53. Y. Li, H.-E. Kim, Y.-H. Koh, Improving the surface hardness of zirconia toughened alumina (ZTA) composites by surface treatment with a boehmite sol, *Ceramics International*, 38(4), (2012) 2889–2892, ISSN 0272-8842, https://doi.org/10.1016/j.ceramint.2011.11.062.
54. G. Pan, Q. Guo, J. Ding, W. Zhang, X. Wang, Tribological behaviors of graphite/epoxy two-phase composite coatings, *Tribology International*, 43(8), (2010) 1318–1325, ISSN 0301-679X, https://doi.org/10.1016/j.triboint.2009.12.068.
55. Rashmi, N.M. Renukappa, B. Suresha, R.M. Devarajaiah, K.N. Shivakumar, Dry sliding wear behaviour of organo-modified montmorillonite filled epoxy nanocomposites using Taguchi's techniques, *Materials & Design*, 32(8–9), (2011) 4528–4536, ISSN 0261-3069, https://doi.org/10.1016/j.matdes.2011.03.028.
56. N. Hassan, P. Habibollah and A. Abolhassan, Properties of the amniotic membrane for potential use in tissue engineering. *European Cells & Materials*, 15 (2008), 88–89.

7 Microstructural and Tribological Behavior of 5083-TiB$_2$ Cast Composites Fabricated by a Flux-Assisted Synthesis Technique

Vikash Gautam, M.J. Pawar, Amar Patnaik, Vikas Kukshal, and Ashiwani Kumar

CONTENTS

7.1 Introduction 107
7.2 Materials and Methods 108
7.3 Microstructural Characterization of 5083-TiB$_2$ Composites 110
7.4 Physical and Mechanical Characterization of 5083-TiB$_2$ Composites 111
7.5 Effect of Sliding Velocity on Tribological Behavior 112
7.6 Influence of Normal Load on Tribological Behavior 113
7.7 Worn-Surface Analysis 114
7.8 Conclusion 115
References 115

7.1 INTRODUCTION

At present, it is a great challenge to develop a lightweight material for various applications with admirable physical and mechanical properties [1, 2]. If we look to the lightweight metals family, we found various elements, and among these elements, aluminum is suitable metals use in industries such as food processing, automobile, aerospace and naval. Pure aluminum alloy has a low weight-to-strength ratio, which causes less appropriate material for high-strength applications. The strength of aluminum is improved by adding alloying elements such as copper, zinc, magnesium, silicon and manganese. After addition of alloying elements, aluminum becomes a suitable alloy for high-strength application. Aluminum alloy is available in seven categories by addition of different alloying elements, and these alloying elements convert aluminum into a better weight-to-strength ratio element [3–12]. Further improvement in aluminum alloy is possible with the help of addition of reinforcement material and convert

DOI: 10.1201/9781003093213-7

aluminum alloy into aluminum-based metal matrix composites. Aluminum metal matrix composites possess superior mechanical properties with better wear-resistance properties, excellent heat treatability, castability, and high strength-to-weight ratio and the like [3–12]. The major elements in aluminum alloy A5083 are aluminum, magnesium and chromium; the combination of magnesium and chromium makes it highly resistant to corrosion that occurs due to seawater and industrial chemicals [7].

Various researchers used various combinations of reinforcement with the 5083 aluminum alloy. Gargatte et al. [13] used silicon carbide (SiC) particles with 5083 aluminum alloy to enhance wear and mechanical characteristics. For the fabrication of composites, stir casting was used with varying the volume percentage of SiC particulates. The experimental results depicted that hardness is enhanced with variation of SiC particles. A wear test was performed, and from the results, it was inferred that the wear rate shows a decrement with variation of SiC particles. Sliding wear behavior of a similar composition with 10% weight fraction of SiC particles was examined by Ravindra et al. [14]. They concluded that the applied normal load was the most significant operating parameter for the wear rate among all operating parameters. Furthermore, Javadi et al. [15] synthesized aluminum nanocomposites by using titanium bromide (TiB_2) as reinforcement material by a flux-assisted synthesis technique and concluded that TiB_2 particles synthesized through a flux-assisted synthesis technique enhance the hardness to composites. Zhao et al. determined the mechanical and sliding wear properties of boron carbide (B_4C) in the aluminum alloy [16]. They fabricated their sample using hot-sintering press technique and experimental results were obtained. The hardness and tribo-resistance properties of the samples were observed superior to the matrix material.

The present research work is emphasizing on fabrication of 5083-TiB_2 composites cast using through a flux-assisted synthesis technique. The effect of TiB_2 particles on the mechanical and tribological performance is analyzed. X-ray powder diffraction (XRD) analysis and scanning electron microscopy (SEM) were used to validate the presence of TiB_2 particles. SEM was also used to examine the wear mechanism obtained during the experiment.

7.2 MATERIALS AND METHODS

In present research, A5083 is used as a base matrix material and TiB_2 as a reinforcement material. The reinforcement particle TiB_2 form in cast composites with the help of halide particles. When halide particles are mixed in molten A536 alloy, TiB_2 particles are formed in cast composites. The chemical element description of 5083 aluminum alloy is presented in Table 7.1.

$$K_2TiF_6 + 13/3\ Al \rightarrow Al_3Ti + 4/3\ AlF_3 + 2\ KF \quad (7.1)$$
$$2\ KBF_4 + 3\ Al \rightarrow AlB_2 + 2\ AlF_3 + 2\ KF \quad (7.2)$$
$$Al_3Ti + AlB_2 \rightarrow TiB_2 + 4\ Al \quad (7.3)\ [15]$$

TABLE 7.1
Chemical Composition of 5083 Aluminum Alloy

Element	Zinc	Iron	Titanium	Copper	Silicon	Lead	Manganese	Magnesium	Chromium	Aluminum
Wt.%	0.03	0.173	0.04	0.0181	0.16	0.014	0.526	5.13	0.097	Balance

Behavior of 5083-TiB$_2$ Cast Composites

A high-temperature vacuum-induction furnace is used for casting purposes. Initially a graphite crucible is preheated to 300°C to prevent oxidation of the base matrix and the proper formation of TiB$_2$ particles. After that AA5083 aluminum alloy pieces are put in the crucible and melted in temperature up to 900 °C. To achieve a homogeneous distribution of reinforcement, a graphite stirrer at 400 rpm was used to mix molten alloy. After melting the A5083 alloy, a preheated mixture of inorganic salts (halide particles) in proper wt.% is poured into the crucible and stirred continuously up to 30 minutes. Table 7.2 gives the measures of the quantities of inorganic salts that were added to obtain the different amounts of TiB$_2$ in AA 5083. Finally, the molten mixture is poured into a preheated mold. An identical procedure was adopted for the preparation of other MMC samples. The schematic diagram of fabrication machines with their component is presented. The schematic diagram for fabrication of TiB$_2$ Particles reinforced 5083 aluminum alloy composites is presented in Figure 7.1.

Specimens were studied using XRD and SEM for an aesthetic examination. With reference to physical properties, the effect of reinforcement material on the density and content of cavities were examined. The theoretical density was determined using the expression stated by Aggarwal and Broggtman [31], and the experimental density was determined through the principle of water dispersion. The void/cavity fraction is the difference fraction between theoretical density and experiment density. Flexural, hardness, impact, and fracture values were determined for analysis.

The flexural test has been performed on the universal testing machine (UTM) as per ASTM standard E290. The testing parameters of crosshead speed = 1 mm/min were used. Throughout the experiments, the chip length was 40 mm, and the sample size was 60 × 10 × 10 mm³. The hardness value is determined using Brinell hardness testing machine according to ASTM (E 10–15) at 250-kgf load for the matrix material and composite samples. The ASTM E-23 standard was followed to estimate the

TABLE 7.2
Measured Weight Amount of Inorganic Salts

TiB$_2$ (wt.%)	KBF$_4$ (g)	K$_2$TiF$_6$ (g)
5	106.011	84.14
7	148.034	117.36
9	188.91	149.97

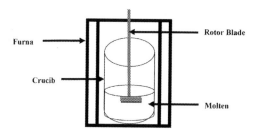

FIGURE 7.1 Schematic Diagram for Flux-Assisted Liquid Stir-Casting Technique

impact resistance. The standard dimension of 64 × 12.7 × 3.2 mm³, and the slot depth is 10.2 mm for each was used.

The sliding wear behavior is determined using a pin-on-disk tribotester. A sample with a cross section of 10 mm and a length 15 mm was used. These specimens slide on the opposite surface of AISI 52100 steel with 61 HRC hardness value. The sliding wear test was performed at a normal temperature, and an average reading of three samples was collected in each experiment. The following parameters were considered for the test: sliding velocity (0.25–1 ms^{-1}) and loads (20–80 N) while sliding distance (250 m) was kept constant. Emery paper of different grit sizes was used to polish both the pin and disc surface to get a uniform roughness of 38 μm. Microstructural examinations of worn surfaces were done using SEM to analyze the applied surface features obtained during the experiment. These salient features were helpful in exploring the dominant wear mechanism.

7.3 MICROSTRUCTURAL CHARACTERIZATION OF 5083-TiB₂ COMPOSITES

The micro structure of the A5083-TiB$_2$ composites cast using flux assisted synthesis technique is necessary to validate the presence of TiB$_2$ particles. XRD graph and SEM images predicted the existence of TiB$_2$ particles in the cast composites using a flux-assisted synthesis technique. The XRD graph of a series of TiB$_2$-particle-reinforced 5083 aluminum alloy composites is presented in Figure 7.2. The variation in TiB$_2$ particles achieved through in situ reaction obtained through the flux-assisted synthesis technique is denoted by x factor.

Figure 7.2 evidently indicates the higher peaks of TiB$_2$ particles due to the flux-assisted synthesis technique. The SEM image of A5083-TiB$_2$ composites cast using the flux-assisted synthesis technique at higher magnification is present in Figure 7.3. The scanning electron microscopic image of A5083-TiB$_2$ composites fabricated through flux-assisted synthesis technique is presented at higher magnification. SEM image of A5083-TiB$_2$ composites cast using a flux-assisted synthesis technique at higher magnification.

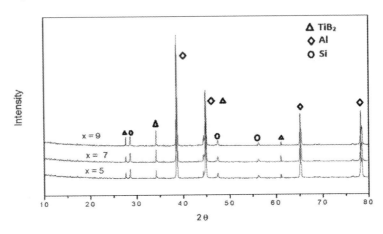

FIGURE 7.2   XRD Patterns Graph of 5083-TiB$_2$ Composites

This SEM image was presented to validate the shape of obtained TiB$_2$ particle through in situ reaction. Various researchers obtained various shapes of TiB$_2$ particles, such as spherical, hexagonal, and cubic, in their research [24–26]. It was clearly observed that the hexagonal shape of TiB$_2$ particle was obtained in the present research.

7.4 PHYSICAL AND MECHANICAL CHARACTERIZATION OF 5083-TiB$_2$ COMPOSITES

From Table 7.3, it is clearly concluded that both theoretical density and experimental density, along with void content, increase with a variation of TiB$_2$ particles. One of the reasons after the increase in void may be an agglomeration of reinforcement particle leads toward the intra-particulate voids in the composites and mistakes in fabrication technology maybe another possible cause [21]. One of the reasons of the enhancement in tensile strength may be the size of the reinforcement particle as well as the bond strength of the constituent materials [14]. A maximum tensile strength of 391 MPa was evaluated for the 5083–9 TiB$_2$ composites, and a minimum tensile strength of 316 MPa was evaluated for 5083–0 TiB$_2$ composites.

It is also observed that the flexural strength of the samples improves with the addition of TiB$_2$ particles. The enhancement in the flexural strength is less compared to tensile strength. The possible reason behind the increase may be the size of the reinforcement particles, as well as the nature of the bonding between the matrix and

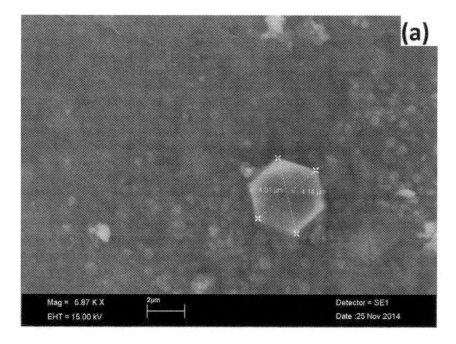

FIGURE 7.3 SEM Micrograph of 5083–5 wt.% TiB$_2$ Composites

TABLE 7.3
Physical and Mechanical Properties of 5083-TiB$_2$ Samples

Designation	Theoretical Density (g/cc)	Experimental Density (g/cc)	Void Fraction (%)	Tensile Strength (MPa)	Flexural Strength (MPa)	Hardness (HB)	Impact Energy (Joule)
5083–0 TiB$_2$	2.7135	2.7135	0	316	103	57.5	46
5083–5 TiB$_2$	2.7654	2.7357	1.07	357	116	63.7	63
5083–7 TiB$_2$	2.7882	2.7131	2.69	376	131	71.3	77
5083–9 TiB$_2$	2.8113	2.6931	4.28	391	149	93.5	83

the reinforcement [14]. A maximum flexural strength of 149 MPa was evaluated for 5083–9 TiB$_2$ composites, and a minimum flexural strength of 103 MPa was evaluated for 5083–0 TiB$_2$ composites.

Hardness value was evaluated through experiment for 5083–0 TiB$_2$ composites and hardness value is present in Table 7.3. From the table it is clearly observed that TiB$_2$ particles synthesis through a flux-assisted liquid stir-casting technique enhanced the hardness of 5083-TiB$_2$ composites. A maximum hardness of 93.5 HB was evaluated for 5083–9 TiB$_2$ composites, and a minimum hardness of 57.5 HB was evaluated for 5083–0 TiB$_2$ composites. The possible reason may be a dispersion effect, which leads to maintaining high hardness in higher environment temperature [26–28].

The impact energy was evaluated through experiment for 5083–0 TiB$_2$ composites are presented in Table 7.3. From the table, it is clear that the impact energy absorption properties were enhanced after the inculcation of TiB$_2$ particles into the virgin matrix. A maximum impact energy of 83 Joules was evaluated for 5083–9 TiB$_2$ composites, and a minimum hardness of 46 Joules was evaluated for 5083–0 TiB$_2$ composites. One of the reasons for the significant improvement in impact energy may be the particle size and the homogeneous distribution of the reinforcement particles [29–30].

7.5 EFFECT OF SLIDING VELOCITY ON TRIBOLOGICAL BEHAVIOR

Figure 7.4 presents the variation in the wear rate with a sliding velocity for 5083-TiB$_2$ composites. For experimentation, the sliding velocity parameter has been varied from 0.25 m/s to 1.00 m/s with the interval of 0.25 m/s while keeping other operating parameters constant. The line plot depicts that the wear rate of 5083-TiB$_2$ composites enhances, with respective to the increase, the sliding velocity and becomes constant at a certain value of velocity. From the graph, it is also found that the tribo-resistance performance of 5083-TiB$_2$ composites enhanced with an increase in TiB$_2$ particles in the base alloy. The possible reason behind the increase of the wear resistance performance of 5083-TiB$_2$ composites may be attributed to high hardness of TiB$_2$ particles and the homogeneous distribution of particles in the base alloy. Similar Experimental Results were obtained for TiB$_2$ filled Aluminum Composites by S. Kumar and C.S. Ramesh [22–23].

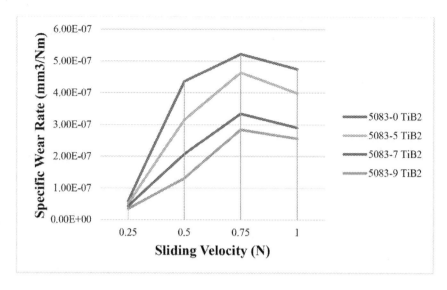

FIGURE 7.4 Variation of the Wear Rate with Sliding Velocity for 5083-TiB$_2$ Composites

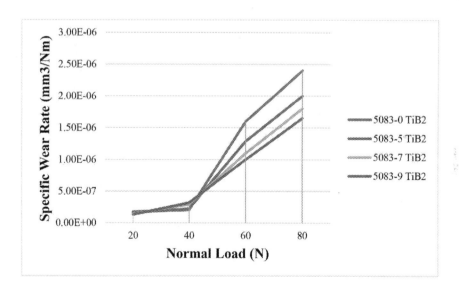

FIGURE 7.5 Variation of the Wear Rate with a Normal Load for 5083-TiB$_2$ Composites

7.6 INFLUENCE OF NORMAL LOAD ON TRIBOLOGICAL BEHAVIOR

The influence of load on the wear rate for the composite samples is presented in Figure 7.5. The experiment condition for the load varies from 20–80 N, with a step size of 20 N. The other control factors remain constant during the experiments. It was concluded from the experimental results that the wear rate is enhanced with the increase in load. The possible reason behind the increase in wear rate is that thermal softening occurs due to the generation of excessive heat. In addition, the tribo-resistance performance of 5083-TiB$_2$ samples was enhanced with increase of TiB$_2$ particles in the base alloy. The possible reason behind the increase in the wear resistance performance of 5083-TiB$_2$ composites may be attributed to the high hardness of TiB$_2$ particles and the homogeneous distribution of the particles in the base alloy.

7.7 WORN-SURFACE ANALYSIS

Worn-surface micrographs were presented to examine the wear mechanism feature obtained during the experimental condition. Worn-surfaces micrographs of 5083-TiB$_2$ composites at different experimental conditions. Figures 7.6a and 7.6b are

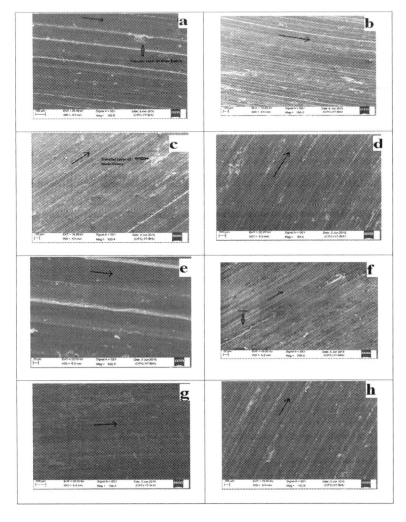

FIGURE 7.6 Worn-Surfaces SEM micrographs of 5083-TiB$_2$ Composites

for base alloy at the minimum and maximum sliding velocity. From the micrographs, a delamination sliding track with shallow grooves and oxidative wear mechanism features were obtained. The oxide wear debris was obtained on the worn surfaced due to adhesion and micro-cutting phenomenon [17]. Figures 7.6c and 7.6d are for 5083-TiB$_2$ composites at the minimum and maximum sliding velocities. The worn-surfaces micrograph shows that fine wear debris particles were pulled out of the base matrix [18].

The worn-surfaces micrograph for the base alloy at the minimum and maximum loads were presented in Figures 7.6e and 7.6f. Both micrograph images show a rich oxidative wear mechanism at the minimum and maximum load condition and adhesive and abrasive wear leading to the formation of a compacted layer, which was the key factor of the third-body abrasive wear due to the formation of abrasive particles and oxidative wear mechanisms. The micrographs of the worn surface also represent the wear creators on the surfaces [19]. The worn-surfaces micrographs for 5083-TiB$_2$ composites at the minimum and maximum loads are presented in Figures 7.6g and 7.6h. From the micrograph of the lower load conditions, shallow grooves were observed due to the presence of void contents, which lead toward the poor interfacial bonding between the particles and the matrix [20]. At the higher load condition micrographs, sliding tracks were obtained by a cutting phenomenon. The presence of hard abrasive TiB$_2$ particles were delaminate from the surfaces due to the excessive load and lead to the formation of a long sliding track with a cutting phenomenon [21].

7.8 CONCLUSION

The major conclusions are formed on the basis of the experimental results:

1. The successful fabrication was of 5083-TiB$_2$ composites through a flux-assisted synthesis technique via chemical reactions between halide particles (K2TiF6 and KBF4).
2. XRD analysis and SEM validate the presence of TiB$_2$ particles in 5083-TiB$_2$ composites cast through a flux-assisted synthesis technique.
3. The physical properties such as density and void content, and mechanical properties such as hardness, fracture strength, flexural strength, and impact strength of 5083-TiB$_2$ composites are enhanced compared to the base alloy.
4. The wear performance of 5083-TiB$_2$ composites is enhanced compared to the base alloy for normal load and sliding velocity.
5. Worn-surfaces micrographs were examined to analyze the wear mechanism obtained at different experimental conditions. The following wear mechanisms were observed such as adhesion, delamination, oxidation, and ploughing were the predominant wear mechanism.

REFERENCES

1. Habisreutinger SN, Leijtens T, Eperon GE, Stranks SD, Nicholas RJ, Snaith HJ, Carbon nanotube/polymer composites as a highly stable hole collection layer in perovskite solar cells, Nano Letters 14 (2014) 5561–5568.

2. Lu H, Wang X, Zhang T, Cheng Z, Fang Q, Design, fabrication, and properties of high damping metal matrix composites—a review, Materials 2 (2009) 958–977.
3. Gautam V, Patnaik A, Baht IK, Thermo-mechanical and fracture characterization of uncoated, single and multilayer (SiN/CrN) coating on granite powder filled metal alloy composites, Silicon 8 (2016) 133–143.
4. Gautam V, Patnaik A, Baht IK, Microstructure and wear behavior of single layer (CrN) and multilayered (SiN/CrN) coatings on particulate filled aluminum alloy composites, Silicon 8 (2016) 417–435.
5. Zhou ZS, Wu GH, Jiang LT, Li RF, Xu ZG, Analysis of morphology and microstructure of B4C/2024Al composites after 7.62 mm ballistic impact, Materials Design 63 (2014) 658–663.
6. Priyan MS, Benjamin MM, James J, Varughese PP, Roy R, In situ wear performance on Al 3102 alloy hybrid composites fabricated by stir casting method. International Research Journal of Modernization in Engineering Technology and Science 2(3) (2020) 466–474.
7. Farayibi PK, Akinnuli BO, Ogu S, Mechanical properties of aluminum—4043/nickel-coated silicon carbide composites produced via stir casting, International Journal of Engineering Technologies 4(1) (2018) 41–46.
8. Alizadeh A, Abdollahi L, Biukani, H, Processing, characterization, room temperature mechanical properties and fracture behavior of hot extruded multi-scale B4C reinforced 5083 aluminum alloy based composites, Transactions of Nonferrous Metals Society of China 27(6) (2017) 1233–1247.
9. Kumar A, Kumar M, Patnaik A, Pawar MJ, Pandey A, Kumar A, Gautam V, Optimization of sliding and mechanical performance Ti/NI metal powder particulate reinforced Al 6061 alloy composite using preference selection index method, Materials Today: Proceedings 44(6) (2021) 4784–4788, https://doi.org/10.1016/j.matpr.2020.10.974.
10. Baradeswaran A, Perumal AE, Influence of B4C on the tribological and mechanical properties of Al 7075–B4C composites, Composites Part B: Engineering 54 (2013) 146–152.
11. Munivenkatappan MSB, Shanmugam S, Veeramani A, Synthesis and characterization of in-situ AA8011-TiB$_2$ composites produced by flux assisted synthesis, International Information and Engineering Technology Association 44(5) (2020) 333–338.
12. Gargatte S, Upadhye RR, Dandagi VS, Desai SR, Waghamode BS, Preparation & characterization of Al-5083 alloy composites, Journal of Minerals and Materials Characterization and Engineering 1 (2013) 8–17.
13. Rana R S, Purohit R, Sharma AK, Rana S, Optimization of wear performance of Aa 5083/10 Wt.% SiCp composites using Taguchi method, Procedia Materials Science 6 (2014) 503–511.
14. Thakur S K, Dhindaw BK, The influence of interfacial characteristics between SiCp and Mg/Al metal matrix on wear, coefficient of friction and microhardness, Wear 247 (2001) 191–201.
15. Javadi A, Cao C, Li X, Manufacturing of Al-TiB 2 nanocomposites by flux-assisted liquid state processing, Procedia Manufacturing 10 (2017) 531–535.
16. Zhao Q, Liang Y, Zhang Z, Li X, Ren L, Microstructure and dry-sliding wear behavior of B4C ceramic particulate reinforced Al 5083 matrix composite, Metals 6 (2016) 227–239.
17. Qing-Ju I, Evaluation of sliding wear behavior of graphite particle-containing magnesium alloy composites, Transaction of Nonferrous Metals Society of China 16 (2006) 1135–1140.
18. Banerjee A, Prasad SV, Surappa MK, Rohatgi PK, Abrasive wear of cast aluminium alloy-zircon particle composites, Wear 82 (1982) 141–151.

19. Cui X, Wu Y, Xiangfa Y, Microstructural characterization and mechanical properties of VB2/A390 composite alloy, Journal of Materials Science & Technology 31 (2015) 1027–1033.
20. Jin K, Qiao Z, Zhu S, Friction and wear properties and mechanism of bronze—Cr—Ag composites under dry-sliding conditions, Tribology International 96 (2016) 132–140.
21. Antony C, Kumar V, Rajadurai JS, Influence of rutile (TiO_2) content on wear and microhardness characteristics of aluminium-based hybrid composites synthesized by powder metallurgy, Transaction of Nonferrous Metals Society of China 26 (2016) 63–73.
22. Kumar S, Chakraborty M, Subramanya Sarma V, Murthy BS, Dry sliding wear behavior of Aluminum 6063 composites reinforced with TiB_2 particles, Wear 265 (2008) 134–142.
23. Ramesh CS, Ahamed A, Channabasappa BH, Keshavamurthy R, Developement of Al6063-TiB_2 in-situ composites, Materials Design 31 (2010) 2230–2236.
24. Ramesh CS, Pramod S, Keshavamurthy R, A study on microstructure and mechanical properties of Al 6061–TiB_2 in-situ composites, Material Science and Engineering A 528 (2011) 4125–4132.
25. Turk A, Durman M, Kayali ES, The effect of manganese on the microstructure and mechanical properties of zinc–aluminium based ZA-8 alloy, Journal of Material Science 42 (2007) 8298–8305.
26. Ergun E, Aslantas K, Tasgetiren S, Effect of crack position on stress intensity factor in particle-reinforced metal-matrix composites, Mechanics Research Communication 35(4) (2008) 209–218.
27. Kari S, Berger H, Ramos RR, Gabber U, Computational evaluation of effective material properties of composites reinforced by randomly distributed spherical particles, Composite Structure 77 (2007) 223–231.
28. Liu FR, Chan KC, Tang CY, Numerical modeling of the thermo-mechanical behavior of particle reinforced metal matrix composites in laser forming by using a multi-particle cell model, Composite Science Technology 68 (2008) 1943–1953.
29. Bindumadhavan PN, Wah HK, Prabhakaran O, Dual particle size (DPS) composites: Effect on wear and mechanical properties of particulate metal matrix composites, Wear 248 (2001) 112–120.
30. Patnaik A, Mamatha TG, Biswas S, Kumar P, Damage assessment of titania filled zinc–aluminum alloy metal matrix composites in erosive environment: A comparative study, Materials Design 36 (2012) 511–521.
31. Agarwal BD, Broutman IJ, Analysis and performance of fiber composites, 2nd edn. Wiley, New York (1990).

8 Study and Prediction of Response Parameters during Oblique Machining for Different Machining Parameters by WEDM of Inconel-HX

I.V. Manoj and S. Narendranath

CONTENTS

8.1 Introduction: Background and Driving Forces 119
8.2 Material 121
8.3 Experimental Technique and Parameters 121
8.4 Analysis of CV and SR 122
    8.5.1 ANOVA and Main Effects Plot 123
8.5 Optimization at Each Angle Using Genetic Algorithm 125
8.6 Influence of WEDM at the Highest CV 127
8.7 Prediction of CV and SR Using ANFIS and ANN 128
8.8 Conclusion 131
References 131

8.1 INTRODUCTION: BACKGROUND AND DRIVING FORCES

Nickel-based alloys have low thermal diffusivity and high-temperature strength. This makes the machining of these alloys very challenging [1–2]. Nontraditional machining like wire electric discharge machining (WEDM) was found to be an ideal process that can machine precision components made of these materials [3]. The complexity of the components demands different shapes in machining with good surface quality. Taper/oblique parts can be produced using WEDM by a unique way of bending the wire during machining. This traditional method causes many problems like inaccuracies, wire break and improper flushing, among others [4–6]. Yan et al. [7], Martowibowo and Wahyudi [8] and Manoj et al. [9–11], among other

DOI: 10.1201/9781003093213-8

researchers, have tried different mechanisms, materials and methods to improvise the traditional tapering in WEDM. Abbasi et al. [12] have explored the effects of pulse-on time, discharge gap, pulse ratio, discharge current and wire speed on surface integrity. Camposeco-Negrete [13] investigated that servo voltage affects on machining time, and the surface was influenced by pulse-off time. There was a decrease of 1.16% in surface roughness and 7.50% in machining time found by multi-objective optimization. Kumar et al. [14] studied surface roughness, material removal rate and kerf width, which was optimized with the aid of the grey relation analysis method. Ishfaq et al. [15] analyzed that 20% of cutting speed can be improved by optimizing the effects of layer thickness, workpiece orientation, dielectric pressure ratio, and wire diameter during the machining of stainless-clad steel. Nain et al. [16] stated that pulse-on time influences the wire–wear ratio, cutting speed and dimensional deviation. Chaubey and Jain [17] stated with the increase in pulse-off time, the surface roughness decreases. The cutting speed at different parameters also affected the surface roughness. Ishfaq et al. [18] employed a genetic algorithm to get the best permutation which derived minimum surface roughness at optimal cutting speed and a kerf width in WEDM of Al6061–7.5% silicon carbide (SiC) composite. Altug et al. [19] explored the outcome of different heat-treated Ti6Al4V by electric discharge machining. It was established that the genetic algorithm optimized parameters generated minimum kerf. Sivaprakasam et al. [20] have used the genetic algorithm optimization technique for optimal conditions, yielding minimum kerf width and surface roughness and maximum material removal rate. Chen et al. [21] employed the genetic algorithm for optimization of machining gap and material removal rate. The results were validated by experiments with prediction errors of 6.70% and 8.38% for the machining gap and material removal rate, respectively. Phate and Kumar et al. [22] have used prediction models devised by an artificial neural network (ANN) and dimensional analysis in WEDM of Al/SiCp metal matrix composite (MMC). ANN models gave precise results in the prediction of surface roughness and material removal rate. Naresh et al. [23] investigated the adaptive neuro-fuzzy system (ANFIS) and ANNs prediction of responses in the machining of Nitinol alloy, it was established that the ANFIS predictor was most accurate. Abhilash and Chakradhar [24] have examined the causes of wire breakage and variation in mean gap voltage instances during machining of Inconel 718 by WEDM. ANFIS was used to foresee the wire-breakage situations, and the model shows the variations between input and output parameters. Riahi-Madvar and Seifi [25] have predicted gravel beds in rivers using ANN and ANFIS, in which it was concluded that ANFIS was superior to ANN in prediction.

The surface roughness (SR) and cutting velocity (CV) are essential factors that control the productivity and quality of machined components. These factors also have to be explored and optimized in the case of the tapering operations in WEDM due to their applications. In the present study, a novel method using the slant type fixture was employed to obtain the taper component without bending the wire, thereby avoiding the disadvantage of traditional tapering. The machining was performed at different oblique/taper angles, namely 0°, 15°, and 30°. The effects of different parameters like wire feed, servo voltage, pulse-off time, and pulse-on time on CV and SR. The oblique angle in machining increases SR and decreases CV in overall aspects, but the effects of the parameters remained unaltered. The genetic algorithm was employed in the optimization of CV and SR during oblique machining at different angles. ANN and ANFIS were used in the prediction of SR and CV.

8.2 MATERIAL

Inconel-HX is also called Hastelloy-X. It is a wrought nickel base alloy known as turbine alloys. Due to its excellent mechanical properties, it can be used in aircraft, furnace, petrochemical, and chemical processing industries. The heat-treatment method of solution annealing was performed at 1175 °C (2150°F) as received material. The energy dispersive spectroscopy (EDS) of Inconel-HX is as shown in Figure 8.1.

8.3 EXPERIMENTAL TECHNIQUE AND PARAMETERS

The WEDM of the make "ELPULS 15 CNC WEDM" from Electronica, Pune was employed in the machining of Inconel-HX. It is hard to machine an alloy that has various applications in the aerospace, chemical, marine, industrial furnace, die and tool areas. The 0.25-mm-diameter wire of zinc-coated copper and deionized water was used in machining. The peak current is 12 A, and the servo feed is 20 mm/min. The slant-type taper fixture made of an aluminum alloy was fixed to the WEDM table as shown in Figure 8.2a. The workpiece was locked in the rotating angular plate, where it was fixed to the necessary position. The angular plate was made to rotate at different angles from 0°, 15° and 30° as shown in Figure 8.2b. The input parameters like servo voltage, wire feed, pulse-on time and pulse-off time were employed for the investigation. The main difficulty in the study was selecting the input parameters that could machine in all three oblique angles. These parameters were chosen based on initial experiments and machining capabilities [11, 26]. An L_{16}

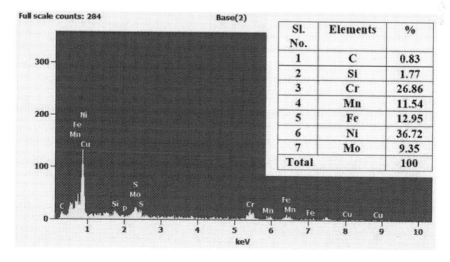

FIGURE 8.1 EDS of Inconel-HX

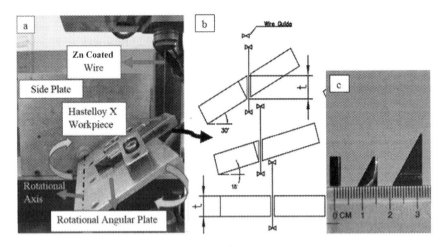

FIGURE 8.2 (a) Fixture Used on WEDM, (b) Oblique Machining, and (c) WEDM-ed Piece

TABLE 8.1
Factors Considered for Experimentation

Factors	Phase 1	Phase 2	Phase 3	Phase 4
Pulse-on time (T_{on}) (μs)	110	115	120	125
Pulse-off time (T_{off}) (μs)	30	40	50	60
Servo gap voltage (SV) (V)	30	40	50	60
Wire feed (WF) (m/min)	5	6	7	8

Taguchi's orthogonal array was employed in which the CV and SR were recorded. The oblique machining was performed on Inconel-HX and Figure 8.2c shows components from the machining. The parametric variations were recorded, the CV was calculated as the average of instantaneous speed and the "Mitutoyo SJ-301" SR testerwas employed for determining SR. Table 8.1 indicates the different steps employed for the parameters in the experimentation.

8.4 ANALYSIS OF CV AND SR

Table 8.2 illustrates the different parametric combinations used to the machine at different angles. It can be noticed that the highest CV obtained was 2.65 mm/min and 0.7 µm; the lowest SR was recorded at 0°. The lowest CV was 0.13 mm/min and the highest SR of 4.27 µm at 30°. It was prominent that the CV decreases and SR increases as the oblique angle increases from 0° to 30°. This was due to the result of the growth in material thickness, which leads to improper flushing [24, 25]. The regression equations were formulated from Equations 8.1 through 8.6 by Minitab software.

TABLE 8.2
L₁₆ Orthogonal Arrays

Experiment No.	T_{on}	T_{off}	SV	WF	0° CV	0° SR	15° CV	15° SR	30° CV	30° SR
1	110	30	30	5	1.58	1.93	1.43	2.31	1.28	2.71
2	110	40	40	6	1.32	1.33	1.17	1.71	1.02	2.11
3	110	50	50	7	0.82	0.89	0.77	1.13	0.62	1.53
4	110	60	60	8	0.79	0.70	0.28	0.92	0.13	1.32
5	115	30	40	7	1.88	2.47	1.73	2.74	1.58	3.14
6	115	40	30	8	1.79	2.76	1.64	3.11	1.49	3.51
7	115	50	60	5	0.90	0.95	0.75	1.22	0.60	1.62
8	115	60	50	6	0.98	1.04	0.83	1.34	0.68	1.69
9	120	30	50	8	2.15	2.63	2.05	2.93	1.90	3.28
10	120	40	60	7	2.14	1.53	2.04	1.73	1.89	2.08
11	120	50	30	6	2.40	3.59	2.30	3.89	2.15	4.24
12	120	60	40	5	1.59	2.22	1.49	2.52	1.34	2.87
13	125	30	60	6	2.22	2.13	2.12	2.43	1.97	2.78
14	125	40	50	5	2.48	3.32	2.38	3.67	2.23	4.02
15	125	50	40	8	2.65	3.19	2.55	3.54	2.40	3.94
16	125	60	30	7	2.54	3.62	2.44	3.87	2.29	4.27

Regression Equations for CV and SR

CV at 0° = −3.14 + 0.0534 T_{on} − 0.02020 T_{off} − 0.00362 SV − 0.0567 WF   (8.1)

CV at 15° = −4.02 + 0.05542 T_{on} − 0.01273 T_{off} − 0.01257 SV − 0.0282 WF   (8.2)

CV at 30° = −5.314 + 0.07062 T_{on} − 0.01712 T_{off} − 0.01423 SV − 0.0271 WF   (8.3)

SR at 0° = −6.34 + 0.0729 T_{on} − 0.0012 T_{off} − 0.0133 SV + 0.044 WF   (8.4)

SR at 15° = −5.06 + 0.0877 T_{on} − 0.01430 T_{off} − 0.00552 SV + 0.0610 WF   (8.5)

SR at 30° = −4.49 + 0.0859 T_{on} − 0.01418 T_{off} − 0.05583 SV + 0.0672 WF   (8.6)

8.5.1 ANOVA AND MAIN EFFECTS PLOT

From Table 8.2, the main effects plot and analysis of variance (ANOVA) were derived by using Minitab software. At higher pulse-on times, the CV also increases, and a contrasting behavior was observed for the SV parameter. With the escalation in pulse-on time, the discharge energy increased across the electrode due to an escalation in spark intensity, and henceforth, this leads to higher melting at the wire–workpiece interface of the material. This increases the CV as observed in the literature [28, 29] and Figure 8.3a. At higher servo voltages, the gap between the job and the electrode is higher. This results in lower ionization, leading to fewer sparks and decreased discharge energy. Therefore, there is a decrease in CV as shown in Figure 8.3a. Regarding SR, it increases as pulse-on time escalates because of large-size craters found on the surfaces machined using WEDM. During higher pulse-on time, the discharge energy increases due to an escalation in the intensity of the spark.

This results in deeper and larger crater formations over the WEDM surface, in turn, increasing the SR as in Figure 8.3b. Similar results have been established by Kumar et al. [28]. But as servo voltage increases, there is a reduction in SR because of the escalation in the spark gap, which results in a lower spark intensity. This forms shallower and smaller craters on WEDM surface [28, 29]. Table 8.3 and Figures 8.3a and 8.3b both show that wire feed had low effects on the CV and SR. As the wire feed increases, the CV also increases, as the molten material formed was flushed out of the wire workpiece interface instantaneously. Due to this flushing action, a large number of the microcavities on the surface were formed, creating a higher surface finish [28]. At higher wire feed, there was a decrease observed in CV due to

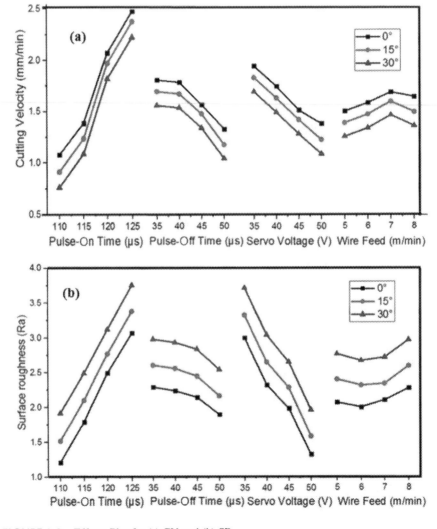

FIGURE 8.3 Effects Plot for (a) CV and (b) SR

TABLE 8.3
ANOVA Indicating Percentage Contribution on Response Parameters

Sl. No.	Source	DF	0° Adj. SS	0° % Contribution	15° Adj SS	15° % Contribution	30° Adj SS	30° % Contribution
CV								
1	T_{on}	3	4.819	72.11	5.065	73.88	5.342	72.62
2	T_{off}	3	0.735	11.00	0.705	10.28	0.843	11.46
3	SV	3	0.883	13.21	0.839	12.24	0.962	13.08
4	WF	3	0.097	1.45	0.120	1.75	0.109	1.48
5	Error	3	0.149	2.23	0.127	1.85	0.099	1.35
6	Total	15	6.683		6.856		7.356	
SR								
1	T_{on}	3	7.846	52.35	7.807	50.57	7.528	49.18
2	T_{off}	3	0.366	2.44	0.467	3.02	0.468	3.06
3	SV	3	5.735	38.26	6.179	40.02	6.302	41.17
4	WF	3	0.195	1.30	0.197	1.28	0.219	1.43
5	Error	3	0.846	5.64	0.791	5.12	0.789	5.15
6	Total	15	14.988		15.439		15.307	

the increase in wire vibration as shown in Figure 8.3a. Therefore, higher wire feed results in wire vibration [29]. In the case of SR, at higher wire feed, the SR increases. This was because of the wire vibration that occurs due to decreased tension at higher wire feeds. This leads to variations in the average spark gap, consequently increasing the SR as established by Chaudhary et al. [30]. In all the cases, the 30° angle shows lower CV and highest SR compared to 0° as stated in Section 8.5. The CV reduces as the oblique angle escalated because of the decrease in the melting of the material as in Figure 8.3. This phenomenon was observed because of the increase in material thickness that leads to better heat distribution as stated in Joy et al. [31]. In the case of the SR, the increased material thickness results in faster cooling of melted material on the WEDM surface as the dielectric fluid cools the material rapidly. So the SR escalates as established in the literature [26, 27] and depicted in Figure 8.3b.

8.5 OPTIMIZATION AT EACH ANGLE USING GENETIC ALGORITHM

This algorithm is a natural selection process whereby the fittest individuals/models/chromosomes are selected for reproduction to produce offspring of the succeeding generation. The genetic algorithm mainly works on three types of operators, namely, reproduction, crossover, and mutation. Reproduction mainly selects the best pair of chromosomes that was formed. This chromosome pair is improved monotonically from one generation to the next generation. The best chromosome pair is allowed for crossover with an assigned probable pair, which is generally good. This crossover introduces diversity in the gene pool by occasional random replacements of other chromosomes. This leads to the creation of new chromosomes caused by the alteration (mutation), based on the crossovers. The chromosomes that possess high fitness in the gene pool among the parents and offspring are selected for the succeeding generation. The whole process is repeated to get the best combination [20]. The genetic algorithm facilitates the simultaneous optimization of two different parameters.

From earlier investigation, it is seen that there should be a trade-off in CV and SR. The conditions were

$$110 \leq T_{on} \leq 125, \qquad (8.7)$$
$$30 \leq T_{off} \leq 60, \qquad (8.8)$$
$$30 \leq V \leq 60, \qquad (8.9)$$
$$5 \leq WF \leq 8. \qquad (8.10)$$

The genetic algorithmic optimization was performed in a MATLAB® environment using the optimization toolbox with parameters as shown in Table 8.4. It was seen that 110 (T_{on}), 56 (T_{off}), 58 (SV) and 6 (WF) were the optimal parameter set at all the three oblique angles. As seen in Table 8.5, the validated experimental results had a less than 5% error compared to the optimal predicted result.

TABLE 8.4
Genetic Algorithm Optimization Parameters

Sl. No.	Parameters	Values
1	Number of variables	4
2	Population	Double vector (size 50)
3	Function	Tournament (size 2)
4	Crossover segment	0.8
5	Probability in mutation	0.2
6	Generations	100
7	Stall generations	50

TABLE 8.5
Prediction and Experimental Validation of Optimized Results

Sl. No.	Oblique Angles (Degrees)	Experimental CV	Experimental SR	Genetic Algorithm Optimization CV	Genetic Algorithm Optimization SR	% Error in CV	% Error in SR
1	0°	1.47	1.59	1.41	1.55	4.08	4.40
2	15°	1.12	1.9	1.10	1.81	1.79	4.74
3	30°	0.9	2.06	0.87	2.12	3.33	2.91

8.6 INFLUENCE OF WEDM AT THE HIGHEST CV

Figure 8.4 shows the scanning electron microscopy (SEM) images highlighting the change in average white-layer thickness (AWT) at the highest CV. During machining, the metal is melted and the molten metal is carried out in form of debris. The molten metal on the parent workpiece resolidifies to form a white layer on the WEDM surface. As the oblique angle rises, the AWT reduces due to better heat distribution. At higher oblique angles, the slant provided by the fixture increases the material thickness. This increase in thickness ensures better heat distribution, in turn, decreasing the thermal degradation [31]. This phenomenon was validated by microhardness (MH) testing at all the oblique angles. Figure 8.5 indications the change in MH at

FIGURE 8.4 SEM Images at Different Oblique Angles: (a) 0°, (b) 15°, and (c) 30°

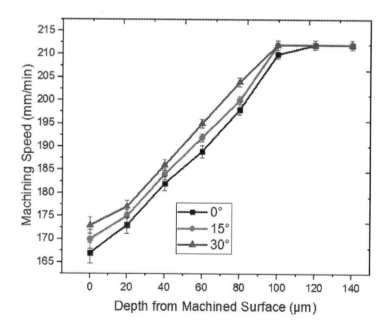

FIGURE 8.5 Variation of MH at Different Oblique Angles

0°, 15° and 30° oblique angles. It was observed that higher oblique angles had lower MH; that is, the 30° oblique-angle WEDM surface showed 167 HV MH. In the case of lower oblique angles, the MH was higher; that is, the 0° oblique-angle WEDM surface showed 173 HV MH. The 15° oblique angle shows an AWT and MH to be between the 0° and 30° oblique angles.

8.7 PREDICTION OF CV AND SR USING ANFIS AND ANN

From the earlier inferences, it was noticed that the oblique angle affects the response characteristics. So this was also considered as a parameter for the prediction of the output parameter. Forty-eight combinations of experiments were performed and used for the formulation of the optimum model in ANN and ANFIS. Figure 8.6 shows the regression plot generated during the training of the ANN model. The ANN and ANFIS models' parameters were set as shown in Table 8.6. Based on different trial experiments and literature [9, 23], these parameters were selected, and the optimal model was derived. From Figure 8.7, it can be noticed that the ANFIS model was better than the ANN model as the percentage error varied from 0–5% for both the response parameters. For the ANN model, it was observed that the error percentage varied from 0–10%. Figure 8.8 shows the actual mapping of input and output parameters. The three-dimensional plots show a better understanding of the actual variation in the output response with the machining parameters compared to the effects plots.

TABLE 8.6
Modeling Parameters for ANN and ANFIS

Sl. No.	Network Parameters	Values
ANN		
1	Network structure	5-8-1-1
2	Training/validation/testing experimental sets	32/8/8 (48 experiments)
3	Algorithm used in network	Feedforward backpropagation
4	Transfer function	Tangential sigmoid
5	Function used for training	TRAINLM
6	Function for learning	LEARNGDM
7	Performance function	MSE
ANFIS		
1	Input parameters	48
2	Output parameters	1
3	Data used for training/testing	38/10
4	Input membership function	Gaussian
5	Output function	Constant
6	Fuzzy rules used in network	32 rules
7	Maximum epochs used	500/output

Parameters during Oblique Machining 129

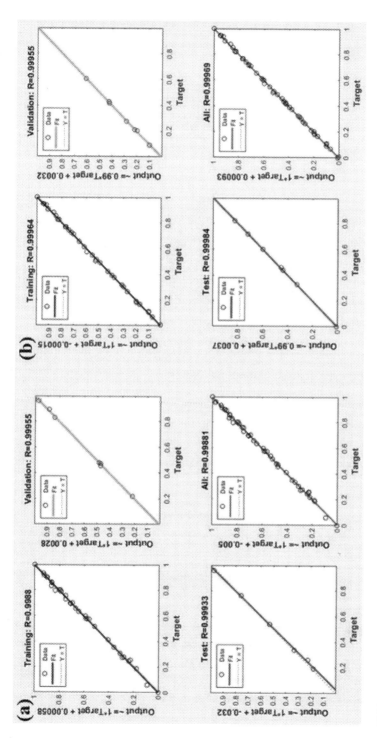

FIGURE 8.6 Regression Plots for (a) CV and (b) SR

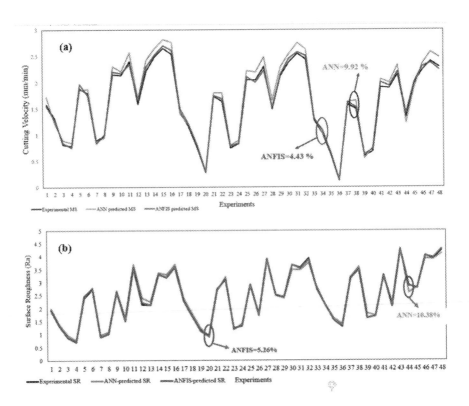

FIGURE 8.7 Experimental, ANN and ANFIS Prediction of (a) CV and (b) SR

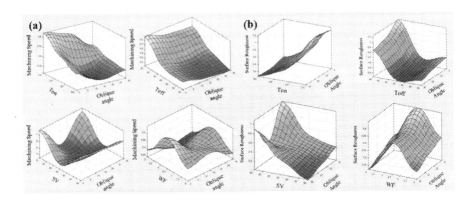

FIGURE 8.8 ANFIS Mapping for (a) CV and (b) SR

8.8 CONCLUSION

This research can be extended in the future to machining circular components without bending the wire as in the traditional process. The wire bending causes wire breakage, guide wear, improper flushing, angular inaccuracies, rough surface and the like. This study shows that oblique/taper machining of Inconel-HX was achieved by a unique angular placement of the workpiece with the aid of a fixture rather than the traditional technique. The effects of machining parameters on different responses for three oblique angles were analyzed, optimized, and predicted. The genetic algorithm used was validated experimentally, giving a less than 5% deviation. The pulse-on time affects both SR and CV. The response parameters increased with a rise in pulse-on time. The SR and CV reduce with an increase in servo voltage. The wire feed increases both SR and CV slightly. As machining is performed from 0° to 30° angles, the highest decrease of 83.54% in the CV and the highest increase of 71.91% in SR were observed. The AWT decreases whereas the MH increases at the WEDM surface at higher oblique angles. The ANFIS model was found to be most accurate in predicting the response parameters with a minimum-errors percentage ranging from 0–5%, and it also shows the variation of both input and output parameters by mapping.

REFERENCES

1. I.A. Choudhury, M.A. El-Baradie (1998) Machinability of nickel-base super alloys: a general review, Journal of Materials Processing Technology, 77, 278–284.
2. R. Arunachalam, M.A. Mannan (2000) Machinability of nickel-based high temperature alloys, Machining Science and Technology, an International Journal, 4, 127–168.
3. R. Maurya, R.K. Porwal (2020) EDM of Hastelloy—An overview, Materials Today: Proceedings, 26, 311–315.
4. N. Kinoshita, M. Fukui, T. Fujii (1987) Study on Wire-EDM: accuracy in taper-cut, CIRP Annals—Manufacturing Technology, 36, 119–122. https://doi.org/10.1016/S0007-8506(07)62567-0.
5. J.A. Sanchez, S. Plaza, L.N. Lopez De Lacalle, A. Lamikiz (2006) Computer simulation of wire-EDM taper-cutting, International Journal of Computer Integrated Manufacturing, 19, 727–735. https://doi.org/10.1080/09511920600628855.
6. S. Plaza, N. Ortega, J. A. Sanchez, I. Pombo, A. Mendikute (2009) Original models for the prediction of angular error in wire-EDM taper-cutting, International Journal of Advanced Manufacturing Technology, 44, 529–538.
7. H. Yan, Z. Liu, L. Li, C. Li, X. He (2017) Large taper mechanism of HS-WEDM, International Journal of Advanced Manufacturing Technology, 90, 2969–2977. https://doi.org/10.1007/s00170-016-9598-9.
8. S.Y. Martowibowo, A. Wahyudi (2012) Taguchi method implementation in taper motion wire EDM process optimization, Journal of the Institution of Engineers (India): Series C, 93, 357–364. https://doi.org/10.1007/s40032-012-0043-z.
9. I.V. Manoj, S. Narendranath (2020) Variation and artificial neural network prediction of profile areas during slant type taper profiling of triangle at different machining parameters on Hastelloy X by wire electric discharge machining, Proceedings of the Institution of Mechanical Engineers, Part E: Journal of Process Mechanical Engineering, 234, 673–683. https://doi.org/10.1177/0954408920938614.

10. I.V. Manoj, S. Narendranath (2021) Influence of machining parameters on taper square areas during slant type taper profiling using wire electric discharge machining, IOP Conference Series: Materials Science and Engineering, 1017, 1–9.
11. I.V. Manoj, R. Joy, S. Narendranath, D. Nedelcu (2019) Investigation of machining parameters on corner accuracies for slant type taper triangle shaped profiles using WEDM on Hastelloy X. IOP Conference Series: Materials Science and Engineering, 591, 1–11.
12. J. Ali Abbasi, M. Jahanzaib, M. Azam, S. Hussain, A. Wasim, M. Abbas (2017) Effects of wire-Cut EDM process parameters on surface roughness of HSLA steel, International Journal of Advanced Manufacturing Technology, 91, 1867–1878. https://doi.org/10.1007/s00170-016-9881-9.
13. C. Camposeco-Negrete (2019) Prediction and optimization of machining time and surface roughness of AISI O1 tool steel in wire-cut EDM using robust design and desirability approach, International Journal of Advanced Manufacturing Technology, 103, 2411–2422. https://doi.org/10.1007/s00170-019-03720-3.
14. A. Kumar, T. Soota, J. Kumar (2018) Optimisation of wire-cut EDM process parameter by grey-based response surface methodology, Journal of Industrial Engineering International, 14, 821–829. https://doi.org/10.1007/s40092-018-0264-8.
15. K. Ishfaq, N.A. Mufti, M.P. Mughal, M.Q. Saleem, N. Ahmed (2018) Investigation of wire electric discharge machining of stainless-clad steel for optimization of cutting speed, International Journal of Advanced Manufacturing Technology, 96, 1429–1443. https://doi.org/10.1007/s00170-018-1630-9.
16. S.S. Nain, D. Garg, S. Kumar (2018) Investigation for obtaining the optimal solution for improving the performance of WEDM of super alloy Udimet-L605 using particle swarm optimization, Engineering Science and Technology, An International Journal, 21(2), 261–273.
17. S.K. Chaubey, N.K. Jain (2018) Investigations on surface quality of WEDM-manufactured meso bevel and helical gears, Materials and Manufacturing Processes, 33, 1568–1577.
18. K. Ishfaq, S. Anwar, M. Asad Ali, M.H. Raza, M.U. Farooq, S. Ahmad, C.I. Pruncu, M. Saleh, B. Salah (2020) Optimization of WEDM for precise machining of novel developed Al6061–7.5% SiC squeeze-casted composite, International Journal of Advanced Manufacturing Technology, 111, 2031–2049.
19. M. Altug, M. Erdem, C. Ozay (2015) Experimental investigation of kerf of Ti6Al4V exposed to different heat treatment processes in WEDM and optimization of parameters using genetic algorithm, International Journal of Advanced Manufacturing Technology, 78, 1573–1583.
20. P. Sivaprakasam, P. Hariharan, S. Gowri (2014) Modeling and analysis of micro-WEDM process of titanium alloy (Tie6Ale4V) using response surface approach, Engineering Science and Technology, An International Journal, 17, 227–235.
21. X. Chen, Z. Wang, Y. Wang, G. Chi (2020) Investigation on MRR and machining gap of micro reciprocated wire-EDM for SKD11, International Journal of Precision Engineering and Manufacturing, 21, 11–22.
22. M.R. Phate, S.B. Toney (2019) Modeling and prediction of WEDM performance parameters for Al/SiCp MMC using dimensional analysis and artificial neural network, Engineering Science and Technology, an International Journal, 22, 468–476.
23. C. Naresh, P.S.C. Bose, C.S.P. Rao (2020) Artificial neural networks and adaptive neuro-fuzzy models for predicting WEDM machining responses of Nitinol alloy: comparative study, SN Applied Sciences, 314, 1–23.
24. P.M. Abhilash, D .Chakradhar (2020) ANFIS modelling of mean gap voltage variation to predict wire breakages during wire EDM of Inconel 718, CIRP Journal of Manufacturing Science and Technology, 31, 153–164.

25. H. Riahi-Madvar, A. Seifi (2018) Uncertainty analysis in bed load transport prediction of gravel bed rivers by ANN and ANFIS, Arabian Journal of Geosciences, 688, 1–20.
26. I.V. Manoj, R. Joy, S. Narendranath (2020) Investigation on the effect of variation in cutting speeds and angle of cut during slant type taper cutting in WEDM of Hastelloy X, Arabian Journal for Science and Engineering, 45(2), 641–651. https://doi.org/10.1007/s13369-019-04111-2.
27. A. Goswami, J. Kumar (2017) Trim cut machining and surface integrity analysis of Nimonic 80A alloy using wire cut EDM, Engineering Science and Technology, An International Journal, 20, 175–186.
28. V. Kumar, V. Kumar, K.K. Jangra (2015) An experimental analysis and optimization of machining rate and surface characteristics in WEDM of Monel-400 using RSM and desirability approach, Journal of Industrial Engineering International, 11, 297–307.
29. P. Sharma, D. Chakradhar, S. Narendranath (2015) Evaluation of WEDM performance characteristics of Inconel 706 for turbine disk application, Materials and Design 88, 558–566.
30. T. Chaudhary, A.N. Siddiquee, A.K. Chanda (2019) Effect of wire tension on different output responses during wire electric discharge machining on AISI 304 stainless steel, Defence Technology, 15, 541–544.
31. R. Joy, I.V. Manoj, S. Narendranath (2020) Investigation of cutting speed, recast layer and micro-hardness in angular machining using slant type taper fixture by WEDM of Hastelloy X, Materials Today: Proceedings, 27, 1943–1946.

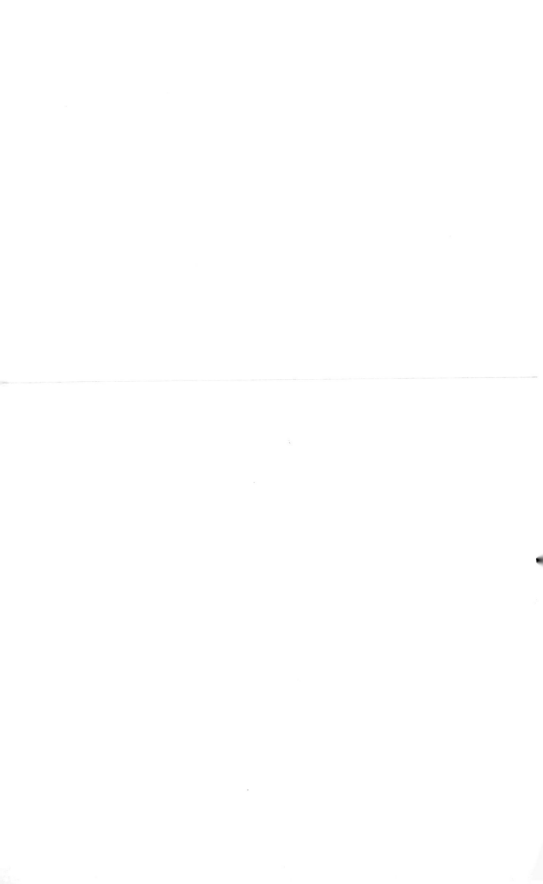

9 A Review on Milestones Achieved in the Additive Manufacturing of Functional Components

T. Sathies and P. Senthil

CONTENTS

9.1 Introduction: Background and Driving Forces 135
9.2 Application of AM in the Aerospace Sector 136
9.3 Application of AM in the Automotive Sector 137
9.4 Application of AM in the Construction Sector 138
9.5 Application of AM in the Electrical and Electronics Sector......................... 139
9.6 Application of AM in the Food Industry 142
9.7 Application of AM in the Medical Sector... 142
    9.7.1 Dental.. 143
    9.7.2 Bioprinting.. 144
    9.7.3 Scaffold Development 144
    9.7.4 Drug Delivery Devices... 145
    9.7.5 Implants 145
    9.7.6 Medical Devices 147
    9.7.7 Prostheses and Orthoses 147
9.8 Other Applications... 148
9.9 Research Scope 149
9.10 Conclusion 149
References.. 150

9.1 INTRODUCTION: BACKGROUND AND DRIVING FORCES

Traditionally, raw materials are converted into components of desired geometry through subtractive manufacturing technology. However, subtractive manufacturing involves huge material wastage, and it is not justifiable to produce customized shapes in low volume. Hence, additive manufacturing (AM) techniques are developed, and they can rapidly respond to change in the product design without wasting material and energy. AM denotes the manufacturing techniques in which the three-dimensional (3D) objects are constructed in a layer-by-layer fashion from different materials based on the computer-aided design (CAD) model. The materials that are possible to process by AM techniques are polymers, photopolymer resins, ceramics,

DOI: 10.1201/9781003093213-9

metals, composites, human tissues, food items, concretes, and the like. The sequence of steps required for the manufacturing of parts through AM is (1) creating a 3D model of an object using CAD software or a 3D scanner, (2) converting a 3D CAD model into an Standard Tessellation Language (STL) file, (3) processing the STL file in AM software and generating a tool path, (4) transferring the G-code to a 3D printer and setting the machine, (5) building the model by depositing the material in a layer-by-layer fashion, and (6) removing the part from the build platform and postprocessing.

The major classifications of AM techniques are (1) material extrusion (fused deposition modeling, contour crafting), (2) powder bed fusion (selective laser melting [SLM], selective laser sintering [SLS], electron beam melting [EBM], direct metal laser sintering [DMLS]), (3) VP (continuous liquid interface production [CLIP], stereolithography [SLA], digital light processing [DLP]), (4) directed energy deposition (laser engineered net shaping [LENS], electron beam freeform fabrication, wire arc AM[WAAM]), (5) sheet lamination (laminated object manufacturing [LOM]), (6) material jetting (MJ), and (7) binder jetting (BJ). During the initial stage of development, AM was widely used for the development of a prototype. Through continuous improvement, the usage of AM has changed from prototype development to functional component manufacturing. The present chapter aims at summarizing the progress in AM of functional components. Technological advancements and obstacles involved in the 3D printing of end-use components are projected in this chapter. Furthermore, the sector-specific future research directions in the additive manufacturing of functional parts are identified and encompassed.

9.2 APPLICATION OF AM IN THE AEROSPACE SECTOR

AM techniques are most preferable for the aerospace field because aerospace involves parts with complex geometry, and they are difficult to machine. Complex parts are needed in small batches but are expected to meet a high buy-to-fly ratio, and AM techniques have the potential to meet these demands. The aerospace sector has a demand for both metallic and nonmetallic AM techniques. Additively manufactured nonmetallic parts are used as prototypes (product development stage), fixture (manufacturing phase) and interiors (end use). Typical nonmetallic parts that are possible to manufacture by AM are tarmac nozzle bezel, airflow ducting, seat backs and entry doors, brackets and door handles, dashboard interfaces, wings and fuselage, and door handle covers. Metal-based AM technique has potential for the fabrication of end-use parts, rapid tooling, and repairing the worn surface. Metal parts that are under study for manufacturing by additive techniques are engine combustion chambers, antenna brackets, propulsion modules, turbine blades, turbine housings, airfoils, fuel nozzles, cabin brackets, exhaust ducts and bracket connectors. Since AM favors fabrication of topologically optimized design and eliminates assembly, a huge material reduction is achieved, and it results in saving of fuel [1, 2]. General electric (GE) additive in collaboration with GE aviation has successfully fabricated several aerospace components and some of them are fuel nozzle tip (cobalt-chrome alloy), low-pressure turbine blades (titanium aluminide), sensor housing (cobalt-chrome alloy), combustion mixer (cobalt-chrome alloy), cyclonic inducer (cobalt-chrome alloy), heat exchanger (aluminum), sump cover (cobalt-chrome alloy), and NACA inlet (Ti6Al4V). A concept laser M2 and an arcam EBM A2X machine were used by the GE additive in the AM of said components [3, 4].

Aircraft have many high-value items, and it is not economical to replace a component upon wearing. For example, integrally bladed rotors (blisks) incur thousands of dollars in their fabrication. Wearing of any single airfoil may lead to the scrapping of the entire blisk. Hence, a repairing stage is unavoidable in the aircraft industry. The geometry of the worn surface can be brought back to its original shape by using a LENS-based repairing system. LENS technology can precisely deposit material on the worn surface without damaging the other features of the component. Optomec used LENS technology for depositing wear-resistant cobalt-based material on worn Ti64 airfoil. The unique advantage of the LENS process is less heat input, a limited heat-affected zone, better mechanical properties, superior quality, the possibility to deposit complex shapes, greater repair access, and near net-shape manufacturing [5]. Factors limiting the usage of AM in the aerospace sector are a lack of consistency, resolution, and build quality. Several metal parts are not available for use immediately after fabrication and need to undergo multiple postprocessing stages. Numerous materials printable by AM are under study and need to be qualified through a safety test for final usage.

9.3 APPLICATION OF AM IN THE AUTOMOTIVE SECTOR

AM acts as a source of product innovation and can transform the supply chain. In the automotive sector, AM is primarily used for assisting the designers in the product development stage, and it helps to accelerate the new-product design phase. Fully functional prototypes are possible to manufacture by AM, and they can be tested for form and fit. As a result, the overall time to market a new product gets reduced drastically. Another use of AM in the automotive field is fabricating spare parts for limited-run and rare vehicles. The concept of reverse engineering is involved in the 3D printing of rare parts [6]. The certain automotive company uses 3D printing for the manufacturing of lighter customized tools. The use of 3D printed customized tools on the shop floor improves workers' comfort and reduces the manufacturing time. Furthermore, AM is also preferred in the luxury car and sports vehicle segment. Additively manufactured topologically optimized structures are used in luxury and sports vehicles for reducing weight and improving comfort. Delphi, an automotive company, adopts the SLM technique for producing an aluminum diesel engine pump. In comparison to aluminum die-casting technique, the quantum of material and energy required for SLM-made pumps found to be lower. Automotive components currently manufactured by AM techniques are bumpers, windbreakers, pumps, valves, and cooling vents. In the near future, AM can be used to manufacture engine components, dashboards, seat frames, hubcaps, tires, and suspension springs [7]. Barriers that can limit the implementation of AM in the automotive sector are the high cost of AM technologies, manufacturing technology limitation, a lack of skilled labors, inadequate management support, limited trust in a technology vendor, and intellectual property right threat in digital inventory. Production technology limitation includes limited material choice, postprocessing requirements, low production speed, and build volume restrictions [8]. By overcoming the previously mentioned barriers, the automotive sector can use AM for fabricating parts on demand, and the necessity of inventory can be reduced in large volume.

9.4 APPLICATION OF AM IN THE CONSTRUCTION SECTOR

The construction sector is one of the major contributors to the global economy. Since a huge human force is involved in construction, the productivity of the sector is lower in comparison to the other industries. Moreover, construction activities involve several hazardous tasks. Hence, the concept of AM is being considered in the construction industry. With AM, it is possible to improve customization, productivity, and human safety in construction. AM has the potential to fabricate entire structures and the necessity of formwork support could be reduced. The elimination of formwork permits the architect to consider novel designs for a building and helps to reduce waste [9]. Among the different AM techniques, extrusion and powder bed–based techniques find wider usage in construction. Different construction AM techniques falling under extrusion and powder bed method are summarized in Figure 9.1. During the initial stage of development, AM techniques were primarily used for developing the architectural model. But now, AM techniques are involved in a variety of construction activities. Research institutes and construction companies across the world have adopted AM techniques for building bridges, villas, apartments, pavilions, and office buildings, among others. Extrusion-based contour crafting and concrete printing techniques have a huge scope for printing

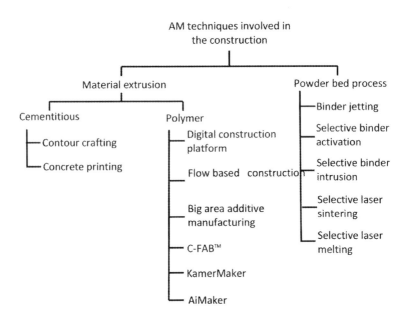

FIGURE 9.1 AM Techniques Involved in the Construction Sector

concrete and finds wider research work. Major parameters to be considered in extrusion-based techniques are extrudability, pumpability, interlayer adhesion, and buildability. To improve the structural performance of 3D-printed concrete, the concept of reinforcement was introduced by researchers. The reinforcement added to the concrete may be external reinforcement, internal reinforcement, fiber reinforcement, and AM reinforcement. Currently, the usage of metal AM in construction is limited because of the high cost of technology, fewer material choices, and dimensional restrictions. WAAM can print larger metal structures, and MX3D has used WAAM for building a dragon bench and a footbridge. Big-area AM is a technology like fused deposition modeling (FDM) developed for printing large-sized polymer structures. D-shape is a BJ technique used to develop large-scale components from sand substrates. The capabilities of AM allow it to be used for construction in harsh environments, war zones, and higher altitude locations. Additional tasks such as repairing and restructuring damaged structures are also possible. Factors limiting the widespread adoption of AM in construction are material-related issues, technology-related issues, and a lack of awareness. For effective implementation of AM in construction, future research can be devoted to multiple nozzle integration, hybrid 3D-printing system, local composition control, new material development, reinforcement study, process parameter optimization, life-cycle analysis, building information modeling, developing new standards, and improving process knowledge [10, 11, 12, 13]. Companies working on AM-based based construction are Apis Cor, BatiPrint, WASP, ICON, Winsun, COBOD, Cybe Construction, XtreeE, SQ4D, BetAbram, Contour Crafting, MudBots, Constructions-3D, and Be More 3D [14, 15].

9.5 APPLICATION OF AM IN THE ELECTRICAL AND ELECTRONICS SECTOR

AM is considered as an alternate manufacturing option in the electronic industry because AM promotes multiple functionalities, faster time to market, design freedom, and favors the production of customized miniature shapes in small batches and waste reduction. 3D-printed parts are tested in the field of embedded electronics, conformal electronics, stretchable electronics, and 3D structural electronics. The concept of embedded electronics helps to decrease the mass and the complexity involved in the assembly of an electronic component. Researchers have successfully developed conformal embedded 3D electronic systems by pausing the printing and embedding the electronic/electrical components inside the 3D structure while printing itself [16]. 3D-printing techniques are focused on the fabrication of traces, interconnects, LED resistors, capacitors, inductors, sensors, EMI shielding, enclosures, energy storage devices, and application-specific electronic devices [17]. Numerous researchers are investigating the AM of electronic components because AM facilitates the printing of substrates and conductive tracks and permits the embedding of electronic components within the print while printing. Noncontact printing methods like the ink-jet and the aerosol-jet printing process help to print conductive tracks within the 3D-printed substrate, and the qualities of the printed line depend on the substrate conditions, ink properties, printing process parameters, and sintering conditions [18]. Generally, conductive tracks printed by ink-based techniques tend to drift, and a nonuniform distribution of conductive nanoparticles may happen. Hence, proper control over the printing process is needed for achieving the required printing uniformity. To fabricate multi-material high-resolution electronic components, researchers have proposed the concept of hybrid manufacturing involving traditional and different AM methods [19].

The demand for complex-shaped low-loss communication devices has led researchers to look for alternate manufacturing strategies, and in that path, the ability of AM techniques to manufacture different communication devices are being investigated by the research community. Communication devices printed through AM techniques and tested for functionality are waveguide-based components, filters, front-end subsystems, antennas, antenna arrays, quasi-optical components, electromagnetic bandgap structures, and frequency-selective surfaces. The functionalities of additively manufactured antenna and other supplementary devices are affected by the resulting dimensional deviations and surface roughness. Dimensional deviation and higher surface roughness may lead to functional band shifting and higher insertion loss [20]. It is possible to fabricate an antenna by using a metallic or nonmetallic AM process. Antennas printed by nonmetallic AM techniques are made functional by providing a metallic coating over the polymer part. The functionalities of the metallic and polymeric antenna coated with metal are tested by researchers in different bands, namely L, C, S, X, K, Ku, Ka, V, and W bands and above. Even though the antenna fabricated by polymeric AM enjoys lightweight and low cost and demonstrates acceptable performance, the entire manufacturing step of the metal-coated polymeric antenna becomes complicated. The thermal stability, physical stability, and performance of metal-coated polymer antennas are inferior to that of metallic antennas. Moreover, the electrostatic discharge problem reported with polymeric antennas is higher than that of 3D-printed metallic antenna [21]. SLA and SLM processes offer better surface finish and may be considered as a suitable choice for fabricating polymeric and metallic antennas. For the successful AM of an antenna, further research work could be devoted to material characteristics improvement, process parameter optimization, surface treatment methods, and the development of hybrid technology [22]. The performance of the additively manufactured photonic and optoelectronic devices has also been discussed in recent years. AM techniques that have the ability to print photonic and optoelectronic devices are stereolithography, multiphoton stereolithography, digital light processing, extrusion-based printing (FDM, direct ink writing), MJ, and hybrid technologies. Examples of the additively manufactured photonic and optoelectronic devices tested for the performance are optical windows, lenses, optical sensors, waveguides, photonic crystals, metamaterials, photodiodes, organic LEDs, solar cells, optical resonators, saturable absorbers, lasers, and color displays. However, the 3D printing of photonic and optoelectronic devices is still in the infancy stage. Spatial resolutions of the most 3D-printed components are less and require postprocessing. The materials involved in the AM of photonic and optoelectronic devices must have low absorption and must exhibit suitable optical properties (optical gain, suitable light-to-current conversion, luminescence, optical nonlinearity) [23].

In recent years, the number of researchers reporting on the AM of sensors has increased steadily. Different types of sensors are 3D-printed, and their sensing ability is experimentally verified. Mostly, the sensing elements are 3D-printed from metal-based materials or carbon-based polymeric nanocomposites. 3D-printed sensors can be categorized into electrical and electronic sensors, mechanical sensors, chemical sensors, biosensors, and the like. 3D-printed sensors are successful in sensing characteristics, such as force, strain, pressure, temperature, gas, viscosity, pH, displacement, and flow [24]. The role of AM in analytical chemistry is inevitable. Examples of electroanalytical components fabricated with 3D printing include microfluidic devices, electrodes, sensors, labware, and support structures. Additively manufactured analytical devices are successful in performing pharmaceutical analysis, chemical analysis, metabolites monitoring, microorganism/biomolecule detection, diagnostics, and more [25]. 3D-printed microfluidic devices and electrodes in combination with other sensing elements are efficient in detecting several biological compounds, and they can be categorized as biomolecule-based sensors, microbial sensors, cell-based sensors, and bionic sensors. Examples of biological compounds detected via 3D-printed microfluidic devices and electrochemical sensors are glucose concentration, lactate concentration, DNA damage, malaria, cyanobacteria gliding movement, influenza virus, bacterial pathogens, cellular aging, urine protein concentration, hemagglutinin, peroxide content in human serum, dopamine in human blood serum, nitrite and uric acid in human urine and saliva, and serotonin overflow, among others [26, 27, 28]. Furthermore, additively manufactured conductive polymer composite parts, conductive polymer composite parts with metallic coating, metallic pats, and metallic parts with different coating exhibit sensitivity toward several compounds, namely, solvents, lead, cadmium, caffeine, paracetamol, copper presence in bioethanol, mercury, dopamine, trinitrotoluene, ferrocene, catechol, hydrogen peroxide presence in milk and mouthwash, picric acid, ascorbic acid, and several environmental pollutants (phenol, p-aminophenol, 1-naphthol, nitrite) [29, 30].

AM is considered a revolutionary process for the fabrication of electrochemistry components because AM facilitates (1) multiscale manufacturing, (2) multi-material manufacturing, (3) on-demand manufacturing, and (4) reinforced material printing [31]. Currently, lightweight 3D structures with well-controlled geometry are preferred for energy storage devices because 3D structures help achieve high energy and power density. AM techniques promote the fabrication of 3D-structured energy storage devices with well-controlled architecture [32]. Published works have discussed the functioning of 3D-printed structural electrodes (employed in batteries and supercapacitors), solid-state electrolyte, electrode supports, electrolyzer devices (current collectors, electrodes, and bipolar plates), electrode spacer, reactor vessels, flow cell components, electrochemical sensors, and microfluidic chips. Postprocessing techniques like electrodeposition, potentiostat polishing, atomic layer deposition, anodization, electrochemical activation, solvent activation, enzyme activation, and thermal activation are adopted by the researchers for improving the functionality of 3D electrochemical devices. The functionality of the 3D-printed electrochemical devices depends on the input material type and the AM process, the structure of the printed structure, and the post-modification steps demanded [33, 34]. Despite the benefits offered by 3D-printing technologies, factors such as lower resolution, higher anisotropy, poor surface finish, lesser material choices, inferior strength, and post-processing requirements limit the widespread adoption of AM for electrochemical device fabrication, and the use of AM is at the laboratory level [35].

9.6 APPLICATION OF AM IN THE FOOD INDUSTRY

Food printing is an emerging application of AM. 3D food printing favors customized food design, personalized nutrition design, food production process automation, and waste reduction. Commonly used AM methodologies in 3D food printing are extrusion, BJ, PBF, and MJ [36]. Materials printable (food inks) by extrusion technique are classified as natively and non natively extrudable by the researchers. Food inks coming under the natively extrudable category has sufficient viscoelastic properties and can be printed directly without the addition of a binding agent. Non natively extrudable food inks contain additives for better printing. For successful printing, the raw material should have suitable printing characteristics and proper printing parameters must be chosen. Furthermore, the printed structure must withstand postprocessing steps [37]. Normally additives (pectin, transglutaminase, methylcellulose, xanthan gum, sodium chloride, polysaccharides) are added with printable food materials because additives favor printability, quality, and stability. In few works, additives are added to provide nutrition and improve flavor. However, proper attention must be given to the selection of printing parameters when additives are added because the additives may change the rheological properties of ink [38]. 3D food printing can be effectively used in military food, space food, elderly food, and the confectionery market. Food printers developed by different companies are CocoJet™ and Chefjet™ by 3D Systems, QiaoKe by 3Dcloud, Foodjet™ by De Grood Innovations, CandyFab-6000 by CandyFab Project, Foodini™ by Natural Machines, Fouche Chocolate printer by Fouche Chocolates, Choc Creator™ by Choc Edge, GumJet 3D printer by GumLab project. For widespread adoption, 3D food printing techniques should be precise, accurate, and productive and must be capable of producing multi flavor, multistructure, and multicolor food items [39]. Furthermore, numerous preprocessing (comminution, gelation) and posttreatment stages (baking, drying, cooling) are involved in 3D food printing, and care must be taken for achieving desired accuracy, flavor, and stability.

9.7 APPLICATION OF AM IN THE MEDICAL SECTOR

Among different sectors, the medical field is one of the biggest beneficiaries of AM. From model development for surgical planning to the 3D printing of the entire organ, AM technologies have found multifold usages in the medical sector. 3D-printed models help doctors to plan an operation and communicate the surgical procedure to patients effectively. Broad classifications of AM applications in the medical sector are provided in Figure 9.2. The use of AM in the different medical fields is discussed in the following sections.

A Review on Milestones Achieved in AM

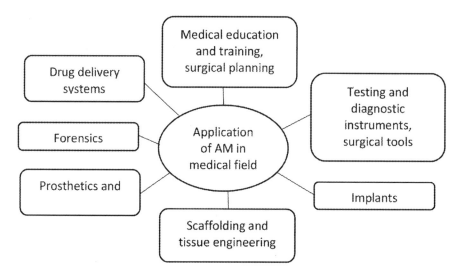

FIGURE 9.2 Classifications of AM Application in the Medical Field

9.7.1 Dental

The different categories of the dental sector that benefit from AM are periodontia, pediatric dentistry, restorative dentistry and endodontics, maxillofacial surgery, implant dentistry, and prosthetic dentistry [40]. AM techniques are capable of fabricating maxillofacial implants, anatomical and training models, dentures and crowns, surgical guides, and so on [41]. In the endodontics field, 3D-printed models developed from patient scan details help dentists to obtain endodontic access, guide drills, and prepare the transplanting site before teeth transplantation. With the 3D-printed space maintainers, it is possible to maintain the existing arc length for children in the case of premature tooth loss. Customized biodegradable and biocompatible scaffold fabricated by AM techniques can meet the requirements of periodontal and alveolar regenerative procedures, such as socket preservation, sinus augmentation, guided bone regeneration, vertical bone augmentation, and guided tissue regeneration [42]. Among different AM techniques, FDM technology is commonly used for printing anatomical and training models due to its low cost. The SLA or the DLP technique offers high resolution and has the potential to print biocompatible material. Hence, the SLA or DLP technique is preferred for printing crowns, bridges, surgical guides and splints, aligners, and retainers, dental satellites, and the like. PBF techniques such as SLM, DMLS, and EBM are preferred for making metallic maxillofacial implants. Printing of ceramic-based dental components could be realized with binder-jet or inkjet printing technology [43]. Although AM techniques are capable of meeting dental requirements, still, in-depth studies are needed on materials, design, quality, and performance for wider adoption of AM in the dental field.

9.7.2 BIOPRINTING

3D bioprinting, an advancement in regenerative medicine, aims at building tissue or organ structures in 3D by depositing the hydrogel-based supporting biomaterials and living cells or cell aggregates into demanded geometries. Bioprinting is explored for use in different applications, namely, the generation and transplantation of several tissues, bone, vascular grafts, heart tissue, tracheal splints, nerves, blood vessels, livers, biodegradable scaffolds, skin grafts, cartilaginous structures, and drug testing. Bio ink is the input material in the case of a bioprinter, and it is normally referred to as the solution or hydrogel form of biomaterials filled with the required cell types. Based on the functionality of structure build, bio inks are classified as structural, fugitive, support, and functional-bio inks. Materials involved in the bioprinting must have better printability, biocompatibility, suitable degradation kinetics and by-products, structural and mechanical properties, and biomimicry [44, 45, 46]. Bioprinters fall into three categories, namely, extrusion-based, inkjet, and laser-assisted, and the companies with expertise in the development of 3D bioprinters are Organovo, CELLINK, Aspect Biosystems, Cyfuse Biomedical, TeVido Biodevices, Digilab, Tissue Regeneration Systems, nScrypt, EnvisionTEC, MedPrin, Nano3D Sciences (n3D), Rokit, Cellbricks, REGEMAT 3D, Allevi, Poietis, RegenHU, Skolkovo (FABION 3D bioprinter), and GeSim's (Bioscaffolder), among others. Currently, most of the bioprinters are limited to preclinical research, and through continuous improvement, implantable organs can be printed in the near future [47, 48].

9.7.3 SCAFFOLD DEVELOPMENT

Scaffold refers to a support framework or structural element that holds cells or tissue together and helps in the formation of new functional tissues and repair bone defects. Both cell-laden and plain scaffolds are printable by AM techniques. Scaffold materials should allow cell adhesion and migration, the diffusion of secreted products and vital cell nutrients, and vascularization and support mechanical and biological functions [49]. Materials printable through AM techniques investigated for the development of scaffold fall into four categories namely, metals (degradable and nondegradable), polymers (natural polymers, synthetic biodegradable, and nondegradable polymers), ceramics, and composites [50]. In the case of scaffold fabrication, AM promotes customization in terms of shape, material type, pore size, and mechanical properties. 3D-printed scaffolds with suitable mechanical properties, porosity, biomimetic property, and surface topography are expected to exhibit biocompatibility, bioresorbability, non-immunogenic characteristics, and bioactive behaviors. In certain cases, the scaffold should have biodegradability, radiolucency, osteoconduction, and osteoinduction characteristics. In the near future, scaffolds printed by AM techniques are believed to promote the growth and regeneration of new bone tissue through the proper interaction of cells, signals, and scaffolds. Numerous in vitro and in vivo studies have been performed to assess the cell adsorption, proliferation, and differentiation characteristics of 3D-printed scaffolds. The usages of AM scaffolds are in the infancy stage because of (1) degradation of material during printing, (2) high processing cost, (3) a lack of demanded mechanical properties, (4) limited medical-grade materials,

A Review on Milestones Achieved in AM

(5) sterilization issues, (6) a lack of bioactivity in many cases, and (7) difficulties in preparing microvascular structures [51, 52].

9.7.4 Drug Delivery Devices

The demand for personalized drug delivery systems is on the rise because the quantity of medication must be designed based on the patient's age, fitness, gender, and health conditions. Currently, the ability of the AM to develop customized drug delivery systems has been widely tested by researchers. There are different ways for administrating drugs, and the ability of the different AM methods have been tested for developing oral solid medications (tailored aesthetics, disintegrated and polypill structures), transdermal patches, surgical meshes, biomedical implants, intravenous devices (eluting catheters), scaffolds, and rectal and vaginal delivery (tailored shapes and drug-eluting devices). Steps involved in the 3D printing of typical drug delivery systems are (1) a laboratory examination, (2) preparing a treatment plan and drug release profile, (3) designing the drug delivery system, and (4) AM of the drug delivery system. AM principles explored for the printing of drug delivery devices (DDD) are material extrusion (ME), BJ, VP, MJ, and PBF. Active pharmaceutical ingredients, namely, paracetamol, isoniazid, progesterone, aripiprazole, dipyridamole, theophylline, 5-ASA, captopril, prednisolone, budesonide, diclofenac sodium, nifedipine, glipizide, domperidone, fenofibrate, carbamazepine, 4-ASA, acetaminophen, naproxen, aspirin, atenolol, hydrochlorothiazide, pravastatin, and ramipril, have been blended with different AM-input materials (polylactic acid, polyvinylpyrrolidone, hydroxypropyl methylcellulose, polycaprolactone, polyvinyl alcohol, polyethylene glycol) and successfully printed [53, 54]. SPRITAM medication developed through ZipDose® Technology (Aprecia Pharmaceuticals) intended for an epilepsy treatment is the first 3D-printed drug that has received approval by the U.S. Food and Drug Administration (FDA) [55]. Researchers from Howard Hughes Medical Institute have developed a molecular 3D printer that can produce protein tyrosine phosphatases. Printlets technology, patented by FabRx, is involved in the development of polypills, fast-dissolving tablets, personalized dosages, and chewable medicines. The Institute of Chemicals Technology (ICT) and Tvasta are working on hybrid AM technology that can produce a tablet with a controlled release rate [56]. The raw materials involved in the 3D printing of a drug must meet the standards set by regulatory agencies, and strong care is required in eliminating the drug degradation during the AM of DDD. The interaction between the drug and AM materials, the distribution of the drug in a matrix, and the drug stability while printing decide the functionality of 3D-printed DDD [57]. By properly optimizing the raw material composition and printing parameters, the DDD of desired functionality could be achieved.

9.7.5 Implants

Owing to many accidents and age-related health issues, the demand for complex profile implants is steadily increasing. Implants placed inside the body should have mechanical properties compatible with human bone. Any mismatch in mechanical properties between human bone and implants may lead to implant failure because of the stress-shielding effect [58]. Furthermore, the implants are expected to manage osseointegration and reduce the risk of infection at the implant interface. To achieve the desired function, AM techniques are considered for the development of metallic, ceramic, and polymeric implants. The mechanical properties, osseointegration, osteoinduction, osteogenesis, and vascularization characteristics of the implants are tailored according to need by altering the implant geometry (porosity, pore size, shape, and permeability), chemical composition, biomechanics, and surface roughness. AM techniques have the potential to address the clinical challenges by fabricating novel implant geometries with the previously mentioned features. Parts printed through the PBF technique meet the requirements of the medical industry, and this process is widely involved in the fabrication of implants. Biocompatible AM materials possessing suitable osseointegration characteristics are Ti-6Al-4V, Co-Cr-Mo, PEEK, and tantalum. Even though the AM techniques offer numerous benefits, issues such as manufacturing defects (pores, lack of fusion, keyholes), reduced fatigue and compressive strength, heterogeneous microstructures, postprocessing requirements, and regulatory challenges inhibit the effectiveness of AM methods in printing implants [59, 60, 61]. Higher surface roughness and the presence of partially melted particles on the surface of the AM implants limit its functionality. Hence, detailed study is demanded regarding postprocessing and surface modification (coating, grafting, and antimicrobial agent immobilization) of AM implants [62]. Table 9.1 lists the details of a few FDA-approved AM implants.

TABLE 9.1
Details of FDA-Approved Implants Printed by AM Techniques [60, 62, 63]

Manufacturer	Device Name	FDA Number
Zavation	Ti3Z Cervical Interbody System	K191354
Stryker	Tritanium X PL Expandable Posterior Lumbar Cage, Tritanium X TL Expandable Curved Posterior Lumbar Cage,	K183249
Nexxt Spine	NEXXT MATRIXX Stand Alone Cervical System	K190546
Kalitec Medical	TiWAVE-L Porous Titanium Lumbar Cage	K182210
Innovasis	AXTi Titanium Stand-Alone ALIF System	K182139
DePuy Synthes	EIT Cellular Titanium Lumbar Cage—T/PLIF	K183447
Choice Spine	Hawkeye Vertebral Body Replacement	K183588
Camber Spine Technologies	SPIRA Open Matrix ALIF and LLIF	K190483
Addivation Medical	Cervical Interbody System	K190291
4 WEB Medical	Cervical Spinal Truss System Stand Alone (CSTS-SA)	K190870
4 WEB Medical	Hammertoe Truss System (HTS)	K190926
Additive Implants	SureMAX Cervical Spacer	K182477
Additive Orthopaedics	Patient Specific 3D Locking Lattice Plates	K183011
Exactech	Equinoxe Stemless Shoulder	K173388
INFINITY	Total Ankle Revision System	K180730
Additive Implants	Sure MAX-X Cervical Spacer	K193359
Osseus Fusion Systems	Aries® Lumbar Interbodies (Intervertebral Fusion Device with Bone Graft)	K181347
Oxford Performance Materials, Inc.	OsteoFab® Patient Specific Cranial Device	K121818

9.7.6 MEDICAL DEVICES

Customization, cost-effectiveness, and accessibility of AM techniques have driven researchers and medical experts toward the AM of medical devices. Besides, several additively manufactured medical devices possess magnetic resonance imaging (MRI) compatibility and biocompatibility. Medical devices of traditional or novel design are manufactured by AM techniques and found to satisfy the demand. Additively manufactured medical devices fall into three major categories, namely, instrumentation (laboratory equipment, diagnostic instrument, assistive tools), implants, and external prostheses and orthoses. 3D-printed general-purpose or specific-purpose instruments are tested in numerous clinical applications, and they are general surgery, intracranial surgery, microsurgery, open surgery, robotic surgery, diagnostics, endoscopic surgery, breast brachytherapy, magnetic resonance (MR)-guided percutaneous procedure, ear nose throat surgery, positron emission tomography–guided biopsy, laparoscopic surgery, endonasal surgery, single-port gastroenterology surgery, percutaneous intervention, neurosurgery, cervical intraepithelial neoplasia, and transrectal brachytherapy. Examples of 3D-printed medical instruments are tweezers, hemostats, forceps, needle drivers, retractors, scalpels, dental elevator, grippers and cutting tools for robotic surgery, reciprocating syringe, trocarcannula, the cap for a conventional colonoscope, laparoscopic graspers (DragonFlex project), smart steerable needles, and the branched overtube system of endoscopes, among others [64]. Customized surgical guides fabricated through AM enjoy benefits such as improved accuracy, reduced operating time, reduced learning curve, enhanced surgeon safety, and reduced blood loss [65]. During the COVID-19 pandemic, owing to supply chain issues, AM techniques have been involved in the development of several medical devices. Parts fabricated by AM techniques finding usage in the medical field during a pandemic are ventilator parts, swabs for testing, face shields, hand sanitizer holders, oxygen valves, lung models, door handle attachments, face masks, medical devices adaptors, and noninvasive positive end-expiratory pressure masks. Most of the reported instruments are printed by FDM, SLA, and SLS techniques [66, 67]. 3D-printed instruments meeting medical regulations and sterilization conditions are the need of the hour. Tight tolerance and appropriate surface finish must be maintained while printing medical devices [68]. Procedures followed in the direct manufacturing of medical implants are the diagnosis, imaging and scanning, data transformation, design and customization, biomechanical simulation, regulatory approval, rapid manufacturing, postprocessing, sterilization, and surgery [69].

9.7.7 PROSTHESES AND ORTHOSES

An orthosis represents an externally applied device aimed at correcting, accommodating, or enhancing the functionality of body parts. The aims of orthosis include (1) controlling biomechanical alignment, (2) correcting or accommodating deformity, (3) protecting and supporting an injury, (4) assisting rehabilitation, (5) reducing pain, and (6) increasing mobility and independence. Some of the commonly prescribed orthoses are ankle orthoses, foot orthoses, and knee orthoses; ankle–foot orthoses and knee–ankle–foot orthoses; upper-limb orthoses; fracture orthoses; and spinal orthoses [70]. Prosthesis denotes an artificial device affixed to the body to replace a missing human part. The major classification of prosthetic devices is joint prosthetics, arm prosthesis (transradial or transhumeral), leg prosthesis (transtibial and transfemoral), and cosmetic prosthesis [71]. In recent years, AM techniques have acted as the preferred choice for developing cost-effective custom-designed orthosis or prosthesis. AM techniques incorporated into the development of prostheses and orthoses favor limitless customization (design and functionality), weight reduction, optimal fit, and comfort. AM has been adopted in developing numerous prosthesis, and some of them are talar prosthesis, sacrum prosthesis, costal cartilage prosthesis, facial prosthesis, ocular prosthesis, below-the-knee prosthesis for snowboarding, prosthetic ankle, scapular prosthesis, prosthetic socket, prosthetic arm, prosthetic hand, obturator prosthesis, foot prosthesis, hand prosthesis, auricular prosthesis, and phalanx toe prosthesis. Factors such as type of actuation, weight, type of control, force distribution, type of extensor, type of flexor, kinematic specifications (degrees of freedom, number of joints, and actuators), range of motion, and grasp type governs the functionality of 3D-printed upper limb prosthesis [72, 73, 74].

Orthosis fabricated through AM techniques can be classified as upper limb orthoses (hand orthoses, elbow orthoses, cock-up splint, wrist orthoses,), spinal orthoses (lumbar orthoses, cervical-thoracic lumbosacral orthoses, thoracic orthoses, cervical orthoses), and lower limb orthoses (foot orthoses, knee orthoses, hip orthoses, insoles). Extrusion, VP, and PBF-based AM techniques find a place in the development of prosthesis and orthosis [75, 76]. POHLIG GmbH, OT4 Othopädietechnik, plus medica OT, Mecuris, Chabloz Orthopédie, Shapeways, ScientiFeet, Invent Medical, Xkelet, RS Print, and HP Fit Station are the leading companies working on the AM of prosthesis and orthosis [77, 78]. Even though 3D printing favors customization and the 3D printed prosthesis and orthosis are lighter, issues such as higher cost, limited mechanical properties, and material degradation over time restrict the wider usage of AM in prosthesis and orthosis development. Support structure requirements and manufacturing constraints must be kept in mind while designing custom prosthesis and orthosis.

9.8 OTHER APPLICATIONS

Recently, the use of AM in the fashion and consumer product industry has been increasing steadily because the AM favors mass customization and improves the ergonomics and sustainability of products. A myriad of consumer products have been produced by AM techniques, and they fall into different categories, namely, consumer electronics, packaging, sporting goods, appliances, musical instruments, lighting, indoor and outdoor furniture, tools, bottles, and toys. Examples of consumer products produced by AM techniques are insoles, eyewear frames, sports shoes, watches, jewelry, dive lights, wheel hubs, camera mounts, helmets, cycle frames, DEEPTIME's audio set, razor handles, violins, showerheads, mascara brushes, dental aligners, and dresses. [79, 80, 81].

The adoption of AM in the education sector is believed to create excitement, give access to knowledge previously unavailable, complement the curriculum, open new possibilities for learning, and promote problem-solving skills. In the medical field, AM helps to create a 3D anatomical model for explaining, understanding, and preparing complex medical procedures. 3D-printed prototypes are found to be effective in helping students to understand topics such as biological molecules, atomic structures, protein structures, material properties, microelectromechanical designs, different geometries, the foundations of engineering, enzymes, and ligand structures, among others. Examples of 3D-printed medical models that help medical practitioners are airway models, bones, femoral artery, heart, limb sections, lungs, oral surgical model, orbital dissections, prosected human cadavers, skeletal tissues, prosthodontic models, and so on [82, 83].

9.9 RESEARCH SCOPE

The application capability of AM techniques has attained a new height in recent years. Certain AM techniques have matured and are widely involved in the fabrication of functional components. However, further research is required for using AM techniques to their fullest potential. Currently, the widespread adoption of AM techniques is limited by constraints such as inferior mechanical properties, poor surface characteristics (certain cases), regulation issues, size restrictions, limited in-house expertise, software challenges, and higher cost. Further research could be devoted to newer material development, process parameter optimization, machine capability improvement, waste minimization, support structure optimization, postprocessing techniques, the standardization of AM techniques, and the integration of artificial intelligence with AM.

9.10 CONCLUSION

The research works performed in the AM of functional components are summarized in the present chapter. Advantages such as limited material wastage, tool-less fabrication, less processing time, the potential to fabricate the lightweight and multi-material structure, lesser inventory requirement, on-demand manufacturing, and design freedom have made the industries consider AM as an alternate manufacturing process for the fabrication of customized shapes in low volume. By improving the range of materials printable and the accuracy of AM techniques, new heights can be achieved in AM of functional components.

Abbreviations

AM	Additive Manufacturing	**SLA**	Stereolithography
3D	Three-Dimensional	**CLIP**	Continuous Liquid Interface Production
CAD	Computer-Aided Design		
STL	Standard Tessellation Language	**SLM**	Selective Laser Melting
SLS	Selective Laser Sintering	**DED**	Directed Energy Deposition
LENS	Laser Engineered Net Shaping	**PBF**	Powder Bed Fusion
WAAM	Wire Arc Additive Manufacturing	**EBM**	Electron Beam Melting
FDM	Fused Deposition Modelling	**GE**	General Electricals
PEEK	Poly Ether Ether Ketone	**LED**	Light-Emitting Diode
MEMS	Micro Electromechanical System	**DDD**	Drug Delivery Devices
EMI	Electromagnetic Interference	**FDA**	Food and Drug Administration
DLP	Digital Light Processing	**DNA**	Deoxyribonucleic Acid
DOF	Degrees of Freedom	**DMLS**	Direct Metal Laser Sintering

REFERENCES

1. Najmon, J.C., S. Raeisi and A. Tovar (2019) Review of additive manufacturing technologies and applications in the aerospace industry. Additive Manufacturing for the Aerospace Industry, 7–31.
2. Liu, R., Z. Wang, T. Sparks, F. Liou and J. Newkirk (2017) Aerospace applications of laser additive manufacturing. In Laser Additive Manufacturing: Materials, Design, Technologies, and Applications, Woodhead Publishing Series in Electronic and Optical Materials, Woodhead Publishing, Cambridge, England, 351–371.
3. GE ADDITIVE (2021) Application of additive manufacturing in aerospace sector. Available at: www.ge.com/additive/sites/default/files/2020-08/GE9X%20Additive%20parts.pdf.
4. GE ADDITIVE (2021) Metal additive parts. Available at: www.ge.com/additive/additive-parts.
5. Optomec (2021) LENS blisk repair solution. Available at: www.optomec.com/wp-content/uploads/2014/04/Optomec_LENS_Blisk_Repair_Datasheet.pdf.
6. Nichols, M.R. (2019) How does the automotive industry benefit from 3D metal printing? Metal Powder Report, 74(5), 257–258.
7. Giff, C.A., B. Gangula and P. Illinda (2014) 3D opportunity in the automotive industry. Available at: www2.deloitte.com/content/dam/insights/us/articles/additive-manufacturing-3d-opportunity-in-automotive/DUP_707-3D-Opportunity-Auto-Industry_MASTER.pdf.
8. Dwivedi, G., S.K. Srivastava and R.K. Srivastava (2017) Analysis of barriers to implement additive manufacturing technology in the Indian automotive sector. International Journal of Physical Distribution & Logistics Management, 47(10), 972–991.
9. Sanjayan, J.G. and B. Nematollahi (2019) 3D concrete printing for construction applications. 3D Concrete Printing Technology, 1–11.
10. Paolini, A., S. Kollmannsberger and E. Rank (2019) Additive manufacturing in construction: A review on processes, applications, and digital planning methods. Additive Manufacturing, 30, 100894.
11. Wu, P., J. Wang and X. Wang (2016) A critical review of the use of 3-D printing in the construction industry. Automation in Construction, 68, 21–31.
12. Zhang, J., J. Wang, S. Dong, X. Yu and B. Han (2019) A review of the current progress and application of 3D printed concrete. Composites Part A, 125, 105533.
13. Bhardwaj, A., S.Z. Jones, N. Kalantar, Z. Pei, J. Vickers, T. Wangler, P. Zavattieri and N. Zou (2019) Additive manufacturing processes for infrastructure construction: A review. Journal of Manufacturing Science and Engineering, 141(9), 091010.
14. Jamie, D. (2020) The manufacturers of 3D printed houses. Available at: www.3dnatives.com/en/3d-printed-house-companies-120220184/#!.
15. Aniwaa (2021) The 13 best construction 3D printers in 2021. Available at: www.aniwaa.com/buyers-guide/3d-printers/house-3d-printer-construction/.
16. Lu, B., H. Lan and H. Liu (2018) Additive manufacturing frontier: 3D printing electronics. Opto-Electronic Advances, 1(1), 170004.
17. Espera, A.H., J.R.C. Dizon, Q. Chen and R.C. Advincula (2019) 3D-printing and advanced manufacturing for electronics. Progress in Additive Manufacturing, 4, 245–267.
18. Zhang, H., S.K. Moon and T.H. Ngo (2020) 3D printed electronics of non-contact ink writing techniques: Status and promise. International Journal of Precision Engineering and Manufacturing-Green Technology, 7, 511–524.
19. Urasinska-Wojcik, B., N. Chilton, P. Todd, C. Elsworthy, M. Bates, G. Roberts and G.J. Gibbons (2019) Integrated manufacture of polymer and conductive tracks for real-world applications. Additive Manufacturing, 29, 100777.

20. Zhang, B., W. Chen, Y. Wu, K. Ding and R. Li (2017) Review of 3D printed millimeter-wave and terahertz passive devices. International Journal of Antennas and Propagation, 2017, 1297931.
21. D. Helena, A. Ramos, T. Varum and J.N. Matos (2020) Antenna design using modern additive manufacturing technology: A review. IEEE Access, 8, 177064–177083.
22. B. Zhang, Y. Guo, H. Zirath and Y.P. Zhang (2017) Investigation on 3-D-printing technologies for millimeter-wave and terahertz applications. Proceedings of the IEEE, 105(4), 723–736.
23. Camposeo, A., L. Persano, M. Farsari and D. Pisignano (2019) Additive manufacturing: Applications and directions in photonics and optoelectronics. Advanced Optical Materials, 7(1), 1800419.
24. Palenzuela, C.L.M. and M. Pumera (2018) (Bio) Analytical chemistry enabled by 3D printing: Sensors and biosensors. Trends in Analytical Chemistry, 103, 110–118.
25. Choudhary, H., D. Vaithiyanathan and H. Kumar (2020) A review on additive manufactured sensors. MAPAN-Journal of Metrology Society of India.
26. Khosravani, M.R. and T. Reinicke (2020) 3D-printed sensors: Current progress and future challenges. Sensors and Actuators A, 305, 111916.
27. Abdalla, A. and B.A. Patel (2020) 3D-printed electrochemical sensors: A new horizon for measurement of biomolecules. Current Opinion in Electrochemistry, 20, 78–81.
28. Ni, Y., R. Ji, K. Long, T. Bu, K. Chen and S. Zhuang (2017) A review of 3D-printed sensors. Applied Spectroscopy Reviews, 52, 7, 623–652.
29. Cardoso, R.M., C. Kalinke, R.G. Rocha, P.L. dos Santos, D.P. Rocha, P.R. Oliveira, B.C. Janegitz, J.A. Bonacin, E.M. Richter and R.A.A. Munoz (2020) Additive-manufactured (3D-printed) electrochemical sensors: A critical review. Analytica Chimica Acta, 1118, 73–91.
30. Muñoz, J. and M. Pumera (2020) Accounts in 3D-printed electrochemical sensors: Towards monitoring of environmental pollutants. Chem Electro Chem, 7(16), 3404–3413.
31. Hashemi, S.M.H., U. Babic, P. Hadikhani and D. Psaltis (2020) The potentials of additive manufacturing for mass production of electrochemical energy systems. Current Opinion in Electrochemistry, 20, 54–59.
32. Cheng, M., R. Deivanayagam and R. Shahbazian-Yassar (2020) 3D printing of electrochemical energy storage devices: A review of printing techniques and electrode/electrolyte architectures. Batteries & Supercaps, 3(2), 130–146.
33. Browne, M.P., E. Redondo and M. Pumera (2020) 3D printing for electrochemical energy applications. Chemical Reviews, 120(5), 2783–2810.
34. Ambrosi, A., R. Rong, S. Shi and R.D. Webster (2020) 3D-printing for electrolytic processes and electrochemical flow systems. Journal of Materials Chemistry A, 8, 21902–21929.
35. Chang, P., H. Mei, S. Zhou, K.G. Dassios and L. Cheng (2019) 3D printed electrochemical energy storage devices. Journal of Materials Chemistry A, 7, 4230–4258.
36. Godoi, F.C., B.R. Bhandari, S. Prakash and M. Zhang (2019) An Introduction to the Principles of 3D-food Printing. In Fundamentals of 3D Food Printing and Applications, Academic Press, Cambridge, MA.
37. Liu, Z., M. Zhang, B. Bhandari and Y. Wang (2017) 3D printing: Printing precision and application in food sector. Trends in Food Science & Technology, 69, 83–94.
38. Voon, S.L., J. Ana, G. Wong, Y. Zhang and C.K. Chua (2019) 3D food printing: A categorised review of inks and their development. Virtual and Physical Prototyping, 14(3), 203–218.
39. Nachal, N., J.A. Moses, P. Karthik and C. Anandharamakrishnan (2019) Applications of 3D printing in food processing. Food Engineering Reviews, 11, 123–141.

40. Vasamsetty, P., T. Pss, D. Kukkala, M. Singamshetty and S. Gajula (2020) 3D printing in dentistry—Exploring the new horizons. Materials Today: Proceedings, 26(2), 838–841.
41. Aishwarya, B., S. Vijayavenkatraman, V. Rosa, L.W. Feng and Y.H. Jerry Fuh (2018) Applications of additive manufacturing in dentistry: A review. Journal of Biomedical Materials Research Part B Applied Biomaterials, 106(5), 2058–2064.
42. Oberoi, G., S. Nitsch, M. Edelmayer, K. Janjic´, A.S. Müller and H. Agis (2018) 3D printing—Encompassing the facets of dentistry. Frontiers in Bioengineering and Biotechnology, 6, 172.
43. Javaida, M. and A. Haleem (2019) Current status and applications of additive manufacturing in dentistry: A literature-based review. Journal of Oral Biology and Craniofacial Research, 9, 179–185.
44. Murphy, S.V. and A. Atala (2014) 3D bioprinting of tissues and organs. Nature Biotechnology, 32(8), 773–785.
45. Javaid, M. and A. Haleem (2020) 3D printed tissue and organ using additive manufacturing: An overview. Clinical Epidemiology and Global Health, 8(2), 586–594.
46. Mataia, I., G. Kaur, A. Seyedsalehi, A. McClinton, C.T. Laurencin (2020) Progress in 3D bioprinting technology for tissue/organ regenerative engineering. Biomaterials, 226, 119536.
47. Medical Futurist (2018) The top bioprinting companies. Available at: https://medicalfuturist.com/top-bioprinting-companies/.
48. Sher, D. (2015) The top 15 bioprinters. Available at: https://3dprintingindustry.com/news/top-10-bioprinters-55699/.
49. Madrid, A.P.M., S.M. Vrech, M.A. Sanchez and A.P. Rodriguez (2019) Advances in additive manufacturing for bone tissue engineering scaffolds. Materials Science & Engineering C, 100, 631–644.
50. Chen, Y., W. Li, C. Zhang, Z. Wu and J. Liu (2020) Recent developments of biomaterials for additive manufacturing of bone scaffolds. Advanced Healthcare Materials, 2000724.
51. Qu, H. (2020) Additive manufacturing for bone tissue engineering scaffolds. Materials Today Communications, 24, 101024.
52. Garot, C., G. Bettega and C. Picart (2020) Additive manufacturing of material scaffolds for bone regeneration: Toward application in the clinics. Advanced Functional Materials, 2006967.
53. Beg, S., W.H. Almalki, A. Malik, M. Farhan, M. Aatif, Z. Rahman, N.K. Alruwaili, M. Alrobaian, M. Tarique and M. Rahman (2020) 3D printing for drug delivery and biomedical applications. Drug Discovery Today, 25(9), 1668–1681.
54. Mohammed, A., A. Elshaer., P. Sareh., M. Elsayed and H. Hassanin (2020) Additive manufacturing technologies for drug delivery applications. International Journal of Pharmaceutics, 580, 119245.
55. West, T.G. and T.J. Bradbury (2019) 3D printing: A case of ZipDose® technology—World's first 3D printing platform to obtain FDA approval for a pharmaceutical product. In 3D and 4D Printing in Biomedical Applications: Process Engineering and Additive Manufacturing, Wiley-VCH Verlag GmbH & Co. KGaA, Weinheim, Germany.
56. Sriram, R. (2019) 3D printing drugs: The latest advancements. Available at: https://all3dp.com/2/3d-printing-drugs-the-latest-advancements-around-the-world/.
57. Wallis, M., S. Al-Dulimi, D.K. Tan, M. Maniruzzaman and A. Nokhodchi (2020) 3D printing for enhanced drug delivery: Current state-of-the-art and challenges. Drug Development and Industrial Pharmacy, 46(9), 1385–1401.
58. Ni, J., H. Ling, S. Zhang, Z. Wang, Z. Peng, C. Benyshek, R. Zan, A.K. Miri, Z. Li, X. Zhang, J. Lee, K.-J. Lee, H.-J. Kim, P. Tebon, T. Hoffman, M.R. Dokmeci, N. Ashammakhi, X. Li and A. Khademhosseini (2019) Three-dimensional printing of metals for biomedical applications. Materials Today Bio, 3, 100024.

59. Burnard, J.L., W. C. H. Parr., W. J. Choy, W. R. Walsh and R. J. Mobbs (2020) 3D-printed spine surgery implants: A systematic review of the efficacy and clinical safety profile of patient-specific and off-the-shelf devices. European Spine Journal, 29, 1248–1260.
60. Lowther, M., S. Louth, A. Davey, A. Hussain, P. Ginestra, L. Carter, N. Eisenstein, L. Grover and S. Cox (2019) Clinical, industrial, and research perspectives on powder bed fusion additively manufactured metal implants. Additive Manufacturing, 28, 565–584.
61. Murra, L.E. (2019) Strategies for creating living, additively manufactured, open-cellular metal and alloy implants by promoting osseointegration, osteoinduction and vascularization: An overview. Journal of Materials Science & Technology, 35, 231–241.
62. Sarker, A., M. Leary and K. Fox (2020) Metallic additive manufacturing for bone-interfacing implants. Biointerphases, 15, 050801.
63. Vetalice, J.A. (2019) FDA 510(k) Review: 2019 additively manufactured product clearances. Available at: www.bonezonepub.com/2503-fda-510-k-review-2019-additively-manufactured-product-clearances.
64. Culmone, C., G. Smit and P. Breedveld (2019) Additive manufacturing of medical instruments: A state-of-the-art review. Additive Manufacturing, 27, 461–473.
65. Yilmaz, A., A.F. Badria, P.Y. Huri, G. Huri (2019) 3D-printed surgical guides. Annals of Joint, 4.
66. Horst, A., F. McDonald and D.W. Hutmacher (2019) A clarion call for understanding regulatory processes for additive manufacturing in the health sector. Expert Review of Medical Devices, 16(5), 405–412.
67. Javaid, M. and A. Haleem (2018) Additive manufacturing applications in medical cases: A literature-based review. Alexandria Journal of Medicine, 54, 411–422.
68. Larrañeta, E., J. Dominguez-Robles and D.A. Lamprou (2020) Additive manufacturing can assist in the fight against COVID-19 and other pandemics and impact on the global supply Chain. 3D Printing and Additive Manufacturing, 7(3), 100–103.
69. Arora, P.K., R. Arora, A. Haleem and H. Kumar (2020) Application of additive manufacturing in challenges posed by COVID-19. Materials Today: Proceedings, In Press.
70. Australian orthotic prosthetic association (AOPA) (2021) Orthoses and prostheses. Available at: www.aopa.org.au/careers/what-are-orthoses-and-prostheses.
71. Bailey, A. (2021) Different types of prosthetics. Available at: https://livehealthy.chron.com/different-types-prosthetics-1244.html.
72. EOS (2021) 3D printing production of orthoses and prostheses. Available at: www.eos.info/en/3d-printing-examples-applications/people-health/medical-3d-printing/orthoses-prostheses.
73. Wang, Y., Q. Tan, F. Pu, D. Boone and M. Zhang (2020) A review of the application of additive manufacturing in prosthetic and orthotic clinics from a biomechanical perspective. Engineering, 6, 1258–1266.
74. Kate, J.T., G. Smit and P. Breedveld (2017) 3D-printed upper limb prostheses: A review. Disability and Rehabilitation: Assistive Technology, 12(3), 300–314.
75. Choo, Y.J., M. Boudier-Revéret and M.C. Chang (2020) 3D printing technology applied to orthosis manufacturing: Narrative review. Annals of Palliative Medicine, 9(6), 4262–4270.
76. Chen, R.K., Yu-an Jin, J. Wensman and A. Shih (2016) Additive manufacturing of custom orthoses and prostheses—A review. Additive Manufacturing, 12, 77–89.
77. Carolo, L. (2020) 3D printed orthotics: 7 most promising projects. Available at: https://all3dp.com/2/3d-printed-orthotics-most-promising-projects/.
78. Carlota, V. (2019) Top 12 3D printed orthoses. Available at: www.3dnatives.com/en/3d-printed-orthoses-110620194/.
79. Protocam (2021) Consumer product prototyping and additive manufacturing, consumer goods product development. Available at: www.protocam.com/markets/consumer-goods/.

80. AMFG (2019) 10 exciting ways 3D printing is being used in the consumer goods industry. available at: https://amfg.ai/2019/03/12/10-exciting-ways-3d-printing-is-being-used-in-the-consumer-goods-industry/.
81. EOS (2021) 3D printing for more diversity in design and engineering. Available at: www.eos.info/en/3d-printing-examples-applications/people-health/sports-lifestyle-consumer-goods.
82. Makerbot (2021) The top 5 benefits of 3D printing in education. Available at: www.makerbot.com/stories/3d-printing-education/5-benefits-of-3d-printing/.
83. Ford, S. and T. Minshall (2019) Invited review article: Where and how 3D printing is used in teaching and education. Additive Manufacturing, 25, 131–150.

10 Fatigue Behavior of Particulate-Reinforced Polymer Composites
A Review

Vijay Verma, Arun Kumar Pandey, and Chaitanya Sharma

CONTENTS

10.1 Introduction 155
    10.1.1 Processing Techniques 156
    10.1.2 The Filler Materials 157
10.2 Fatigue Properties of Particulate Polymer Composites 158
    10.2.1 Effect of Particle Size on the Fatigue Properties of
            Particulate Polymer Composites 159
    10.2.2 Effect of Particle Shape on the Fatigue Properties of
            Particulate Polymer Composites 159
    10.2.3 Effect of Particle Content on the Fatigue Properties of
            Particulate Polymer Composites 162
    10.2.4 Effect of Particle and Matrix Interface on the Fatigue
            Properties of Particulate Polymer Composites 163
10.3 Overall Summary 166
10.4 Future Scope 167
References 168

10.1 INTRODUCTION

The relatively low cost, low density, and high strength-to-weight ratio of polymers make them good material for engineering applications. Polymers have added advantage of high resistance toward environmental and chemical corrosion. Even then, the use of polymers is limited to low-load application and noncritical places. Low resistance toward crack initiation and propagation, lower stiffness, and distortion temperature are the limiting criteria for the use of polymers in engineering applications. Polymers are viscoelastic material, so their use is further restricted to low-temperature applications. Due to these limitations, the application of polymers in their pure form is limited. To overcome these limitations and to widen the scope of application, both rigid and elastomeric materials are used to reinforce polymer material.

DOI: 10.1201/9781003093213-10

Long and short fibers as fillers have been an area of research for a long time. But the fiber-reinforced composites have the disadvantage of having directional properties. The properties in the longitudinal direction are highly enhanced by these fibers, whereas, in the transverse direction, the properties still remain very poor. Due to these directional properties of the materials, the fibers are considered as single-dimensional bodies. To cope up with these limitations of fibers, the use of particulate fillers has increased. The particulate composites have an advantage of having a uniform improvement of properties in all the direction. Thus, to enhance the mechanical properties of polymers, different types of micro-particulates [1–15] and nano-particulates [16–37], such as glass beads [4–6, 11], alumina particles (Al_2O_3) [9, 21–24], silica particles [15, 31–32, 37], elastomeric particles (both natural and synthetic) [38–40], clay [41–52] and the like are used to blend with the polymer matrix. The size of particulate fillers may be in the micrometer range or in the nanometer range. When the fillers are in the micrometer range, the number of particles of reinforcement is less than that of fillers of nano size [22]. Reducing the size of reinforced particles from the micrometer to the nanometer range, conventional fabrication methods of composites with micro-particles cannot be used for composites with nanofiller. The nano-sized fillers have entirely different characteristics compared to micro-fillers [1–7, 16–25]. And for the fabrication of nano-sized filler composites, different methods should be adopted.

Both rigid and elastomeric particles modify the fatigue properties of polymer composites by different mechanisms. The inclusion of rigid fillers of micrometer size improves Young's modulus and the fracture toughness of composites while strength and failure strain are adversely affected [9, 26, 32]. The inclusion of elastomeric fillers enhances the fracture toughness and failure strain of composites with a simultaneous decrease in Young's modulus and the tensile strength of composites [38–40]. The inclusion of these fillers also imparts a brittle to tough transition to the composite by increasing ductility in the brittle thermoset epoxy matrix due to the massive plastic deformation of the reinforced fillers.

10.1.1 PROCESSING TECHNIQUES

Depending on type of fillers and the polymer matrix, the method of processing may vary from sol-gel [53–54], solution mixing/in situ polymerization [43–44, 47–48] or melt blending [18, 46, 49–50]. Several other variants of the previously mentioned techniques are also being developed and have been reported in the literature [20–21]. These techniques have their own merits and demerits.

Sol-gel is a chemical process with the advantage of controlling the properties at a molecular level. In this method, maximum homogeneity may be achieved by control at a molecular level, but the total densification of composites is not possible by this process. The sol-gel process starts from precursors, and an interpenetrating network of organic and inorganic phases are grown simultaneously. A low volume of batch size and the incomplete removal of the pores even after densification are the main limitations of the sol-gel process [53].

In situ polymerization technique, generally used for thermosetting polymers and involves polymerization with reinforcement already in place [43–44, 47–48]. Fillers are initially dispersed in liquid resin by mechanical mixing techniques such as ultra-sonication, high-speed shear mixing, or ball milling. No chemical reaction takes place during the dispersion of the filler material. The polymerization process is then initiated by using a catalyst called a hardener, which initiates and accelerates the crosslinking process. The in situ polymerization process consumes less time compared to the sol-gel method. The homogeneous dispersion of fillers requires a high shear energy wave, which limits the batch size to be processed by this method.

The melt-blending technique is a general process used for processing thermoplastics. Micro-granules/platelets of polymer are mixed with fillers using mechanical mixing methods. The temperature during mechanical mixing of polymer and filler is controlled to prevent overheating the polymer. After the mixing, the mixture then melted either in a single or doublescrew extruder to form a filament or pellets [18, 46, 49–50]. The filament or pellets so formed are reprocessed either in an injection molding machine or a compression molding machine to form the composites. The mechanical properties of composites are governed by the type of filler material reinforced and the method used for the fabrication of composites.

10.1.2 THE FILLER MATERIALS

There are large varieties of materials available that may be used as filler materials for the processing of composites. Clay is one of the most widely used filler materials reported in the literature. Both thermoplastics [41–42, 45–46, 49] and thermosets [43, 47–48] are reinforced with clay. Clay, the layered silica, may be dispersed in bunches with 1–2-nm interlayer spacing as micro-sized tactoids or an intercalated or exfoliated structure with individual layer randomly oriented [45]. The advantage of the high stiffness of ceramics makes them a good choice for filler material. The reinforcement of polymer with a low volume fraction of ceramic nanofillers, such as alumina oxide, silicon carbide, and silicon dioxide, has shown improvement in various mechanical properties [17, 19–20, 23, 26].

Carbon nanotubes (CNTs) are also of great use as filler material due to their characteristics of high aspect ratio. CNTs have a diameter in the range of few nanometers and a length up to few micro-meters. CNTs have a structure similar to highly crystalline fiber. CNTs are basically the cylindrical folding of graphene sheets. CNTs can be single-layered, double-layered, or multilayered. The inclusion of CNTs imparts an enhancement in the strength and modulus of composites [55–58]. Thermoplastics and elastomers are another class of reinforcement [8, 38–40]. Both the thermoplastics and elastomers have a similar type of reinforcing effect. Fracture toughness of the composites increases, but the tensile properties are reduced by the reinforcement of these materials.

The shape of reinforced particles varies as spherical, cylindrical or any regular or irregular shapes. The researchers are using micro as well as nanoparticles for the reinforcement of polymer matrix composites. Aluminum oxide (Al_2O_3) platelets of 200 nm thickness and 5–10 μm of plane dimension have been used to reinforce the epoxy matrix [24, 26]. The platelets had an added advantage of high contact-surface area; thus, the interfacial area is increased and has a high aspect ratio compared to spherical or rod-shaped reinforcement. A higher improvement in mechanical properties of composites is reported by using platelet-type reinforcement even at a low volume fraction of filler material due to the larger surface area.

The size and shapes of the reinforcement particles have a great effect on the properties of polymer matrix composites. The different mechanical properties are adversely affected by the size and shape of the reinforced materials. The present chapter focuses on the effect of morphology and volume fraction of the particulate reinforcement on fatigue properties of polymer composites.

10.2 FATIGUE PROPERTIES OF PARTICULATE POLYMER COMPOSITES

The cumulative damage of materials against the fluctuating stress is termed as fatigue. All engineering materials have entirely different behavior under fatigue loading compared to their behavior under static loading. Even materials that have highly ductile nature under the static loading condition will fracture in a brittle manner under the influence of fatigue loading, whereas materials that are brittle in nature are very poor inhibitors of fatigue crack nucleation and propagation. Epoxy, a thermoset plastic, has a very low resistance to fatigue crack initiation and propagation which limits the use of such material in engineering applications. Various attempts have been made to improve the fatigue life of epoxy by reinforcing it with fibers and particulates.

The fatigue property of the brittle thermoset epoxy is enhanced by reinforcement with both hard ceramics and soft rubber (natural and synthetic) particles [10, 13, 15, 25, 33–35, 51–52, 62–63]. The smaller particles (0.2 µm) have resulted in greater improvement in the resistance to fatigue crack propagation (FCP) than larger particles (1.5 µm). It has also been shown that V_f.% of particles has no influence on the threshold fatigue (ΔK_{th}) properties, whereas the near-threshold (ΔK_{th}) fatigue properties are strongly affected by the blend morphology [10].

The temperature rise during cyclic loading also has a great effect on the fatigue life of polymer. As the stress level during cyclic loading increases, the temperature will also increase, resulting in a rapid decay of strength and modulus. By the addition of nanoclay as reinforcement, there will be less thermal softening and modulus drop of polyamide-6 during cyclic loading [28]. Furthermore, good interfacial bonding between the matrix and the reinforcement improves the FCP resistance of the composite [27]. Particles debonding, plastic void creation, crack deflection and crack pinning are the major contributors to the increase in the resistance to FCP of polymers [27, 55]. Less matrix splitting was observed during cyclic loading, as a result of strong interfacial bonding in single-walled carbon nanotube (SWCNT) epoxy composites [55].

The measurement of change in residual resistance of SWCNT network composites is a better way to predict the location of failure of a composite than the measurement of residual strain [56]. A rapid increase in residual resistance during initial cycles that was followed by the slower and consistent increase in residual resistance in consecutive cycles was also reported.

Nanoparticles have a tendency to form agglomerates in a polymer matrix if not processed properly. To get enhancement in long-term mechanical behavior, such as fatigue and creep properties, the homogeneous dispersion of second-phase particles in the composite has to be achieved. The use of nanostructured epoxidized triblock copolymer or functionalization with silane coating can reduce the agglomeration of nanoparticles and results in a homogeneous dispersion of nanoparticles [26, 30, 56–57].

The reinforcement of low modulus fibers with high modulus fibers will enhance the fatigue resistance of fiber composites [62]. During fatigue loading, when the fracture starts in high modulus fibers with partial matrix damage, the low modulus fiber will stabilize the fatigue damage. Percent volume fraction and the type of reinforcement have a great effect on the fatigue behavior of the polymer nanocomposites [63].

10.2.1 Effect of Particle Size on the Fatigue Properties of Particulate Polymer Composites

The size of the reinforcement particles greatly influences the fatigue life of polymer nanocomposites. One order higher FCP resistance at the near-threshold regime was reported when the epoxy was reinforced with 0.2-μm rubber particles compared to 1.5-μm rubber particles. The V_f.% of the rubber particles had no effect on FCP resistance at ΔK levels below $ΔK_{th}$, whereas at higher ΔK (ΔK > $ΔK_{th}$), the FCP resistance had a strong dependency on the volume fraction of rubber particles [10].

The morphology of reinforced particles affects the fatigue properties of polymer composites. High-density polyethylene (HDPE) was reinforced with hydroxyapatite (HA) at 20% volume fraction and 40% volume fraction [13]. Micro-whiskers and micro-particles of HA (1.3 μm) were used as reinforcement. The fatigue life of the composites was enhanced more by the micro-whiskers reinforcement compared to micro-particles reinforcement [12–13]. The fatigue life of composites with micro-whiskers has been found 4 to 5 times higher than that of composites with particles at low volume fraction (20%) of particles. The increase in fatigue life was due to the higher aspect ratio and higher surface area of the whiskers [13].

The number of smaller size particles will be greater in comparison to that of bigger size particles at the same V_f.% of fillers in composite. In a 50,000-μm³ reference volume composite having 3% volume fraction of titanium oxide (TiO_2) particles will have only 3 particles of 10 μm size, whereas the number of particles will increase to 3×10^6 if the particle size is decreased to 100 nm [16, 22]. In the case of smaller-sized particles, a greater number of hindering particles is present, retarding the FCP rate and enhancing the fatigue performance of the composite. Achieving uniform spread is an important governing factor [64–65]. Both these important factors govern the fatigue property of composite.

10.2.2 Effect of Particle Shape on the Fatigue Properties of Particulate Polymer Composites

The size of the particle is also a very important factor controlling the fatigue life of polymer-based composites. Researchers have used different shapes of reinforcement particles and found varying effect of these particles on fatigue life. Verma and Sharma [67] suggested that spherical alumina particles are more efficient in improving the fatigue life of an epoxy composite compared to rod-shaped Al_2O_3 particles of similar dimensions. The variation of fatigue life of composite reinforced with

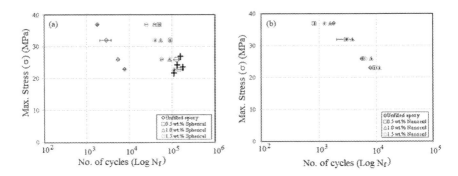

FIGURE 10.1 S-N Fatigue Life of an Alumina Epoxy Composite Filler by (a) Spherical and (b) Rod-Shaped Alumina Nanoparticles (Adopted from [67])

spherical and rod-shaped alumina nanoparticles is shown in Figure 10.1. This figure elaborates that the fatigue life of a spherical particle–reinforced composite is 10–28 times more in comparison to an unfilled epoxy and a rod-shaped Al_2O_3 nanoparticle–reinforced epoxy composite [67]. The increase in the fatigue life of spherical alumina composites at all weight fractions of reinforcement considered under study is more relative to a composite reinforced with rod-shaped alumina nanoparticles. Among spherical alumina nanoparticle composites, the increase in fatigue life is more in a high cyclic fatigue regime, whereby the applied maximum stress is low in comparison to a low cyclic fatigue regime (higher σ_{max}).

Orientation is independent of the fabrication and dispersion method used, thus increasing the fatigue life of spherical Al_2O_3 particulate composites. The spherical particles support the mobility and alignment of polymer chains. It may be the other reason for increasing the fatigue behavior of spherical Al_2O_3 particulate composites. The rod-shaped particles can have different orientations and dispersion behaviors depending on the fabrication technique and the method of dispersion. The rod-shaped particles offer more resistance to mobility and arrangement of the polymer chain, thus resulting in less fatigue behavior of the epoxy composite [66–67].

The FCP rate will depend on the weight fraction as well as on the morphology of the particulate filler. Verma and Sharma [67] reported in their study that spherical shape alumina nanoparticles are more influential in improving the FCP resistance compared to rod shape alumina nanoparticles. Figure 10.2 shows the FCP behavior da/dN (m/cycle) with an applied stress-intensity factor ΔK (MPa√m) for rod-shaped and spherical alumina nanoparticle–reinforced epoxy composite. From Figure 10.2, it is clearly evident that as the shape of the reinforced nanoparticles changes from rod-shaped to spherical, the FCP resistance of the epoxy composite improves. The improvement in the FCP resistance is due to a higher surface area and lower stress concentration caused by spherical nanoparticles compared to rod-shaped nanoparticles.

The higher surface area will have more surface binding force, so propagating crack requires more energy to overcome the obstacle caused by the reinforced particles. Thus, higher surface area nanoparticles can efficiently cause crack pinning, bowing and will delay the crack propagation rate. Figure 10.3 shows

Particulate-Reinforced Polymer Composites 161

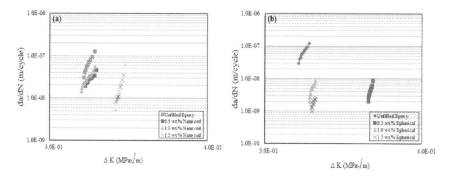

FIGURE 10.2 FCP Rate (da/dN) with ΔK of an Epoxy Alumina Composite with (a) Rod-Shaped and (b) Spherical Alumina Nanoparticles [Adopted from 67]

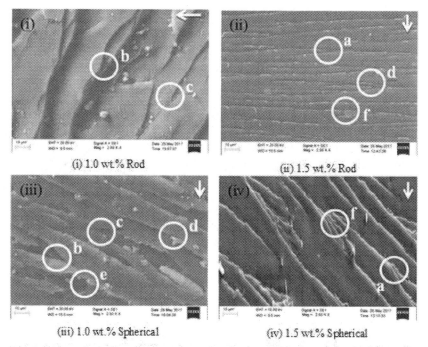

((a) crack bifurcation, (b) crack bifurcation and crack plane shift, (c) crack bowing, (d) matrix spalling, (e) deep craters and (f) secondary striation)

FIGURE 10.3 Scanning Electron Microscopy (SEM) Images of the Fatigue Fracture Surface of Nanoparticle-Reinforced Composite (Adopted from [67])

the fractured surface of a composite having 1.0 and 1.5 wt.% of rod-shaped shape (Figures 10.3a and 10.3b) and spherical (Figures 10.3c and 10.3d) alumina nanoparticles. Figure 10.3 clearly shows evidence that different energy decapitation mechanisms occurred during crack advancement in the case of rod-shaped and spherical alumina nanoparticle reinforced epoxy composites.

10.2.3 Effect of Particle Content on the Fatigue Properties of Particulate Polymer Composites

The fatigue life of HDPE HA composites decreased with an increase in the volume fraction of HA from 20% to 40% [13]. The decrease in the fatigue life was more severe in the case of composites reinforced with HA micro-particles compared to that of composite of HA micro-whiskers [13]. The flexural fatigue response of the SiC/epoxy composites was influenced by particle wt.%. At a lower level of applied load, 1.5 wt.% of SiC nanoparticle composites performed better (increased fatigue life) compared to that of neat epoxy. The threshold level (ΔK_{th}) was approximately 60%. Above the threshold load level, nanocomposites with the 1.5 wt.% of SiC nanoparticulate performed better (increased fatigue life) in comparison to nanocomposites having 3.0 wt.% of SiC, whereas, at high levels of applied load (i.e., 85% and 90%), the performance of neat epoxy is better than both (1.5 wt.% and 3 wt.%) composites [19].

Loading of composites with hard (silica 20 nm), as well as soft (rubber 100 nm), particles also influenced the fatigue property of the composites [29]. Resistance to FCP was dominant with 10 wt.% of silica in comparison to its counterpart of 2 wt.% rubber-reinforced composites. At a higher stress level (40 MPa), the increase in fatigue life was greater for composites having 2 wt.% of silica nanoparticles compared to 10 wt.% of silica nanocomposites. At a low stress level (20 MPa), the rubber/epoxy nanocomposite showed a reduction in fatigue life at all wt.% of reinforcement considered in the study.

In a similar work, a decrease in Paris-Erdogan constants "C" and "m" was more with silica/epoxy nanocomposites relative to rubber/epoxy or silica-rubber/epoxy nanocomposites [35]. At a constant wt.% (12%) of reinforcement, the decrease in the FCP rate was higher for composites having silica nanoparticles (20 nm) compared to that of composites of rubber nanoparticles (100 nm) or of hybrid nanoparticles (6.0 wt.% each of rubber and silica nanoparticles). Crack deflection, debonding of silica particles and void growth were the main mechanisms reported for the increases in the fatigue life of composites with silica nanoparticles while shear yielding and rubber cavitation were the main mechanisms in the case of rubber/epoxy composites.

At constant wt.% reinforcement of crushed silica micro-particles (average particles size 20–30 µm), the fatigue life (da/dN) and the threshold-stress intensity factor (ΔK_{th}) depend on stress (R) ratio (defined as the ratio of minimum applied cyclic stress and maximum applied cyclic stress) [15]. Furthermore, the value of ΔK_{th} decreases with an increase in the R-ratio. The decrease in the ΔK_{th} value for both neat epoxy and epoxy composite reinforced with crushed silica was higher at a higher R-ratio. At an R-ratio of 0.1, the ΔK_{th} of the composite was increased by 26% relative to the unfilled epoxy. The increased micro-cracking ahead of the main crack and the deflection of crack were the main mechanisms reported for the improvement in FCP resistance of silica epoxy composites.

Reinforcing epoxy with alumina nanoparticles had positive impact on the FCP behavior of a composite [22]. At a constant applied stress, the FCP rate decreased with increase in the volume fraction of alumina nanoparticles.

A reduction of a thousandfold in FCP rate has been reported when the volume fraction of alumina was increased from 1% to 10%. The applied threshold-stress intensity, ΔK_{th}, increased with the reinforcement of silica nanoparticles. The fatigue life of the nanocomposite was also increased. Furthermore, the whole da/dN vs ΔK curve was shifted to higher limits of the applied stress-intensity factor range [25].

Carbon nanofibers (CNFs) also increased the fatigue life of epoxy composites [33–34]. At low wt.% (0.5%) of CNFs, the fatigue life of the composite increased by 15% compared to unfilled epoxy in a low cyclic fatigue regime [32–34]. The increase in fatigue life depends on the R-ratio. At a constant wt.% (1%) of CNFs, the fatigue life of composites at R = 0.1 was increased by 150% to 580% over the entire loading range (40 to 110 MPa maximum applied stress), whereas a 340%–670% increase in the fatigue life of the composites was reported over the entire loading range (40 to 110 MPa maximum applied stress) consider for study at R = −0.5 [34]. Similar behavior of enhanced fatigue life was also observed at a low wt.% of fullerene (C60) for fullerene-epoxy composites [36].

Increasing the wt.% of nano silica, 20 nm in size, the crack growth rate versus the applied stress-intensity curve was moved to a higher ΔK_{th} value. At a constant ΔK_{th}, the fatigue crack growth rate decreases with an increase in wt.% of silica nanoparticles. At a constant ΔK of 0.4, the crack growth rate decreased from 2.3×10^{-4} m/cycle to 6.6×10^{-6} m/cycle by including 20 wt.% of silica nanoparticles in epoxy. Paris-Erdogan constant "m" decreased with an increase in wt.% of silica.

10.2.4 Effect of Particle and Matrix Interface on the Fatigue Properties of Particulate Polymer Composites

Interfacial bonding plays an important role in the fatigue properties of the polymer composites [14, 26–27]. In case of untreated fillers, there is lack of chemical bonding between the matrix and the reinforcement and thus making weak interfacial bonds. The use of coupling agents or silane to modify the contact surface of the second-phase particles provides a strong chemical bond between the fillers and matrix [26–27]. The Paris-Erdogan equation has been widely used for predicting the FCP rate during fatigue loading [22, 27, 59–61].

Functionalized and nonfunctionalized alumina/epoxy nanocomposites (having untreated as well as 3-aminopropyltriethoxysilane [APTES]–treated nanoparticles) resulted in a decrease in the FCP rate relative to that of unfilled epoxy. At higher stress-intensity range (ΔK), the nanocomposite reinforced with functionalized alumina particles showed a greater reduction in FCP rate relative to nanocomposites of untreated alumina [27].

Figure 10.4 presents the transmission electron microscope view of epoxy composite reinforced with spherical and rod-shaped alumina nanoparticles. From Figures 10.4c and 10.4d, we can see that at constant weight fraction of reinforcement, the dispersion of spherical nanoparticles is quite homogeneous throughout the epoxy matrix, whereas at the same weight fraction of reinforcement, the rod-shaped nanoparticles form bunches (Figures 10.4a and 10.4b). The ability to form a

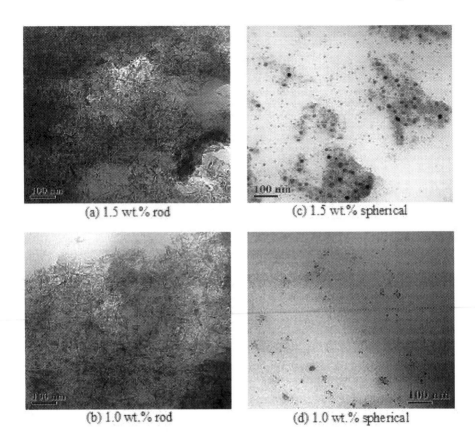

FIGURE 10.4 Transmission Electron Microscopy Images of an Epoxy Alumina Composite Reinforced with Different Weight Fraction of Rod-Shaped and Spherical Alumina Nanoparticles (Adopted from [67])

more homogeneous dispersion of spherical particles in an epoxy matrix will render a higher fatigue life compared to the rod-shaped alumina nanoparticles, although the overall dimensions of both types of particles remain the same, that is, less than 50 nm. Thus, the morphology of the particles also has a great influence on the dispersion ability and fatigue properties epoxy composite [67–69].

Table 10.1 summarizes the effect of rigid and elastomeric reinforcement on the stress-based fatigue life of polymer composites. Table 10.2 summarizes the effect of reinforcing rigid and elastomeric reinforcement on the fracture-based fatigue properties (FCP rate) of polymer composites. The maximum increase of 365% in the fatigue life of epoxy was in the case of 1.0 wt.% of CNFs [33]. The FCP resistance of epoxy was enhanced by both rubber and ceramic (alumina, silica, etc.) particles. The maximum (67%) decrease in the FCP rate was at 3.1 wt.% of alumina in epoxy [27].

TABLE 10.1
Influence of Nanoparticles on the Fatigue Life of Composites

Material System		Treatment	% Change in Fatigue Life (% Particle wt.%)	Remarks	Ref.
Matrix	Reinforcement				
HDPE	HA Whiskers	–	+4–5 times (20 wt.%)	Percentage change is in comparison to HA powder filler	20
Epoxy	Carbon + SiC	–	+62% (1.5 wt.%)	when applied stress is 50% of ultimate tensile strength (UTS) maximum improvement is 62%, while reduction of 97% at 90% of UTS.	35
Epoxy	Nano silica Nano rubber	– –	+145% (10 wt.%) −43% (10 wt.%)	Nano-silica/epoxy in low stress regime and nano-rubber/epoxy decreases in low as well as high stress regime.	43
Epoxy	Carbon nanofiber	–	+365% (1.0 wt.%)	Higher increases at low applied stress (20 MPa), marginal increases at high applied (40 MPa) stress.	49
Epoxy	Carbon nanofiber	–	150–670% (1.0 wt.%)	Increase in entire range from fully compression to tensile–tensile loading.	50
Epoxy	CNT	Yes	+2400% (0.5 wt.%)	Number of fatigue cycles increases from 20,000 to 5,00,000 (0.5 wt.%) Triton X-100 treated composites.	95
Polyurethane	MWCNT	–	+248% (0.3 wt.%)	248% in high-stress low-cycle fatigue regime, while no significant increase in low-stress high-cycle fatigue regime.	97

TABLE 10.2
Influence of Micro- and Nanoparticles on Fracture-Based Fatigue Parameters of Polymer Composites

Material System		Treatment	% Change in Slope "m" (% Particle wt.%)	Remarks	Ref.
Matrix	Reinforcement				
Epoxy	Rubber (0.2 μm)	No	−57% (10 wt%)	FCP rate decreased from 10^{-5} to 10^{-6} (10 wt.%).	10
Epoxy	Micro silica	No	—	R-ratio is a governing parameter for FCP rate and threshold ΔK_{th}.	15
Epoxy	Al_2O_3 (30 nm)	No	—	1000 times decrease in FCP rate (10% [Vol.]).	22
Epoxy	Silica (20 nm)	Yes	—	ΔK_{th} increased by 80% at 7.8 wt.%.	25
Epoxy	Al_2O_3 (45 nm)	No Yes	−30% (3.1 wt%) −50% (3.1 wt%)	67% decrease in FCP rate (3.1 wt.%) of treated Al_2O_3 epoxy composite.	27
Epoxy	Nano silica Nano rubber	No No	−36% (12 wt%) −53% (12 wt%)	The fatigue life and ΔK_{th} increased with silica and decreased with reinforcement of rubber.	78
Epoxy	Carbon fullerene	No	−17% (1.0 wt%)	The increase in fatigue life was due to higher strength of C60 fullerene due to sp2 bond.	36
Epoxy	Silica (20 nm)	Yes	—	FCP resistance increases by 97% at 20 wt.% of silica. The Threshold ΔK_{th} increases by 50% at 20 wt.% of silica.	37

10.3 OVERALL SUMMARY

The experimental results of various polymer matrix particulate composites with micro and nano-sized fillers were critically evaluated under the influence of fatigue load. The following points can be summarized based on the discussion made in this chapter:

- Reinforcement was either soft elastomer or rigid particles of a micro or nano size.
- It is easy to disperse a micro-sized filler, while a nano-sized filler requires high energy dispersion techniques such as ultrasonication, high-speed shear mixer, and the like.
- Micro-fillers improve the fracture toughness of a composite whereas the strength and failure strain decrease.

- The fatigue life of polymer nanocomposites increases with the volume fraction of the filler. The increase in fatigue life is higher in a high-cyclic fatigue regime compared to a low-cyclic fatigue regime.
- FCP resistance of polymer nanocomposites increases with the filler volume fraction. The applied threshold stress intensity (ΔK_{th}) increases with the volume fraction of filler.
- Functionalized nanofillers disperse more uniformly in the composite and provide better fatigue life behavior compared to unfilled polymer and composites having agglomerated nanofillers.
- The morphology of the reinforced particles also governs the dispersion quality as well as the fatigue behavior of composites.

10.4 FUTURE SCOPE

Following areas may be the future scope of the present chapter:

1. Mathematical correlation and simulation on the fatigue behavior with variable loading and overloading conditions can be conducted for better understanding and application of epoxy-alumina composite in engineering applications.
2. Other types of fillers, especially carbon fillers with different shapes, can be studied.

List of Abbreviations

S. No.	Abbreviation	Full Form
1	Al_2O_3	Alumina oxide
2	SiC	Silicon carbide
3	SiO_2	Silicon dioxide
4	CNT	Carbon nanotubes
5	SWCNT	Single-walled carbon nanotubes
6	MWCNT	Multiwalled carbon nanotubes
7	CNF	Carbon nanofiber
8	ΔK	Range of applied stress intensity factor
9	ΔK_{th}	Threshold applied stress intensity factor
10	Da/Dn	Crack growth rate per cycle
11	FCP	Fatigue crack propagation rate
12	HA	Hydroxyapatite
13	APTES	3-Aminopropyl-triethoxysilane
14	TiO_2	Titanium dioxide
15	UTS	Universal tensile strength
16	HDPE	High-density polyethylene
17	R-ratio	Stress ratio
18	$V_f.\%$	Volume fraction percentage

REFERENCES

1. Radford K.C., "The mechanical properties of an epoxy resin with a second phase dispersion", J. Mater. Sci., 1971; 6: 1286–1291.
2. Young R.J. and Beaumont P.W.R., "Failure of brittle polymers by slow crack growth-Part 3 effect of composition upon fracture of silica particle-filled epoxy resin composites", J. Mater. Sci., 1977; 12: 684–692.
3. Faber K.T. and Evans A.G., "Crack deflection processes-II. Experiment", Acta. Metall., 1983; 31 (4): 577–584.
4. Spanoudakis J. and Young R.J., "Crack-propagation in a glass particle filled epoxy-resin: Part 1 effect of particle volume fraction and size", J. Mater. Sci., 1984; 19: 473–486.
5. Spanoudakis J. and Young R.J., "Crack-propagation in a glass particle filled epoxy-resin: Part 2 effect of particle-matrix adhesion", J. Mater. Sci., 1984; 19: 487–496.
6. Kinloch A.J., Maxwell D.L. and Young R.J., "The fracture of hybrid—particulate composites", J. Mater. Sci., 1985; 20: 4169–4184.
7. Moloney A.C., Kausch H.H., Kaiser T. and Beer H.R., "Review—parameters determining the strength and toughness of particulate filled epoxide resins", J. Mater. Sci., 1987; 22: 381–393.
8. Pearson R.A. and Yee A.F., "Toughening mechanisms in thermoplastic modified epoxies: 1-Modification using poly (phenylene oxide)", Polymer, 1993; 34 (17): 3658–3670.
9. Hussain M., Oku Y., Nakahira A. and Niihara K., "Effects of wet ball milling on particle dispersion and mechanical properties of particulate epoxy composites", Mater. Lett., 1996; 26: 177–184.
10. Azimi H.R., Pearson R.A. and Hertzberg R.W., "Fatigue of rubber-modified epoxies: Effect of particle size and volume fraction", J. Mater. Sci., 1996; 31: 3777–3789.
11. Lee J. and Yee A.F., "Effect of rubber interlayers on the fracture of glass bead/epoxy composites", J. Mater. Sci., 2001; 36: 7–20.
12. Roeder R.K., Sproul M.M. and Turner C.H., "Hydroxyapatite whiskers provide improved mechanical properties in reinforced polymer composites", J. Biomed. Mater. Res., 2003: 67: 801–812.
13. Kane R.J., Concerse G.L. and Roeder R.K., "Effect of the reinforcement morphology on the fatigue properties of Hydroxyapatite reinforced polymers", J. Mech. Behav. Biomed. Mater. 2008; I: 261–268.
14. Mallarino S., Chailan J.F. and Vernet J.L., "Interphase study in cyanate/glass fiber composites using thermomechanical analysis and micro-thermal analysis", Compos. Sci. Technol., 2009; 69: 28–32.
15. Boonyapookana A., Nagata K. and Mutoh Y., "Fatigue crack growth behavior of silica particulate reinforced epoxy resin composite", Compos. Sci. Technol., 2011; 71; 1124–1131.
16. Wetzel B., Haupert F., Friedrich K., Zhang M.Q. and Rong M.Z., "Impact and wear resistance of polymer nanocomposites at low filler content", Polym. Eng. Sci., 2002; 42: 1919–1927.
17. Zheng Y., Zheng Y. and Ning R., "Effects of nanoparticles SiO_2 on the performance of nanocomposites", Mater. Lett., 2003; 57: 2940–2944.
18. Mahfuz H., Adnan A., Rangari V.K., Jeelani S. and Jang B.Z., "Carbon nanoparticles/whiskers reinforced composites and their tensile response", Compos. Part A., 2004; 35: 519–527.
19. Chisholm N., Mahfuz H., Rangari V.K., Ashfaq A. and Jeelani S., "Fabrication and mechanical characterization of carbon/SiC-epoxy nanocomposites", Compos. Struct., 2005; 67: 115–124.

20. Kuo M.C., Tsai C.M., Huang J.C. and Chen M., "PEEK composites reinforced by nano-sized SiO_2 and Al_2O_3 particulates", Mater. Chem. Phys., 2005; 90: 185–195.
21. Zheng H., Zhang J., Lu S., Wang G. and Xu Z., "Effect of core—shell composite particles on the sintering behavior and properties of nano-Al_2O_3/polystyrene composite prepared by SLS", Mater. Lett., 2006; 60: 1219–1223.
22. Wetzel B., Rosso P., Haupert F., Friedrich K., "Epoxy composites—fracture and toughening mechanisms", Eng. Fract. Mech., 2006; 73: 2375–2398.
23. Zunjarrao S.C. and Singh R.P., "Characterization of the fracture behavior of epoxy reinforced with nanometer and micrometer sized aluminum particles", Compos. Sci. Technol., 2006; 66: 2296–2305.
24. Shukla DK., Parameswaran V., "Epoxy composites with 200 nm thick alumina platelets as reinforcements", J. Mater. Sci., 2007, 42: 5964–5972.
25. Blackman B.R.K., Kinloch A.J., Lee J.S., Taylor A.C., Agarwal R., Schueneman G. and Sprenger S., "The fracture and fatigue behaviour of nano-modified epoxy polymer", J. Mater. Sci., 2007; 42: 7049–7051.
26. Shukla D.K., Kasisomayajula S.V. and Parameswaran V., "Epoxy composites using functionalized alumina platelets as reinforcements", Compos. Sci. Technol., 2008; 68: 3055–3063.
27. Zhao S., Schadler L.S., Hillborg H. and Auletta T., "Improvements and mechanisms of fracture and fatigue properties of well-dispersed alumina/epoxy nanocomposites", Compos. Sci. Technol., 2008; 68: 2976–2982.
28. Ramkumar A. and Gnanamoorthy R., "Axial fatigue behaviour of polyamide-6 and polyamide-6 nanocomposites at room temperature", Compos. Sci. Technol., 2008; 68: 3401–3405.
29. Wang G.T., Liu H.Y., Saintier N. and Mai Y.W., "Cyclic fatigue of polymer nanocomposites", Eng. Fail. Anal., 2009; 16: 2635–2645.
30. Ocando C., Tercjak A. and Mondragon I., "Nanostructured systems based on SBS epoxidized triblock copolymers and well dispersed alumina/epoxy matrix composites", Compos. Sci. Technol., 2010; 70: 1106–1112.
31. Hsieh T.H., Kinloch A.J., Masania K., Taylor A.C. and Sprenger S., "The mechanisms and mechanics of the toughening of epoxy polymer modified with silica nanoparticles", Polymer, 2010; 51: 6284–6294.
32. Uddin M.F. and Sun C.T., "Improved dispersion and mechanical properties of hybrid nanocomposites", Compos. Sci. Technol., 2010; 70: 223–230.
33. Bortz D.R., Merino C. and Gullon I.M., "Carbon nanofibers enhance the fracture toughness and fatigue performance of a structural epoxy system", Compos. Sci. Technol., 2011; 71: 31–38.
34. Bortz D.R., Merino C. and Gullon I.M., "Augmented fatigue performance and constant life diagram of hierarchical carbon fiber/nanofiber epoxy composites", Compos. Sci. Tech., 2012; 72: 446–452.
35. Liu H.Y., Wang G. and Mai Y.W., "Cyclic fatigue crack propagation of nanoparticle modified epoxy", Compos. Sci. Technol., 2012; 72: 1530–1538.
36. Rafiee M.A., Yavari F., Rafiee J. and Koratkar N., "Fullerene-epoxy nanocomposites-enhanced mechanical properties at low nanofiller loading", J. Nanopart. Res., 2011; 13: 733–737.
37. Kothmann M.H., Zeiler R., Anda A.R.D., Bruckner A. and Altstadt V., "Fatigue crack propagation behaviour of epoxy resins modified with silica nanoparticles", Polymer, 2015; 60: 157–163.
38. Huang Y. and Kinloch A.J., "The toughness of epoxy polymers containing microvoids", Polymer, 1992; 33 (6): 1330–1332.

39. Bagheri R. and Pearson R.A., "The use of microvoids to toughen polymers", Polymer, 1995; 36 (25): 4883–4885.
40. Bagheri R. and Pearson R.A., "Role of particle cavitation in rubber-toughened epoxies: II. Iner-particle distance", Polymer, 2000; 41: 269–276.
41. Kojima Y., Usuki A., Kawasumi M., Okada A., Fukushima Y., Kurauchi T. and Kamigaito O., "Mechanical properties of nylon 6-clay hybrid", J. Mater. Res., 1993; 8 (5): 1185–1189.
42. Kojima Y., Usuki A., Kawasumi M., Okada A., Kurauchi T. and Kamigaito O., "Sorption of water in nylon 6-clay hybrid", J. Appl. Polym. Sci., 1993; 49: 1259–1264.
43. Chin I.J., Albrechta T.T., Kima H.C., Russella T.P. and Wang J., "On exfoliation of montmorillonite in epoxy", Polymer, 2001; 42: 5947–5952.
44. Kornmann X., Lindberg H. and Berglund L.A., "Synthesis of epoxy-clay nanocomposites. Influence of the nature of the curing agent on structure", Polymer, 2001; 42: 4493–4499.
45. Daniel I.M., Miyagawa H., Gdoutos E.E. and Luo J.J., "Processing and characterization of epoxy/clay nanocomposites", Society Exp. Mech., 2003; 43(3): 348–354.
46. Modesti M., Lorenzetti A., Bon D. and Besco S., "Effect of processing conditions on morphology and mechanical properties of compatibilized polypropylene nanocomposites", Polymer, 2005; 46: 10237–10245.
47. Mohan T.P., Kumar M.R. and Velmurugan R., "Mechanical and barrier properties of epoxy polymer filled with nanolayered silicate clay particles", J. Mater. Sci., 2006; 41: 2929–2937.
48. Mohan T.P., Kumar M.R. and Velmurugan R., "Thermal, mechanical and vibration characteristics of epoxy-clay nanocomposites", J. Mater. Sci., 2006; 41: 5915–5925.
49. Deshmane C., Yuan Q. and Misra R.D.K., "High strength—toughness combination of melt intercalated nanoclay-reinforced thermoplastic olefins", Mater. Sci. Eng., A, 2007; 460–461: 277–287.
50. Wilkinson A.N., Man Z., Stanford J.L., Matikainen P., Clemens M.L., Lees G.C. and Liauw C.M., "Tensile properties of melt intercalated polyamide 6—Montmorillonite nanocomposites", Compos. Sci. Technol., 2007; 67: 3360–3368.
51. Mallick P.K. and Zhou Y., "Yield and fatigue behavior of polypropylene and polyamide-6 nanocmposites", J. Mater. Sci., 2013; 38: 3183–3190.
52. Ferreira J.A.M., Borrego L.P., Costa J.D.M. and Capela C., "Fatigue behaviour of nanoclay reinforced epoxy resin composites", Compos. Part B, 2013; 52: 286–291.
53. Livage J., "Sol-gel processes", Curr. Opin. Solid Mater. Sci., 1997; 2: 132–138.
54. Qian Z., Hu G., Zhang S. and Yang M., "Preparation and characterization of montmorillonite—silica nanocomposites: A sol—gel approach to modifying clay surfaces", Physica B, 2008; 403: 3231–3238.
55. Ren Y., Li F., Cheng H.M. and Liao K., "Tension-tension fatigue behavior of unidirectional single-walled carbon nanotubes reinforced epoxy composites", Letters to Editor/Carbon, 2003; 41: 2177–2179.
56. Nofar M., Hoa S.V. and Pugh M.D., "Failure detection and monitoring in polymer matrix composites subjected to static and dynamic loads using carbon nanotubes network", Compos. Sci. Technol., 2009; 69: 1599–1606.
57. Jen Y.M. and Yang Y.H., "A study of two-stage cumulative fatigue behavior for CNT/epoxy composites", Procedia. Eng., 2010; 2: 2111–2120.
58. Loos M.R., Yang J., Feke D.L., Zloczower I.M., Unal S. and Younes U., "Enhancement of fatigue life of polyurethane composites containing carbon nanotubes", Compos. Part B, 2013; 44: 740–744.

59. Kumar P, Element of fracture mechanics, Wheeler Publication, 2014, pp. 188–218, Chapter 9.
60. Agrawal B.D. and Broutman L.J., Analysis and performance of fiber composites, John Wiley & Sons, 3rd edition, 2015, pp. 324–367, Chapter 8.
61. Friedrich K., Fakirov S. and Zhang Z., Polymer composites from nano to Micro scale, Springer Science and Business, 2005, pp. 91–106, Chapter 6.
62. Wu Z., Wang X., Iwashita K., Sasaki T. and Hamaguchi Y., "Tensile fatigue behaviour of FRP and hybrid FRP sheets", Compos. Part B, 2010; 41: 396–402.
63. Manjunatha C.M., Chandra A.R.A. and Jagannathan N., "Fracture and fatigue behavior of polymer nanocomposites—A review", J. I. I. Sc., 2015; 95 (3): 249–266.
64. Fu S.Y., Feng X.Q., Lauke B. and Mai T.W., "Effect of particle size, particle/matrix interface adhesion and particle loading on mechanical properties of particulate-polymer composites", Compos. Part B, 2008; 39: 933–961.
65. PippanR. And HohenwarterA., "Fatigue crack closure: A review of the physical phenomena, fatigue", Fract. Eng. M., 2017; 40: 471–495. https://doi.org/10.1111/ffe.12578.
66. Verma V., Shukla D.K. and Kumar V., "Estimation of fatigue life of epoxy-alumina polymer nanocomposites", Procedia. Mater. Sci., 2014; 5: 669–678.
67. Verma V. and Sharma C., "Fatigue behavior of epoxy alumina nanocomposite—role of particle morphology", Theor. Appl. Fract. Mec., 2020; 110: 102807.
68. Verma V., Sayyed A.H.M., Sharma C. and Shukla D.K., "Tensile and fracture properties of epoxy alumina composite: Role of particle size and morphology", J. Polym. Res., 2020; 27 (12): 1–14.
69. Verma V. and Tiwari H., "Role of filler morphology on friction and dry sliding wear behavior of epoxy alumina nanocomposites", P. I. Mech. Eng. J.-J. Eng., 2020, https://doi.org/10.1177/1350650120970433.

11 Characterization of Surface Engineering and Coatings

*K. Ashish Chandran, A. Inbaoli,
C.S. Sujith Kumar, and S. Jayaraj*

CONTENTS

11.1 Introduction to Surface Engineering and Coating 173
11.2 Necessity and Relevance of Surface Engineering and Coating 174
11.3 Characterization in Surface Engineering and Coatings 175
11.4 Different Areas of Characterization 175
    11.4.1 Electron Microscopy 175
        11.4.1.1 Scanning Electron Microscopy 176
        11.4.1.2 Transmission Electron Microscopy 177
    11.4.2 Scanning Probe Microscopy 179
        11.4.2.1 Atomic Force Microscopy (AFM) 179
        11.4.2.2 Scanning Tunneling Microscope (STM) 180
        11.4.2.3 Electrostatic Force Microscopy (EFM) 181
        11.4.2.4 Piezo Force Microscopy (PFM) 182
        11.4.2.5 Current AFM 182
        11.4.2.6 Scanning Capacitance Microscope (SCM) 182
    11.4.3 Spectroscopic Techniques 182
        11.4.3.1 X-Ray Diffraction (XRD) 183
        11.4.3.2 Energy-Dispersive X-Ray Spectroscopy (EDX, EDS) 184
        11.4.3.3 Wavelength Dispersive X-ray Spectroscopy (WDX, WDS) 184
        11.4.3.4 Electron Energy Loss Spectroscopy (EELS) 185
        11.4.3.5 Auger Electron Spectroscopy (AES) 185
        11.4.3.6 Raman Spectroscopy 186
11.5 Conclusion 187
11.6 Future Possibilities 187
References 188

11.1 INTRODUCTION TO SURFACE ENGINEERING AND COATING

Surface engineering is a process to alter the material surface. Surface engineering deals with tailoring the surface properties to exhibit enhanced functionality than the underlying substrate material. It emphasizes enhancing surface properties

DOI: 10.1201/9781003093213-11

independently from that of the underlying substrate material [1]. Enhanced surface characteristics or properties widen the application of substrate material in the areas such as optical, wettability, tactile properties, corrosion resistance, appearance, tribology, and the like [2]. The enhancements in properties or altering surfaces can be achieved by different methods such as modifying the surface that exists with no change in composition, modifying the surface that exists by changing the composition of the surface layer, coating processes in which foreign material is deposited on the substrate.

The method for modifying the existing surface with no change in composition can be realized by processes such as surface texturing, surface melting, transformation hardening, and so on. A significant feature like surface layer composition is being changed by modifying the existing surface by tailoring the composition of the surface layer [3]. By forming a solid solution or by lattice disruption, the existing crystal structure is modified or leads to changes to the transformation behavior. New phases differing from those of the substrate material may also be formed due to the reaction between substrate elements and those introduced by the process [4, 5]. Although the newly created layer has properties similar to coating, the process differs from the coating as a new phase is formed from the substrate and not added to the substrate like coating.

The coating is the process of modifying the surface in which a foreign material is added to the surface as a covering in which the changes occur to the original surface. The surface of an object to which coating is applied is usually referred to as the substrate [6]. These processes do not involve the substrate material constituents at the surface, and it will add new material as a coating to the surface. The behavior or properties of the modified surface engineered layer depends on the process employed for coating. The need for coating may be decorative, functional, or both.

11.2 NECESSITY AND RELEVANCE OF SURFACE ENGINEERING AND COATING

There are intended properties of the surface of materials in which conventional alloying process can't be achieved, and they can be achieved only by surface engineering or treatment. The desired surface properties of a material are gained or improved by this, and it will improve component performance, service lifetime, economics etc.. Failure can occur in an engineering component when its surface cannot adequately withstand the external forces or environment subjected to it, such as optical, thermal, magnetic, and electrical wear or corrosion. Further improvement in manufacturing efficiency and seamless technological progress demands surface modifications.

Coating plays a major role in the modification of surfaces as well as enhancement of surface properties [7]. The change in substrate surface properties such as wear and corrosion resistance, wettability, adhesion, and so on occurs when the coatings are applied. In making a semiconductor device, a property that is unique, such as a magnetic response or electrical conductivity, is also applied to the coating [8, 9]. The significantly increased demand in recent years is due to the developments in

Surface Engineering and Coatings 175

paint and coating production methods. By the end of 2026, it's not a tedious task to reach the size of $236.1 billion for the global paints and coatings market as per Fortune Business Insights.

11.3 CHARACTERIZATION IN SURFACE ENGINEERING AND COATINGS

Large varieties of materials are involved in engineering applications, and they should be characterized for understanding their metallurgical, and mechanical properties. Usually, the materials to be characterized are handled in situ or/and ex situ using a possible range of techniques [10]. The most powerful tool that helps us change the surfaces that exist, create new coatings, formulations, understand the mechanism of surface degradation and its improvement is surface characterization.

While there are various techniques for characterizing bulk solids, surface coatings' characterization occupies the main portion of the techniques available. Various physical and analytical methods such as X-ray optical inspection analysis are used for investigating the structure and composition of the coating. The major motives in the characterization of surface engineering and coating are determining physical characteristics such as strength parameters (hardness, bond strength), shape and size (thickness, roughness), and chemical characteristics such as bulk analysis (constituent compounds and elements) and local analysis (distribution of elements/compounds) [11]. The techniques used for material characterization are electron microscopy (EM), optical microscopy, scanning probe microscopy (SPM), imaging techniques, and X-ray methods. In surface engineering and coatings, characterization plays a significant role as it provides in-depth information about the effects created by modified surfaces. There have been significant developments in the field of materials in recent years.

This chapter discusses different existing characterization techniques, majorly analyzing topographical, morphological, structural, and compositional analysis. The techniques mainly focused on in the chapter include experimental techniques like EM, SPM, and spectroscopic analysis. This chapter also provides an idea about the recent advancements and future possibilities in the field of characterization of surfaces.

11.4 DIFFERENT AREAS OF CHARACTERIZATION

11.4.1 Electron Microscopy

For almost 1000 years, microscopes have existed in different forms. The light focused through lenses produced a magnifying power of 6–10× in the earliest days of technology. At the optimum viewing distance of 25 cm, the resolution of the unaided eye is about 0.1 mm, and the resolution limit for optical microscopy is about 2,000 Å. The latest precise optical microscopes can achieve magnification between 500–1000×. The properties of light limit the power of optical microscopes. Scientists began working with electron microscopes in the 1930s to exceed such primitive limits. In this field, the two pioneers were M. Manfred von Ardenne and Knoll. EM was established with the invention of the deflection of electrons by the magnetic field. It is obtained by replacing the light source with a high-energy electron beam [12]. EM is used to get high-end resolution pictures of biological and non-biological specimens. It is used in engineering and biomedical research to vividly study the structure of different specimens in detail. Very short-wavelength electrons are used to illuminate radiation for obtaining high-resolution images instead of the ambient light source as in a conventional microscope. These images provide crucial information regarding the structural details of the specimens. To make electrons behave like light, electron microscopes use a vacuum. In EM, glass lenses are replaced by electromagnetic lenses. Instead of the lens being shifted by shifting the current through the lens coil, the focal length can be changed. A fluorescent screen or a digital camera can be used to visualize the sample. EM can be categorized into (1) Scanning electron microscopy (SEM) and (2) Transmission electron microscopy (TEM).

11.4.1.1 Scanning Electron Microscopy

One of the most sophisticated methods used for studying and analyzing microstructure morphology and chemical composition characteristics is SEM. Credit for the first true SEM goes to Zworykin (1942) [13]. He showed that topographic contrast is provided by secondary electrons by biasing the collector positively relative to the specimen. A raster scan pattern is introduced to the sample surface with a high-energy beam of electrons. The sample surface is imaged by the SEM in this way. By interacting with the atoms that constitute the sample, electrons generate information containing signals, such as electrical conductivity, surface topography, composition, and other characteristics of the specimen [14]. The accelerating voltage used is 0.5–30 KV.

An SEM sample applies a finely focused beam of electrons to produce a high-resolution image. The three-dimensional presence of SEM images examines the sample surface structure. Rather than using a ray of light, an electron beam is used by the SEM and is guided into the sample under observation [15]. An electron gun shoots out a beam of highly focused electrons centered on the top of the unit. When the sample is struck by an electron beam known as the incident beam, three categories of electrons and X-rays are released. These includes Auger electrons, backscattered electrons, and secondary electrons. In the SEM, backscatter and secondary electrons are used. The rebounding electrons are picked up, and an electron recorder logs their imprint. These data are translated on a computer and help the three-dimensional dimensions be clearly represented [16].

The capability of SEM to reproduce textual information consistently and simply is one of the greatest advantages. Since the electrons provide limited data about the topography and sample size, it was impossible to produce color images for traditional SEMs. Recently, researchers have been allowed to test the sample's reaction energy signatures to the impact of the incident laser. Color coding of these energy signatures with each distinct element produces their particular signature. With reference to those colors, it is possible for scientists to collect in extreme detail of each element in the sample. Figures 11.1a and 11.1b show the schematic of SEM and the matter electron interaction, respectively.

Surface Engineering and Coatings

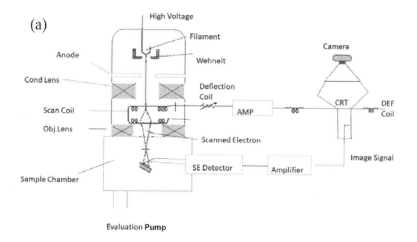

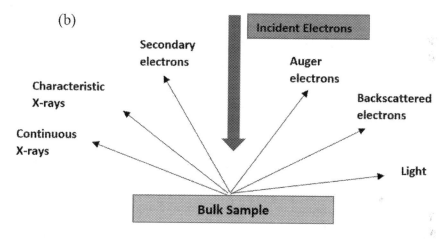

FIGURE 11.1 (a) Schematic Illustration of SEM; (b) the Matter–Electron Interaction

SEMs also opened doors in areas that range from chemistry to engineering, which allows researchers to work on various projects to gain access, with macroscopic implications, to modern, accessible microscopic process information. Energy-Dispersive X-Ray Spectroscopy (EDS) is applied by SEM is used to develop elemental maps [17]. This is correctly described by the distribution of elements within samples. Elemental analysis, mineral orientation, morphology, and comparisons are the most common applications. For accurate visual details, researchers around the world today rely on SEMs.

11.4.1.2 Transmission Electron Microscopy

For material science, the TEM is a very efficient method. A thin sample is exposed to a high-energy beam of electrons, and sample upon interacting with the accelerated electrons, reveals various features such as the crystallinity, crystal defects

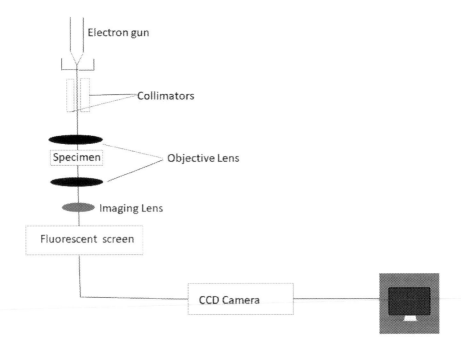

FIGURE 11.2 Schematic Diagram of TEM

(line defect), and grain boundaries. It is possible to conduct chemical analysis as well. By using TEM, the composition, defects, and growth of layers in semiconductors can be researched. For the analysis of the consistency, form, size, and so on, high resolution may be used [18].

TEMs magnify items up to 2 million times. To produce an image, a high-voltage electron beam is employed. An electron gun mounted at the top of the TEM releases electrons that pass through the vacuum tube of the microscope. TEM focuses the electrons through a very fine beam by employing an electromagnetic lens instead of a glass lens focusing on the rays (as in light microscopes). Figure 11.2 schematic of TEM and its basic parts.

The specimen passes through this beam, and the specimen is very thin. At the lower part of the microscope, the electrons hit a fluorescent screen or scatter. The specimen picture with different colors for its assorted parts appears on the screen depending on its density. This image can then be photographed or directly analyzed inside the TEM. For both bright-field images and dark-field images, exact particle size will be determined by TEM. It also provides descriptions regarding nanoparticle as energetic electrons are used to provide specific morphological details, crystallographic and compositional information data [19]. Imaging, spectroscopy, and diffraction techniques are the three primary techniques used in TEM techniques. Table 11.1 gives an overall comparison of characteristics of optical microscope and electron microscope.

TABLE 11.1
Overall Comparison of Characteristics of Optical Microscope, SEM, and TEM

Characteristic	Optical Microscope	SEM	TEM
Atmosphere	Air	Vacuum	Vacuum
Source	Light	Electron	Electron
High Voltage	—	0.5–30 kV	25–300 kV
Lens	Glass	Pole piece	Pole piece
Magnification	X 1K	X 800K	X 1000K
Resolution	5–0.1 μm	7–0.6 nm	0.5–0.1 nm
Focal depth	Shallow	Deep	Deep
Color	Color	B/W	B/W
EDX	N/A	Possible	Possible
Field	Wide	Wide	Small
Sampling	Easy	Easy	Difficult
Sample size	Large	Large	Small
Image	Transmitted/reflected	Surface	Transmitted

11.4.2 Scanning Probe Microscopy

This probe technique reveals the morphology of surfaces by scanning the specimen through employing a physical probe [20]. The picture is obtained by moving the probe mechanically, line by line on the surface, and thereby recording the probe-surface interaction in a raster scan of the specimen as a function of location [21]. Compared to existing microscopes, SPM have advantages such as high resolution (2~3 Å lateral, 0.1 Å vertical), provides quantitative three-dimensional information, applicable to nonconductors as well as conductors and semiconductors, can be operated in air, liquid, and vacuum, and electrical, magnetic, optical, and physical properties can be measured, with atomic-scale manipulations and lithography [22]. The two main SPM techniques are described as follows.

11.4.2.1 Atomic Force Microscopy (AFM)

AFM was invented at Stanford University in 1985 and applicable for nonconductors as well as conducting samples. It is utilized for topographical imaging to gauge and control matter on a nanometer-scale by feeling the surface by employing a probe (mechanical). In AFM, the probe (sharp-tip cantilever) scans the sample surface. The curvature of cantilever tip radius is in the range of nanometers, made up of silicon nitride or silicon. As per Hooke's law, when the sample surface is very near to the tip the cantilever deflects more and this deflection due to attraction or repulsion forces prevailing at the vicinity of tip and sample surface [23]. Figure 11.3 shows a schematic of AFM.

It comprises of a cantilever, a tip, the surface, a laser, and a multi-segment photo detector. The cantilever can bend or deflect due to the forces between the tip and the sample surface. A detector will measure the cantilever distance as the tip is tested over the sample. The deflection of an AFM cantilever is made of significant

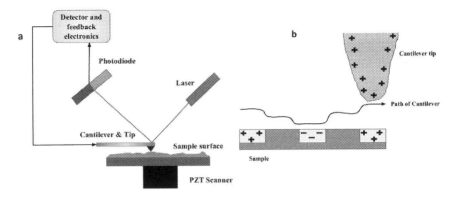

FIGURE 11.3 (a) Illustration of the AFM Setup and (b) Cantilever Tip and Sample Surface Enlarged View

contribution by several forces. AFM can calculate different forces, such as contact forces, the forces of Van der Waals, electrostatic forces, capillary forces, magnetic forces, and so on. The force most frequently related to AFM is an interatomic force called the Van der Waals force. In touch AFM mode, an AFM tip makes gentle physical contact with the surface of sample (repulsive mode). To house changes in topography, the contact force causes the cantilever to bend. The Position-Sensitive Photodetector (PSPD) tests light displacements as low as 10 times [24].

Centered at the end of a cantilever, a sharp tip, rasters across an area and tip force on the surface are analyzed with a laser and photodiode. Between the photodiode and the piezo crystal, there exists a feedback loop that preserves a uniform force for constant amplitude and contact mode imaging during periodic contact mode imaging.

Imaging Modes in AFM are Contact Mode, with damage to the sample, high resolution, and measures of the frictional forces; Non-Contact Mode, with no damage to the sample and lower resolution; and Tapping Mode, with minimal damage to sample and better resolution.

11.4.2.2 Scanning Tunneling Microscope (STM)

STM was invented by Binnig and Roher in 1981, for which they were awarded the Nobel Prize in Physics in 1986. In STM, an atomic scale probe mechanically tests the surface of the sample. The fundamental principle of STM is quantum tunneling. When a conducting tip is brought very close to a metallic or semiconducting surface, a bias is formed, facilitating the electrons to tunnel from the tip to sample [25]. The important components of STM are scanning tip, piezoelectric electrodes, tunneling current amplifier (nA), Bi-potentiostat (bias), and feedback circuit. Figure 11.4 presents a schematic of STM.

A very fine metallic tip scans the sample, and the scanning can be managed in three-dimensions by a piezo-scanner mounted to the tip or connected to the sample point. The tip of the STM should be performing (metals, like platinum). When scanning, a piezo-electric crystal is used to change the height and the piezo voltage is

Surface Engineering and Coatings

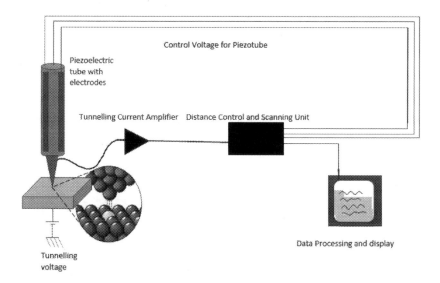

FIGURE 11.4 Schematic of STM

regulated, maintaining a steady tunneling current allows a surface under electric crystal to be imaged [26].

The "tunneling current" streams on the off chance that the tip is in near vicinity to the test as the test is emphatically or adversely one-sided. The low tunnel current is amplified and measured. The majority of the tunneling current passes across an individual projecting molecule on the tip, allowing for sub-angstrom z determination with a sharp tip on a clean surface. The determination of x–y is marginally more prominent. The STM sensors state the thickness of a fabric employing a tunnel current. Utilizing the tunnel current, the input hardware keeps up the distance between the conclusion and the test consistently. In case the tunnel current surpasses its predefined esteem, the distance between the tip and the test is expanded. In the event that it is lower than that esteem, the input diminishes the distance. The tip is filtered line by line over the test surface, taken after the geography of the test.

The steady height mode pictures the diverse surface atoms uncovering the composition of the surface. Constant current mode pictures the surface geology at the nuclear determination in case the same atoms constitute the surface, that is, the as-it-were figure influencing the tunneling current is the distance.

11.4.2.3 Electrostatic Force Microscopy (EFM)

By measuring the electrostatic drive between the surface and a one-sided AFM cantilever, EFM tests the electric properties on a test surface. As the gadget is switched on and off, EFM will outline the electrostatic areas of an electronic circuit. A voltage between the tip and the test is applied by EFM as the cantilever floats over the surface, not touching it. EFM images provide information regarding the electric properties of test specimen such as surface potential and charge distribution of a sample surface.

11.4.2.4 Piezo Force Microscopy (PFM)

PFM tests the mechanical reaction when an AFM conductive tip is attached to the test surface by an electrical voltage. The test at that stage stretches or contracts locally in reaction to the electrical boost. When the spaces have a vertical polarization pointed downward, and a positive voltage is connected to the tip, the test will locally extend. On the off chance that the polarization is pointed up, the test will locally contract. This, in turn, causes the cantilever to divert, which can at that point be measured and deciphered in terms of the piezoelectric properties of the test.

11.4.2.5 Current AFM

In I-AFM, between the tip and the test, a tilt voltage is associated. A conductive AFM tip in the I-AFM mode filters the surface when in contact. Geology and surface conductivity can also be measured at the same time. The current streams between the tip and the test allow the surface conductivity to be graded by the I-AFM mode.

11.4.2.6 Scanning Capacitance Microscope (SCM)

In SCM, a metal/oxide/semiconductor (MOS) capacitor is formed when the SCM tip is fetched near the test surface. An AC predisposition is associated when the test and surface are in contact and produce capacitance varieties within the test, which are detected using an ultra-high recurrence thunderous capacitance sensor. The tip is, at that point, to check over the surface. By applying a rotating inclination to the test, carriers then again collect and exhaust inside the semiconductor's surface layers, varying the tip-sample capacitance.

11.4.3 SPECTROSCOPIC TECHNIQUES

Spectroscopy deals with the interaction of electromagnetic radiation with matter. Electromagnetic radiation could be a basic consonant wave of electric and magnetic fields fluctuating orthogonal to each other. This bunch of strategies utilizes an extend of standards to uncover the chemical composition, composition variety, precious stone structure, and photoelectric properties of materials [27].

Ultraviolet (UV)–visible spectroscopy alludes to retention spectroscopy or reflectance spectroscopy in the portion of the ultraviolet and the total, adjoining obvious unearthly regions. The retention or reflectance within the obvious run specifically influences the seen color of the chemicals included. In this locale of the electromagnetic range, atoms and molecules experience electronic moves. UV and visible assimilation (UV-Vis) spectroscopy tests the constriction of a ray of light from a test surface as it passes through a test or after reflection. Retention estimations can be at a single wavelength or over an amplified unearthly extend [28]. It is utilized for the location of useful bunches, location of debasements, subjective investigation, and quantitative examination, among others. It makes a difference to show the relationship between diverse bunches and is valuable to distinguish the conjugation of the compounds.

UV absorption spectra emerge from the movement of the electron in a particle from the lower level to the next level. The atom retains bright frequency radiation

Surface Engineering and Coatings 183

(Δ), the electron in the experience of the atom moves from a lower energy to a higher energy level. This energy is determined by equation

$$E = h\vartheta. \quad (11.1)$$

When a light energy test matching the energy contrast between conceivable electronic movements inside the particle is discovered, the atom will retain a division of the light energy, and the electrons would be advanced to the orbital state of higher energy [29]. The more efficiently the electrons are energized (i.e., the lower energy hole), the longer it can hold the wavelength of light. A spectrometer records the degree of absorption at various wavelengths through a test, and a range is known as the resulting plot of absorbance (A) versus wavelength (λ). Absorbance estimates are regularly used to evaluate the concentration of an obscure sample by exploiting the Beer-Lambert Law that depicts how light is weakened based on the materials through which it passes. The transmittance and thus the absorbance corresponds directly to the concentration of the sample, the molar absorption, and the length of the cuvette path. In addition, UV-V is spectroscopy can determine the reflectance of a test aside from transmission and retention or how viable a surface is in reflecting the entire sum of light that occurs.

11.4.3.1 X-Ray Diffraction (XRD)

This is a novel and nondestructive chemical analysis technique, and it is the phenomenon in which a crystal's nuclear planes cause an incident beam of X-rays to interact with each other when they take off the crystal. The power of X-rays to be diffracted by a crystalline lattice is applied to X-ray diffraction since the wavelength of X-rays is equivalent to interplanar dispersion. The form of the diffraction pattern is peculiar to a specific crystalline structure. Every crystalline material gives a pattern, and the same substance continuously gives the same design, and each creates the pattern freely in a mix of substances [30]. As a result, the X-ray diffraction pattern of an unadulterated material is like a particular finger impression of the substance. It is based on the diffusing of X-rays by crystals. XRD is especially valuable for inorganic chemical investigation and is less compelling for natural compounds. This technique will usually clearly identify oxides, nitrides, carbides, and sulfides of metals.

XRD procedure involves use of a diffractometer for generation of X-rays, diffraction, location, and elucidation. A diffractometer is a measuring device that uses the scrambling pattern generated when particles such as X-rays or neutrons or a beam of radiation interacts with fabric to analyze its structure. An X-ray diffractometer comprises an X-ray source, a test holder able of rotation, and an arrangement of the counter to screen the diffracted intensity. Unlike imaging, the X-rays utilized are monochromatic. The discovery of the diffraction pattern is made conceivable by the revolution of the test and maybe an arrangement of dim and light groups that indicate damaging and valuable impedances between scattered X-rays.

The preferences of XRD are nondestructive; quick, simple test arrangement; high accuracy for d-spacing calculations; can be done in situ; reasonable for single crystal, poly, and shapeless materials; and guidelines are accessible for thousands of material frameworks.

11.4.3.2 Energy-Dispersive X-Ray Spectroscopy (EDX, EDS)

For the natural inspection or chemical characterization of a specimen, EDX or X-ray microanalysis is an explanatory technique. It refers to the interactions of a few sources of X-ray excitement and a test. When electrons of suitable energy encroach on a test, they cause the emanation of X-rays whose energies and relative abundance depend upon the composition of the test [31]. Its ability to characterize is largely due to the basic concept that each part contains a special nuclear structure that enables its electromagnetic outflow range to provide a one-of-a-kind set of crests.

The excitation source, X-ray locator, pulse processor, and analyzer are four critical components of the EDS setup. Spectrometers for X-ray fluorescence (XRF) uses excitation by X-ray beam. A locator is used to transform the vitality of X-rays into voltage signals, and this information is sent to a pulse processor that analyzes the signals and passes them on to them onto an analyzer for information show and investigation. The predominant locator utilized to be Si (Li) detector cooled to cryogenic temperatures with fluid nitrogen. Presently, with silicon float finders (SDD) with Peltier cooling systems, more existing systems are often prepared. An energy-dispersive spectrometer can calculate the number and vitality of X-rays radiated from an example. The EDS allows the basic composition of the sample to be determined since the energy of the X-rays is function of the energy distinction among the two shells and the nuclear structure of the emitting portion.

11.4.3.3 Wavelength Dispersive X-ray Spectroscopy (WDX, WDS)

WDX or WDS may be a nondestructive analysis technique used by measuring characteristic X-rays within a limited wavelength range to collect basic data regarding a range of materials. WDS isolates characteristic X-rays transmitted from a test utilizing their distinctive wavelengths [32]. This approach produces a spectrum in which the crests can be easily differentiated in contrast to individual X-ray lines and components. WDS is used mainly in chemical analysis, spectrometry of wavelength dispersive X-ray fluorescence (WDXRF), electron microprobes, electron magnifying lens power, and high-precision tests for nuclear and plasma material science research.

The main components of a WDS spectrometer are a diffracting crystal and a locator or counter. The WDS method measures the wavelength of characteristic X-ray outflow to distinguish constituent components. From the number of X-rays collected, the sums of these components can be decided. X-rays characteristic of distinctive components is separated by X-ray diffraction employing a crystal of known lattice spacing d. The point of diffraction required to gather X-rays of a specific wavelength is predicted utilizing the Bragg Law ($n\lambda = 2d\sin\theta$), where n could be an entire number that demonstrates the diffraction arrange, λ is the X-ray wavelength that will be diffracted, d is the interbank dividing for the diffracting precious stone, and θ the point at which diffraction is happening (the point between the gem planes and the diffracted X-ray). WDS offers micro-analysis around 10× more delicate than

EDS, follows component examination (underneath 0.01%), the highest sensitivity for light element detection, and improved X-ray mapping.

WDS is the strategy of choice for best chemical microanalysis than filtering electron magnifying lens because it may be a helpful and delicate strategy for deciding the chemical constituents and composition of stages on the small scale, great for "mapping" chemical varieties, simple test planning, appropriate for most SEMs, works with most sorts of tests, can be utilized in combination with the quicker but less touchy EDS investigation method [33].

11.4.3.4 Electron Energy Loss Spectroscopy (EELS)

In EELS, substance is revealed to electron beam with a known, minimal run of active energies, out of which some electrons will diffuse inelastically, indicating their possibility for energy loss with their paths deflected moderately and randomly. The lost amount of energy is calculated and translated into terms of what caused the energy loss using an electron spectrometer. [34]. Shell ionizations, inter-and intra-band transitions, Cherenkov radiation, and phonon and plasmon excitations are integrated into inelastic interactions. For identifying the natural components of a substance, inner-shell ionization is particularly important. It is possible to specify the kinds of particles, and the numbers of atoms of each kind that are hit by the beam are determined [35]. In addition, the scattering angle can be calculated, giving data nearly the scattering relation of any material excitation that induced the inelastic diffusion.

11.4.3.5 Auger Electron Spectroscopy (AES)

AES may be broadly utilized explanatory strategy to explore the composition of surface layers of solids. It was, to begin with, found in 1923 by Lise Meitner and afterward independently found once again in 1925 by Pierre Twist drill. It can also be utilized for the consideration of oxidation catalysis, the chemistry of broken interfacing, and grain boundaries. Until the early 1950s, Twist drill moves were considered annoyance impacts by spectroscopists. Since 1953, in any case in testing chemical and compositional surface situations, AES has become a practical and clear characterization method and has found applications in metallurgy, gas-phase chemistry, and throughout the microelectronics industry [36].

The fundamental components of an AES spectrometer are UHV (ultra-high vacuum) environment, electron weapon, electron vitality analyzer, electron locator, information recording, handling, and yield framework. Auger spectroscopy can be considered as including three fundamental steps, such as nuclear ionization by evacuation of a center electron, electron emanation (the Auger prepare), and the examination of the transmitted Auger electrons. A fine-centered electron beam besieges the test and launches an electron of the inward shell of the particle. This opportunity must be refilled by an electron from a higher vitality level. When the higher energy electron fills the gap, the discharge of energy is exchanged to an electron in an external circle electron. That electron has adequate energy to overcome the binding energy, and the work function to be shot out with characteristic active energy [37]. The shot-out electron is alluded to as a Twist drill electron after Pierre Auger.

The Auger prepare require three electrons. In this way, we cannot identify hydrogen and helium but can find all other elements from lithium on up.

The preferences of AES are surface touchy, essential, and chemical composition investigation by comparison with standard tests of known composition, the discovery of components heavier than lithium, exceptionally great affectability for light components, and profundity profiling examination.

11.4.3.6 Raman Spectroscopy

It is a spectroscopic method used in a system to monitor vibration, cyclic, and other low-frequency modes found by Indian physicist C. V. Raman in 1928. He found that the visible wavelength of a little division of the radiation scattered by certain atoms varies from that of the occurrence beam and, moreover, that the shifts in wavelength depend on the chemical structure of the particles capable of diffusing [38]. Raman spectroscopy is commonly utilized in chemistry to supply a unique finger impression by which atoms can be recognized. When the radiation passes through the simple medium, the species shows a division of the beam in all courses to diffuse. Figure 11.5 shows schematic of instrumentation of Raman spectroscopy.

Raman spectra are produced by illuminating a test with visible or near-infrared monochromatic radiation with a capable laser source. The test is illuminated with a coherent source, regularly a laser. When monochromatic radiation is occurrence upon a test at that point, this light will be associated with the test in a few molds. It may be reflected, retained, or scattered in a few ways [39]. Most of the radiation is dispersed flexibly (Rayleigh scramble). A small portion is dispersed inelastically (Raman scramble with feeds and anti-Stokes portions). It is the scrambling of the radiation that happens that gives data around the atomic structure. This last-mentioned parcel is taken into consideration since it contains the specified data. The transmitted radiation is of three sorts, that is, stirs scrambling, anti-Stokes scrambling, and Rayleigh scrambling. Amid light, the range of the scattered radiation is measured at a few points (frequently 90°) with a reasonable spectrometer. The range is measured with the laser line as a reference and, the peaks are determined as the shift from the laser line.

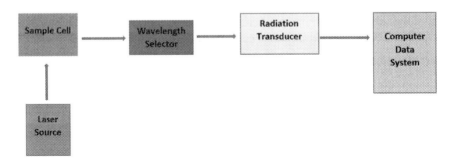

FIGURE 11.5 Schematic of Instrumentation of Raman Spectroscopy

A critical advantage of Raman spectra over infrared lies within the truth that water does not cause obstructions; undoubtedly, Raman spectra can be gotten from aqueous solutions. Water could be a powerless Raman scattered but a solid safeguard of infrared radiation. Hence, aqueous solutions can be considered by Raman spectroscopy but not by infrared. The sources utilized in advanced Raman spectrometry are continuous lasers since their high concentration is necessary to create Raman scrambling of adequate intensity [40]. Common laser sources utilized for Raman spectroscopy are Argon particles, Krypton particles, diode, and others.

11.5 CONCLUSION

The developments in surface engineering have paved the way for enhancement of surface characteristics or properties leading to wide range of applications in the areas such as optical, wettability, tactile properties, corrosion resistance, appearance, tribology, and so on. The enhancements in properties or altering surfaces can be achieved by modify the existing surface with or without change in composition layer or coating. Introduction of characterization to materials in engineering applications, enables understanding their physical, chemical, and mechanical properties. The major characterization techniques were discussed in the present chapter.

11.6 FUTURE POSSIBILITIES

More than ever, sometimes recently, materials-driven item developments in the industry and shorter time-to-market for unused items require high development rates and near participation between research, advancement, and fabricating. This approach, in which researchers and engineers from industry and investigate organizing work together, incorporates supported advancements in materials science and designing, and especially for materials and prepare characterization. Concurring with the European Materials Characterization Committee, explanatory procedures and individual instruments are considered to be crucial drivers for advancement in European industry. In recent years, Ceramic and Teflon coating has played a major role in the car detailing industries.

In the thought of the brief supply of raw materials, the rise in their costs, and other unfavorable cost-related conditions, examination innovation is anticipated to serve successfully for the advancement of modern items for eras to come and their fabricating forms. As per the UK conference on "Future Patterns on Material Science and Nanotechnology" held in 2020, the world commercial center for material science and designing is anticipated to be a solid development all through next decade.

List of Abbreviations

Å	Angstrom
EM	Electron Microscopy
SEM	Scanning Electron Microscopy
TEM	Transmission Electron Microscopy
SPM	Scanning Probe Microscopy

AFM	Atomic Force Microscopy
PSPD	Position-Sensitive Photo Detector
STM	Scanning Tunneling Microscopy
EDS	Energy-Dispersive X-Ray Spectroscopy
EFM	Electrostatic Force Microscopy
PFM	Piezo Force Microscopy
I-AFM	Current Atomic Force Microscopy
SCM	Scanning Capacitance Microscope
MOS	Metal/Oxide/Semiconductor
UV-V	Ultraviolet and Visible assimilation
A	Absorbance
XRD	X-ray Diffraction
EDX,	EDS Energy-Dispersive X-ray Spectroscopy
XRF	X-ray fluorescence
WDX,	WDS Wavelength Dispersive X-ray Spectroscopy
WDXRF	Wavelength Dispersive X-ray Fluorescence
EELS	Electron Energy Loss Spectroscopy
AES	Auger Electron Spectroscopy
UHV	Ultra-High Vacuum

REFERENCES

1. Hutchings, I. and Shipway, P., 2017. *Tribology: Friction and Wear of Engineering Materials.* Butterworth-Heinemann.
2. Davis, J.R., 2014. *Surface Engineering for Corrosion and Wear Resistance.* CRC Press.
3. Woodruff, D.P., 2016. *Modern Techniques of Surface Science.* Cambridge University Press.
4. Arumugaprabu, V., Ko, T.J., Kumaran, S.T., Kurniawan, R. and Uthayakumar, M., 2018. A brief review on importance of surface texturing in materials to improve the tribological performance. *Reviews on Advanced Materials Science*, 53(1).
5. Mao, B., Siddaiah, A., Liao, Y. and Menezes, P.L., 2020. Laser surface texturing and related techniques for enhancing tribological performance of engineering materials: A review. *Journal of Manufacturing Processes*, 53, pp. 153–173.
6. Zhang, S., 2010. *Nano Structured Thin Films and Coatings.* CRC Press.
7. Fotovvati, B., Namdari, N. and Dehghanghadikolaei, A., 2019. On coating techniques for surface protection: A review. *Journal of Manufacturing and Materials processing*, 3(1), p. 28.
8. Bag, P.P. and Roymahapatra, G., 2021. Surface engineering for coating: A smart technique. In *Advanced Surface Coating Techniques for Modern Industrial Applications* (pp. 261–282). IGI Global.
9. Babu, C.V., Yaradi, M. and Thammineni, D.T., 2018, July. A review on recent trends in surface coatings. In *IOP Conference Series: Materials Science and Engineering* (Vol. 390, No. 1, p. 012079). IOP Publishing.
10. Panwar, A.S., Singh, A. and Sehgal, S., 2020. Material characterization techniques in engineering applications: A review. *Materials Today: Proceedings*, 28, pp. 1932–1937.
11. Batchelor, A.W., Lam, L.N. and Chandrasekaran, M., 2011. *Materials Degradation and Its Control by Surface Engineering.* World Scientific.
12. Goodhew, P.J. and Humphreys, J., 2000. *Electron Microscopy and Analysis.* CRC Press.

13. Bogner, A., Jouneau, P.H., Thollet, G., Basset, D. and Gauthier, C., 2007. A history of scanning electron microscopy developments: Towards "wet-STEM" imaging. *Micron*, *38*(4), pp. 390–401.
14. Ul-Hamid, A., 2018. *A Beginners' Guide to Scanning Electron Microscopy* (Vol. 1, p. 402). Springer International Publishing.
15. Goldstein, J.I., Newbury, D.E., Michael, J.R., Ritchie, N.W., Scott, J.H.J. and Joy, D.C., 2017. *Scanning Electron Microscopy and X-ray Microanalysis*. Springer.
16. Akhtar, K., Khan, S.A., Khan, S.B. and Asiri, A.M., 2018. Scanning electron microscopy: Principle and applications in nanomaterials characterization. In *Handbook of Materials Characterization* (pp. 113–145). Springer.
17. Abd Mutalib, M., Rahman, M.A., Othman, M.H.D., Ismail, A.F. and Jaafar, J., 2017. Scanning electron microscopy (SEM) and energy-dispersive X-ray (EDX) spectroscopy. In *Membrane Characterization* (pp. 161–179). Elsevier.
18. Fan, Z., Zhang, L., Baumann, D., Mei, L., Yao, Y., Duan, X., Shi, Y., Huang, J., Huang, Y. and Duan, X., 2019. In situ transmission electron microscopy for energy materials and devices. *Advanced Materials*, *31*(33), p. 1900608.
19. Zuo, J.M. and Spence, J.C., 2017. Advanced transmission electron microscopy. In *Advanced Transmission Electron Microscopy*, ISBN 978-1-4939-6605-9. Springer Science+ Business Media New York.
20. Salapaka, S.M. and Salapaka, M.V., 2008. Scanning probe microscopy. *IEEE Control Systems Magazine*, *28*(2), pp. 65–83.
21. Bhushan, B. and Marti, O., 2017. Scanning probe microscopy—principle of operation, instrumentation, and probes. In *Nanotribology and Nanomechanics* (pp. 33–93). Springer.
22. Takahashi, Y., Kumatani, A., Shiku, H. and Matsue, T., 2017. Scanning probe microscopy for nanoscale electrochemical imaging. *Analytical Chemistry*, *89*(1), pp. 342–357.
23. Lanza, M. ed., 2017. *Conductive Atomic Force Microscopy: Applications in Nanomaterials*. John Wiley & Sons.
24. Wang, D. and Russell, T.P., 2018. Advances in atomic force microscopy for probing polymer structure and properties. *Macromolecules*, *51*(1), pp. 3–24.
25. Battisti, I., Verdoes, G., van Oosten, K., Bastiaans, K.M. and Allan, M.P., 2018. Definition of design guidelines, construction, and performance of an ultra-stable scanning tunneling microscope for spectroscopic imaging. *Review of Scientific Instruments*, *89*(12), p. 123705.
26. Ko, W., Ma, C., Nguyen, G.D., Kolmer, M. and Li, A.P., 2019. Atomic-scale manipulation and in situ characterization with scanning tunneling microscopy. *Advanced Functional Materials*, *29*(52), p. 1903770.
27. Robinson, J.W., 2017. *Practical Handbook of Spectroscopy*. Routledge.
28. Alessio, P., Aoki, P.H., Furini, L.N., Aliaga, A.E. and Constantino, C.J.L., 2017. Spectroscopic techniques for characterization of nanomaterials. In *Nanocharacterization Techniques* (pp. 65–98). William Andrew Publishing.
29. Polman, A., Kociak, M. and de Abajo, F.J.G., 2019. Electron-beam spectroscopy for nanophotonics. *Nature Materials*, *18*(11), pp. 1158–1171.
30. Bunaciu, A.A., Udriştioiu, E.G. and Aboul-Enein, H.Y., 2015. X-ray diffraction: Instrumentation and applications. *Critical Reviews in Analytical Chemistry*, *45*(4), pp. 289–299.
31. Mishra, R.K., Zachariah, A.K. and Thomas, S., 2017. Energy-dispersive X-ray spectroscopy techniques for nanomaterial. In *Microscopy Methods in Nanomaterials Characterization* (pp. 383–405). Elsevier.

32. Hall, M., 2017. X-ray fluorescence-energy dispersive (ED-XRF) and wavelength dispersive (WD-XRF) spectrometry. In *The Oxford Handbook of Archaeological Ceramic Analysis* (pp. 343–381). Oxford University Press.
33. Mei, B., Gu, S., Du, X., Li, Z., Cao, H., Song, F., Huang, Y. and Jiang, Z., 2020. A wavelength-dispersive X-ray spectrometer for in/ex situ resonant inelastic X-ray scattering studies. *X-Ray Spectrometry*, 49(1), pp. 251–259.
34. Hofer, F., Schmidt, F.P., Grogger, W. and Kothleitner, G., 2016. Fundamentals of electron energy-loss spectroscopy. In *IOP Conference Series: Materials Science and Engineering* (Vol. 109, No. 1, p. 012007). IOP Publishing.
35. Brydson, R., 2020. *Electron Energy Loss Spectroscopy*. Garland Science.
36. Unger, W.E., Wirth, T. and Hodoroaba, V.D., 2020. Auger electron spectroscopy. In *Characterization of Nanoparticles* (pp. 373–395). Elsevier.
37. Wang, H. and Linford, M.R., 2015. X-ray photoelectron spectroscopy and auger electron spectroscopy. *Vacuum Technology & Coating*, pp. 2–7.
38. Rostron, P., Gaber, S. and Gaber, D., 2016. Raman spectroscopy, review. *Laser*, 21, p. 24.
39. Larkin, P., 2017. *Infrared and Raman Spectroscopy: Principles and Spectral Interpretation*. Elsevier.
40. Shipp, D.W., Sinjab, F. and Notingher, I., 2017. Raman spectroscopy: Techniques and applications in the life sciences. *Advances in Optics and Photonics*, 9(2), pp. 315–428.

12 Review on the Influence of Retained Austenite on the Mechanical Properties of Carbide-Free Bainite

Siddharth Sharma, Ravi Kumar Dwivedi, and Rajan Kumar

CONTENTS

12.1 Introduction 191
12.2 Influence of RA in the Microstructure of CFB on Strength 192
12.3 Current Trends in the Mechanical Properties of CFB 194
12.4 Conclusion 199
References 200

12.1 INTRODUCTION

When it comes to the commercial use of a material, the desired properties are to be kept in mind. Usually, the high-strength steels that are used for commercial purposes are heat-treated in two ways: one is by austenitizing and cooling them slowly, and the other one is by quenching them to martensite [1]. These carbide-free bainites (CBFs)/martensites, shown in Figure 12.1, are a focus of researchers due to their excellent mechanical properties. With the addition of silicon (Si), the formation of carbide is suppressed while during the transformation of the bainite, the austenitic films are retained, increasing the plasticity, hence providing the combination of the ductility and strength at the same time [2].

It has been observed that CFB possesses better resistance to rolling contact fatigue and thus has huge potential for applications in rails, where more severe service conditions exist [3]. In this present work, the influence that retained austenite (RA) has over the different properties of CFB is analyzed as studied by various researchers who have performed different tests with specimens.

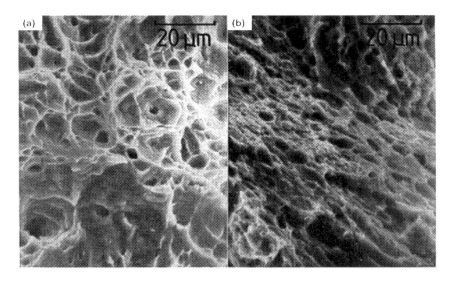

FIGURE 12.1 Scanning Electron Microscopy (SEM) Images of Samples That Were Heat-Treated to Form Bainite [1]

12.2 INFLUENCE OF RA IN THE MICROSTRUCTURE OF CFB ON STRENGTH

In a study performed by Balaji [4], it was found that when the transformation temperature of the sample was decreased, a significant change in the refinement of the bainitic microstructure is obtained (from 300 nm at 350 °C to 60 nm at 200 °C); also, the RA morphology changes a to fully film type instead of the combination of the filmy and blocky–type RA, resulting in the increase in wear resistance of the microstructure of nanobainite transformed at low temperature. The blocky-type RA can cause the transformation-induced plasticity (TRIP) phenomenon while abrasion leads to the formation of coarse martensite, which helps in crack initiation and propagation. The findings of Guhui Gao [5] also support that the improved mechanical properties are the aftereffect of different types of the RA in the Q-P-T–treated triplex microstructure having CFB.

Binggang Liu [6] studied how the impact abrasion wear resistance is affected by the stability of RA. It has been shown by him that if the stability of RA is greater, then the wear resistance shown by the bainite steels was much better while the worsened wear resistance was observed when RA was sufficient to suppress the martensite transformation. It has also been shown that RA improves the hardness of the contact surface along with a resistance to crack openings and propagation.

Tao Jiang [7] performed the tests that resulted in data supporting that bainite transformation process is accelerated when the austenite gets refined from 53 micrometers to 3 micro-meters, but approximately half of the bainite was not nanostructured and a large amount of the blocky RA, shown in Figure 12.3, is obtained when reducing the toughness of the specimen. The best impact toughness, along with good strength and ductility, was obtained when the dimensions were kept moderate.

The Influence of RA on CFB 193

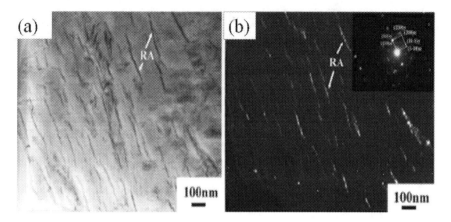

FIGURE 12.2 Microstructures Showing RA in Slowly Quenched and Tempered (SQ&T) Samples [5]

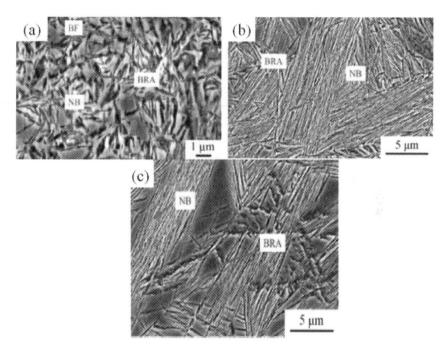

FIGURE 12.3 SEM Images of Samples Showing Blocky RA [7]

Ran Diang enhanced the mechanical properties by increasing the volume fraction of RA by proposing a new method. Before the annealing process, he preserved some interlath RA at a temperature that was relatively higher, creating a new inter critical austenite between the laths. Thus, the blocky RA formation was suppressed, enhancing the RA content and improving the mechanical properties [8].

F. Hu analyzed the effect of RA on the stirring wear resistance; the results showed that enhanced hardness was seen in Quenched and Partitioned (Q&P) martensitic steel in comparison to nanostructured bainitic steel, and the reason for this is the higher stability of RA in Q&P martensitic steel, which provides a better TRIP effect that results in an increase in the hardness of surface and a decrease in the wear rate [9].

12.3 CURRENT TRENDS IN THE MECHANICAL PROPERTIES OF CFB

The recent trends of the mechanical properties of different samples of CFB used by different researchers are enlisted in Table 12.1. It gives us a broad idea of the capabilities of this material under different treatment routes.

Table 12.1 shows a variation of the yield stress from 796 MPa to 2019 MPa, a variation of ultimate tensile stress from 553 MPa to 2329 MPa as well as an elongation variation from 0.30% to 30%, as found by the investigators. Although the treatment route plays an important role in obtaining the combination of strength and ductility, at the same time, the different constituents of the various sample, as shown in Table 12.1, also matter a lot. As the constituents of the material change, the way with which they react to the heat treatment also changes because of the different amount of formation of RA, which ultimately influences mechanical behavior [21–22].

TABLE 12.1
Mechanical Properties of CFB Obtained in Chronological Order

S. No.	Author (Year)	Material	Yield Strength (MPa)	Ultimate Tensile Strength (MPa)	% Elongation	Treatment Route	Reference
1.	S. Sharma et al. (2011)	0.47C-1.22Si-1.07Mn-0.04Ni-0.01Mo-0.7Cr-0.01S-0.01Al-0.01P-Fe	1166	1313	11.5	Soaked at 900 °C for 20 min and tempering at 400 °C for 60 min	[10]
		0.30C-1.76Si-1.57Mn-0.045Ni-0.025Mo-0.144Cr-0.016S-0.04Al-0.022P-Fe	1441	1560	12	Soaked at 900 °C for 20 min and tempering at 350 °C for 90 min	

S. No.	Author (Year)	Material	Yield Strength (MPa)	Ultimate Tensile Strength (MPa)	% Elongation	Treatment Route	Reference
2.	T. Sourmail (2013)	0.99C-1.58Si-0.76Mn-0.46Cr-0.02Mo	1834	2205	11.20	Bainitized at 250 °C for 16 h	[11]
		1.01C-1.51Si-0.82Mn-0.46Mn-0.10Mo	1852	2164	8.30	Bainitized at 250 °C for 16 h	
		1.01C-1.51Si-0.82Mn-0.46Mn-0.10Mo	1883	....	0.30	Bainitized at 220 °C for 22 h	
		1.01C-1.51Si-0.82Mn-0.46Mn-0.10Mo	2019	....	0.40	Bainitized at 200 °C for 64 h	
		1.02C-1.53Si-0.76Mn-0.46Cr-0.02Mo-0.02Nb	1798	....	0.90	Bainitized at 220 °C for 22 h	
		1.02C-1.53Si-0.76Mn-0.46Cr-0.02Mo-0.02Nb	1866	2278	7.20	Bainitized at 240 °C for 16 h	
		0.98C-C-2.90Si-0.77Mn-0.45Cr-0.01Mo	1698	2068	21.30	Bainitized at 250 °C for 16 h	
		0.98C-C-2.90Si-0.77Mn-0.45Cr-0.01Mo	1704	2287	7.40	Bainitized at 220 °C for 22 h	
		0.68C-1.60Si-1.25Mn-1.50Cr	1484	2023	14.20	Bainitized at 250 °C for 12 h	
		0.58C-1.58Si-1.25Mn-1.43Cr-0.02Mo	1443	2022	19.10	Bainitized at 250 °C for 12 h	
		0.64C-1.60Si-1.27Mn-1.50Cr-0.01Mo-0.04Nb	1358	2234	5.10	Bainitized at 220 °C for 22 h	
		0.64C-1.60Si-1.27Mn-1.50Cr-0.01Mo-0.04Nb	1480	2017	19.10	Bainitized at 250 °C for 12 h	
		0.61C-1.45Si-0.76Mn-2.42Cr	1582	2030	14.20	Bainitized at 250 °C for 12 h	
		0.9C-1.65Si-0.79Mn-0.48Cr-0.01Mo	1931	2329	4.10	Bainitized at 220 °C for 40 h	
		0.9C-1.65Si-0.79Mn-0.48Cr-0.01Mo	1910	2213	11.90	Bainitized at 250 °C for 22 h	
		0.9C-1.65Si-0.79Mn-0.48Cr-0.01Mo	1701	2036	12.60	Bainitized at 270 °C for 7 h	

(*Continued*)

TABLE 12.1
Mechanical Properties of CFB Obtained in Chronological Order (Continued)

S. No.	Author (Year)	Material	Yield Strength (MPa)	Ultimate Tensile Strength (MPa)	% Elongation	Treatment Route	Reference
3.	Durbadal Mandal et al. (2014)	0.13C-1.03Mn-1.2Si-0.08V-0.042P-0.05S-Fe-Bal.	....	553	21%	Cast blocks were homogenized at a temperature of 1000 °C for 6 h	[12]
			....	663	27	Austenitization at 900 °C for 30 min and austempering at 400 °C for 10 min	
			....	668	26	Austenitization at 900 °C for 30 min and austempering at 400 °C with holding time from 30 min	
			....	675	19	Austenitization at 900 °C for 30 min and austempering at 400 °C for 60 min	
4.	F. Hu et al. (2016)	Fe-0.95C-0.91Si-1.30Mn-2.30Cr-0.99Mo-0.17Ti	....	1975±20	5.8±1.0	Austenitization at 1000 °C for 30 min with isothermal transformation at a temperature of 200 °C for 10 days	[13]
			....	1910±20	15.0±1.0	Quenched at 0 °C for 2 min in water after which partitioning at 450 °C for 60 s in a molten salt bath	
5.	J. Kang et al. (2016)	0.34C-1.52Mn-1.48Si-1.15Cr-0.71Al-0.93Ni-0.40Mo	1100	1498	16.0	Austenitizaton at 930 °C for 45 min with austempering at 320 °C for 1 h	[14]

S. No.	Author (Year)	Material	Yield Strength (MPa)	Ultimate Tensile Strength (MPa)	% Elongation	Treatment Route	Reference
			1032	1495	17.3	Austenitizaton at 930 °C for 45 min with austempering at 395 °C for 2 h	
6.	Behzad Avishan (2017)	0.88C-1.62Si-1Cr-0.27Mo-2.55Ni-1.20Mn-0.89Al-1.22Co-0.1V-Fe	....	1600	30%	Austenitization at 950 °C for 30 min isothermally austempering at 300 °C for 10 h	[15]
			....	1890	17%	Austenitization at 950 °C for 30 min isothermally austempering at 250 °C for 16 h	
			....	2180	14%	Austenitization at 950 °C for 30 min isothermally austempering at 200 °C for 72 h	
7.	C. Garcia-Mateo and F. G. Caballero (2017)	0.8C-1.5Si-0.75Mn-0.50Cr	1931	2329	4.1	Bainitized at 220 °C	[16]
		0.8C-1.5Si-0.75Mn-0.50Cr	1701	2036	12.64	Bainitized at 220 °C	
		1C-1.50Si-0.75Mn-0.50Cr-0.1Mo	2019	2091	0.38	Bainitized at 270 °C	
		1C-1.50Si-0.75Mn-0.50Cr-0.1Mo	1852	2164	8.29	Bainitized at 250 °C	
		1C-2.9Si-0.75Mn-0.50Cr	1704	2287	7.37	Bainitized at 220 °C	
		1C-2.9Si-0.75Mn-0.50Cr	1698	2068	21.32	Bainitized at 250 °C	
8.	Avanish Kumar and Aparna Singh (2018)	0.85C-1.30Si-1.92Mn-0.44Al-2.05Co-0.29Mo	1560±32	1807±156	7.2±0.16	Austenitization at 950 °C for 40 min and austempering at 250 °C for 40 h	[17]

(*Continued*)

TABLE 12.1
Mechanical Properties of CFB Obtained in Chronological Order (Continued)

S. No.	Author (Year)	Material	Yield Strength (MPa)	Ultimate Tensile Strength (MPa)	% Elongation	Treatment Route	Reference
			1382±20	1676±7	14.1±2	Austenitization at 950 °C for 40 min and austempering at 300 °C for 30 h	
			1028±52	1285±27	25.7±3.65	Austenitization at 950 °C for 40 min and austempering at 350 °C for 20 h	
9.	Binggang Liu et al. (2018)	Fe-0.95C-2.90Si-0.75Mn-0.52Cr-0.25Mo-0.03Nb	1833	2135	21.2%	Austenitized at 950 °C for 30 minutes austempering at 250 °C for 24 h	[18]
10.	Binggang Liu et al. (2019)	Fe-0.95C-2.90Si-0.75Mn-0.52Cr-0.25Mo-0.03Nb	740±10	1773±10	19.0±2	Austenitized at a temperature of 950 °C for 30 min after which austempering at 350 °C for 3 h before quenching to water	[19]
		Fe-0.95C-2.90Si-0.75Mn-0.52Cr-0.25Mo-0.03Nb	1414±9	1845±9	23.0±2	Austenitized at a temperature of 950 °C for 30 min and then austempering 300 °C for 7 h before quenching to water	
		Fe-0.95C-2.90Si-0.75Mn-0.52Cr-0.25Mo-0.03Nb	1689±12	2052±12	15.0±2	Austenitized at a temperature of 950 °C for 30 min and then austempering 250 °C for 24 h before quenching to water	

S. No.	Author (Year)	Material	Yield Strength (MPa)	Ultimate Tensile Strength (MPa)	% Elongation	Treatment Route	Reference
		Fe-0.95C-2.90Si-0.75Mn-0.52Cr-0.25Mo-0.03Nb	1931±10	2174±10	3.1±2	Austenitized at a temperature of 950 °C for 30 min and then austempering at 220 °C for 72 h before quenching to water	
		Fe-0.62C-1.44Si-1.26Mn-1.47Cr-0.26Mo-0.02Nb	796±10	1475±15	8.04±1	Austenitized at a temperature of 950 °C for 30 min and then austempering at 350 °C for 3 h before quenching to water	
		Fe-0.62C-1.44Si-1.26Mn-1.47Cr-0.26Mo-0.02Nb	1166±10	1677±15	16.2±2	Austenitized at a temperature of 950 °C for 30 min and then austempering at 300 °C for 9 h before quenching to water	
11.	A.B. Rezende et al. (2020)	0.71C-0.43Si-0.84Mn-0.27Cr-0.21Cu-0.05V-0.22(Mo+Nb)	1351±47	1600±22	11±1	Austenitization at 900 °C and austempered at 350 °C for 50 min	[20]

12.4 CONCLUSION

From this study, it can be concluded that CFB can give us greater strengths up to 2329 MPa as well as the combination of high ductility and high yield strength, with a proper heat treatment route (in this case bainitization at 220 °C), which creates a higher amount of stable RA, thus enhancing the mechanical behavior of material. Different routes of heat treatment lead to different microstructures of CFB, affecting the desired properties.

Abbreviation

CFB Carbide free bainite
RA Retained austenite
SQ&T Slowly quenched and tempered
(Units- h: hour, min: minutes, s: seconds, MPa: Mega Pascal)

REFERENCES

1. Bhadeshia, H. K. D. H., & Edmonds, D. V. (1983). Bainite in silicon steels: new composition—property approach part 1. *Metal Science*, *17*(9), 411–419.
2. Gao, G., Zhang, H., Tan, Z., Liu, W., & Bai, B. (2013). A carbide-free bainite/martensite/austenite triplex steel with enhanced mechanical properties treated by a novel quenching-partitioning-tempering process. *Materials Science and Engineering A*, *559*, 165–169.
3. Liu, J. P., Li, Y. Q., Zhou, Q. Y., Zhang, Y. H., Hu, Y., Shi, L. B., Wang, W. J., Liu, F. S., Zhou, S. B., & Tian, C. H. (2019). New insight into the dry rolling-sliding wear mechanism of carbide-free bainitic and pearlitic steel. *Wear*, 432–433 (June).
4. Narayanaswamy, B., Hodgson, P., Timokhina, I., & Beladi, H. (2016). The impact of retained austenite characteristics on the two-body abrasive wear behavior of ultra-high strength bainitic steels. *Metallurgical and Materials Transactions A: Physical Metallurgy and Materials Science*, *47*(10), 4883–4895.
5. Gao, G., Zhang, H., Tan, Z., Liu, W., & Bai, B. (2013). A carbide-free bainite/martensite/austenite triplex steel with enhanced mechanical properties treated by a novel quenching–partitioning–tempering process. *Materials Science and Engineering: A*, *559*, 165–169.
6. Liu, B., Li, W., Lu, X., Jia, X., & Jin, X. (2019). The effect of retained austenite stability on impact-abrasion wear resistance in carbide-free bainitic steels. *Wear*, 428–429 (February), 127–136.
7. Jiang, T., Liu, H., Sun, J., Guo, S., & Liu, Y. (2016). Effect of austenite grain size on transformation of nanobainite and its mechanical properties. *Materials Science and Engineering A*, *666*, 207–213.
8. Ding, R., Tang, D., & Zhao, A. (2014). A novel design to enhance the amount of retained austenite and mechanical properties in low-alloyed steel. *Scripta Materialia*, *88*, 21–24.
9. Hu, F., Wu, K. M., & Hodgson, P. D. (2016). Effect of retained austenite on wear resistance of nanostructured dual phase steels. *Materials Science and Technology (United Kingdom)*, *32*(1), 40–48.
10. Sharma, S., Sangal, S., & Mondal, K. (2011). Development of new high-strength carbide-free bainitic steels. *Metallurgical and Materials transactions A*, *42*(13), 3921–3933.
11. Sourmail, T., Caballero, F. G., García-Mateo, C., Smanio, V., Ziegler, C., Kuntz, M., . . . & Teeri, T. (2013). Evaluation of potential of high Si high C steel nanostructured bainite for wear and fatigue applications. *Materials Science and Technology*, *29*(10), 1166–1173.
12. Mandal, D., Ghosh, M., Pal, J., Chowdhury, S. G., Das, G., Das, S. K., & Ghosh, S. (2014). Evolution of microstructure and mechanical properties under different austempering holding time of cast Fe—1.5 Si—1.5 Mn—V steels. *Materials & Design (1980–2015)*, *54*, 831–837.
13. Hu, F., Wu, K. M., & Hodgson, P. D. (2016). Effect of retained austenite on wear resistance of nanostructured dual phase steels. *Materials Science and Technology*, *32*(1), 40–48.
14. Kang, J., Zhang, F. C., Long, X. Y., & Lv, B. (2016). Low cycle fatigue behavior in a medium-carbon carbide-free bainitic steel. *Materials Science and Engineering: A*, *666*, 88–93.
15. Sharma, S., Sangal, S., & Mondal, K. (2011). Development of new high-strength carbide-free bainitic steels. *Metallurgical and Materials Transactions A*, *42*(13), 3921–3933.
16. Garcia-Mateo, C., & Caballero, F. G. (2017). Nanocrystalline bainitic steels for industrial applications. *Nanotechnology for Energy Sustainability*, 707–724.
17. Kumar, A., & Singh, A. (2018). Toughness dependence of nano-bainite on phase fraction and morphology. *Materials Science and Engineering: A*, *729*, 439–443.

18. Liu, B., Lu, X., Li, W., & Jin, X. (2018). Enhanced wear resistance of nano twinned austenite in higher Si nanostructured bainitic steel. *Wear, 398*, 22–28.
19. Liu, B., Li, W., Lu, X., Jia, X., & Jin, X. (2019). An integrated model of impact-abrasive wear in bainitic steels containing retained austenite. *Wear, 440*, 203088.
20. Rezende, A. B., Fonseca, S. T., Fernandes, F. M., Miranda, R. S., Grijalba, F. A. F., Farina, P. F. S., & Mei, P. R. (2020). Wear behavior of bainitic and pearlitic microstructures from microalloyed railway wheel steel. *Wear, 456*, 203377.
21. Kumar, R., Dwivedi, R. K., & Ahmed, S. (2020). Stability of retained austenite in carbide free bainite during the austempering temperature and its influence on sliding wear of high silicon steel. *Silicon, 13* (2113), 1–11.
22. de Moor, E., Lacroix, S., Clarke, A. J., Penning, J., & Speer, J. G. (2008). Effect of retained austenite stabilized via quench and partitioning on the strain hardening of martensitic steels. *Metallurgical and Materials Transactions A: Physical Metallurgy and Materials Science, 39*(11), 2586–2595.

13 Tribological Behavior of Carbide-Free Bainite in High Silicon Steel
A Critical Review

Rajan Kumar, Ravi Kumar Dwivedi, and Siraj Ahmed

CONTENTS

13.1 Introduction 203
13.2 Mechanism of Bainite Formation during Treatment 204
     13.2.1 Upper Bainite (UB) 205
     13.2.2 LB 206
13.3 Research Progress in Wear Rate of CFB Steel 206
13.4 Conclusion 208
References 209

13.1 INTRODUCTION

Finding an efficient and sound material is quite a difficult task to perform, and therefore, it's an area of keen interest to researchers. The manufacturing industries has huge demands and applications of such a material, which has a combination of properties like being corrosion-resistant and long-lasting, having better strength and ductility, and being lightweight. Due to the astonishing mechanical properties, carbide-free bainitic (CFB) steels are also called super bainite; their microstructural constituents, whose thickness is less than 100 nm, makes them useful and therefore have a wide range of applications in engineering; due to the low cost of production and the better mechanical properties of CFB steel, various steel industries have a keen interest in it [1].

    The production of these nanocrystalline materials does not require any alloying elements or any thermomechanical process; it can be obtained by single-step austempering only [2–10]. Bainitic ferrite and austenite films are nanometer-sized constituents that are somewhat related to better mechanical properties. Therefore, the handling of the stability of the retained austenite (RA) is very important [11, 12, 49]. For enhanced mechanical properties, the austenite-to-martensitic transformation in the deformation stage must be gradual [1]. The strength of these microstructures containing nanoscale bainitic laths that are immersed in film-type RA comes from the dislocation movement interruption of the boundaries of the nanoscale bainite

DOI: 10.1201/9781003093213-13

laths [13]. According to the report, depending on isothermal temperature of transformation of the bainite (900 °C), the elongation of 10–40%, along with the strength of 1.5–2.3 GPa, was found [14–20].

The mechanical properties of CFB steels are influenced by the transformation temperature used during the austempering process. At higher temperatures, the bainitic laths are coarse, and the steel has very good toughness. As the transformation temperature is lowered, the strength is increased at the cost of some toughness. However, because the strengthening is due to grain refinement, the reduction of toughness is not so steep.

In CFB steels, alloying is done mainly to improve hardenability and to reduce the martensite start (Ms) temperature of the steel. This allows bainitic transformation at lower temperatures. Currently, commercially available CFB steels can be treated as low as 180 °C. However, at these low temperatures, the rate of transformation is very slow. Therefore, these steels usually take up to 10 days of austempering at the aforementioned temperature to complete the transformation. While alloying is low, so is the cost of these steels; because of the long transformation, times their use is somewhat limited.

Newer studies (among them one related publication not appended in this thesis) have tried to improve the transformation rates of these materials in order to reduce their processing time. These studies have shown that it is possible to achieve a nearly complete transformation in more reasonable times (10–24 h) to obtain a microstructure consisting of ferrite laths that can be as thin as 20 nm.

Most of the scientific interest in CFB steels has been directed toward studying conventional mechanical properties and the mechanisms of transformation of this new microstructure. However, their wear and fatigue properties have not received as much attention. The current work is an attempt to increase understanding and knowledge about CFB in these two contexts.

13.2 MECHANISM OF BAINITE FORMATION DURING TREATMENT

In the range of 400–650 °C, when the austenite transforms, some acicular structures are formed; the term *bainite* refers to these acicular structures, shown in Figure 13.1. Usually, the lath structure also contains martensite, which is the shear transformation product, but as with pearlite and ferrite reactions, a diffusion component is in the bainitic transformation. Since the bainitic reaction is masked by these pearlite and ferrite reactions, therefore, obtaining a homogeneous bainite transformation is quite difficult. To retard the formation of pearlite and ferrite, 0.5 wt.% Mo–0.002 wt.% B combination is used [21] to achieve the bainitic structure over a wide range of cooling rates. Usually, strength and ductility are inverse to each other, but here, some metallic alloying elements like manganese, nickel and chromium are used to control the austenite-to-bainite transformation, which helps in the reduction of the start transformation temperature of bainite; due to the lowering of this temperature, the strength increases, and no significant loss of ductility occurs [22].

CFB in High-Silicon Steel

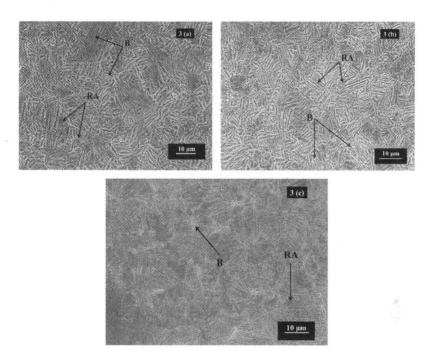

FIGURE 13.1 FE-SEM Micrographs of High Silicon Steel after Austempering for 10 Minutes at: (a) 300 °C, (b) 350 °C, and (c) 400 °C [23]

13.2.1 Upper Bainite (UB)

The bainite transformation is one of the many complex decomposition reactions of the austenitic phase in steels at high temperatures [24]. Its diverse microstructural appearance and the complexity of the mechanism of formation, along with its kinetics, do not agree with creating the right definition [25, 26]. The basic and well-known difference is in the distribution of carbide in the bainite formed at higher temperatures, called intralath, and formed at low temperatures, called interlath, which exist in most steels, and thus, they are the basis of the classification as upper bainite (UB) and lower bainite (LB), distinguishing the appearance and the mechanism of the overall reaction [27–33].

UB can be defined as an aggregate of high-carbon RA and bainitic ferrite because of the effect of a larger amount of silicon (Si) in the alloy without any carbide precipitation. FEF-SEM images of UB are shown in Figure 13.1. The categorization of the UB based on its crystallographic features are as follows [34]: Type (A): two variants of laths with a small misorientation of about 10° contained by a packet.

Type (B): a packet that is further divided by three blocks whose orientations are completely unaligned, and they also contain some sub-blocks. Type (C): six variants of a block are contained in a packet, and every block has one variant of the laths. Preferentially, variant pairs that are formed are twin-related.

13.2.2 LB

The definition of LB is the bainitic ferrite that is aggregated with a small amount of RA the intralath precipitation of the sheets, which are like the particles of cementite in the ferrite.

Initially, the nomenclature of them comes from the high-range and low-range variants of bainite, and they are termed as the UB and LB, respectively [35], and it is useful even today. The schematic in Figure 13.1 shows the structure of UB and LB. Another less frequently used term for UB, used by Smith and Mehl [36], is feathery bainite because of its formation in the forms of bundles of plates at the austenite grain boundaries at high reaction temperatures only. Instead, the aggregates of plates are included by both the UB and LB, which were later designated as the sheaves of bainite [37]. The fraction of volume of the RA that is enclosed between the bainite platelets is found lower for LB [38].

13.3 RESEARCH PROGRESS IN WEAR RATE OF CFB STEEL

TABLE 13.1
Remarks and Findings in Chronological Order of the Wear Behavior of CFB in High-Si Steel

S. No.	Authors (Year)	Materials	Wear Parameters	Tribological Outcomes	References
1.	H. Yokoyama et al. (2002)	Head hardened rail steel, standard carbon rail steel	60 to 2400 rpm, 5 to 40 kN	The head hardened pearlitic rail steel showed lower rolling contact fatigue than the low strength bainitic rail steel.	[39]
2.	Ki Myung Lee and Andreas A. Polycarpou (2005)	0.26C-2.00Mn-1.81Si-1.93Cr, 0.79C-0.91Mn-0.66Si-0.47Cr	0–1000 rpm and contact load of 0.45–45N	The initially harder bainitic rail steel was work hardened less when compared to the initially softer pearlitic rail steel.	[40]
3.	A. Leiro et al. (2013)	0.99C-1.5Si-0.76Mn-0.46Cr, 0.98C-2.9Si-0.77Mn-0.46Cr, 0.9C-1.65Si-0.79Mn-0.48Cr, 0.68C-1.6Si-1.25Mn-1.5Cr, 100Cr6	100 rpm while 5% slip was kept	With similar hardness values, the wear resistance of nanostructured steels in dry rolling-sliding is significantly superior to that of bainitic steels, which are transformed at high temperature.	[41]

S. No.	Authors (Year)	Materials	Wear Parameters	Tribological Outcomes	References
4.	S. Das Bakshi et al. (2014)	0.83C-2.28Mn-0.011P-0.008S-1.9Si-0.044Al-0.12Cu-1.44Cr-0.24Mo-0.11V-1.55Co-0.019Sn-0.023Nb	30,000 cycles at 100 rpm, 95 rpm with load of 300 N which is applied externally	During the contact, the maximum shear stresses generated are below the contact surface thus the role of sliding is minimal.	[42]
5.	R. Rementeria et al. (2015)	0.30C—1.48Si—2.06Mn—0.43Cr—0.27Mo, 0.99C—2.47Si—0.74Mn—0.12Ni—0.97Cr—0.03Mo—0.17Cu	Contact load of 0.5 N with average velocity of 10 mm/s	The nano-scaled bainites were improved as shown by the results with ultrafine and similar hardness values.	[43]
6.	F. Hu et al. (2015)	0.95C-0.91Si-1.30Mn-2.30Cr-0.99Mo-0.17Ti-Fe	Rotational speed of and 2150±20 rpm and time of 30 min × 5 cycles	The hardness of the worn surface of the Q&P martensitic steel significantly enhanced from 674 to 762 HV1 and the thickness of the deformed layer is increased 3.3 mm, as compared to the nanostructured bainitic steel.	[44]
7.	A.M. Gola (2017)	0.32C-1.64Si-0.85Mn-0.20Cr-1.17Ni-0.01V-0.004B, 0.39C-1.6Si-0.61Mn-0.72Cr-0.33Mn-1.66Ni-0.06V	Load 1 kgf-30 kgf and rpm of 39	Heat-treated to CFB microstructure contains higher volume fraction of RA as suggested by the results.	[48]
8.	Sk Md Hasan et al. (2018)	Wheel steel, 880 pearlitic steel, bainitic steel	Slip ratio = 10% and speed and load are 400 rpm and 100 g-30 kg respectively	The wear resistance shown by bainitic rail steels was higher than the pearlitic rail steel, also with the increasing starting hardness the wear resistance increases.	[45]

(Continued)

TABLE 13.1
Remarks and Findings in Chronological Order of the Wear Behavior of CFB in High-Si Steel (Continued)

S. No.	Authors (Year)	Materials	Wear Parameters	Tribological Outcomes	References
9.	Changle Zhang et al. (2019)	0.45C-0.81Cr-0.30Mo-0.004B-1.00Si-2.19Mn-Fe	Test load of 294 N and 200 rpm	When the content of silicon increases the content of retained austenite is reduced, the bainitic lath is refined, and the hardness also tends to increase.	[46]
10.	Rajan Kumar et al. (2020)	0.55C-1.75Si-0.82Mn-0.13Cr-0.03S-0.03Al-0.03P-Fe	Speed 300 RPM and applied load is 50 N	Before austempering when CC time in air is increased, a decrement is observed in the phases of P, B, and RA while an increase is observed in the F.	[47]

13.4 CONCLUSION

On the basis of the preceding review of the influence of the current trends of RA on tribological behavior of CFB, the following conclusions are made:

1. The percentage of carbon in the material composition is found out to be 1% for the highest number of samples. As the carbon percentage increases, the hardness of the steel also increases; hence, higher percentages of carbon are needed so that the wear resistance can also be increased.
2. Silicon plays a vital role in balancing the strength and ductility of the material, which is one of the major advantages of CFB. The percentage of Si in the material composition was 1.4% while higher percentages of Si, up to 2%, were rarely taken, which could have enhanced the mechanical properties of the material.
3. In high-silicon steel, as per the literature survey, better wear resistance of CFB steel was found as compared to other microstructures.

List of Abbreviations

Abbreviations	Full Forms
UB	Upper Bainite
LB	Lower Bainite

RA   Retained Austenite
RCF   Rolling Contact Fatigue
CFB   Carbide Free Bainite
P   Pearlite
F   Ferrite

REFERENCES

1. Avishan, B., Sefidgar, A., & Yazdani, S. (2019). High strain rate deformation of nanostructured super bainite. *Journal of Materials Science, 54*(4), 3455–3468.
2. Gong, W., Tomota, Y., Adachi, Y., Paradowska, A. M., Kelleher, J. F., & Zhang, S. Y. (2013). Effects of ausforming temperature on bainite transformation, microstructure and variant selection in nanobainite steel. *Acta Materialia, 61*(11), 4142–4154.
3. Gong, W., Tomota, Y., Koo, M. S., & Adachi, Y. (2010). Effect of ausforming on nanobainite steel. *Scripta Materialia, 63*(8), 819–822.
4. Garcia-Mateo, C., Fg, C., & Hkdh, B. (2003). Development of hard bainite. *ISIJ International, 43*(8), 1238–1243.
5. Bhadeshia, H. K. D. H., Brown, P., & Garcia-Mateo, C. (2010). Bainite steel and methods of manufacture thereof, patent no. *GB2462197.*
6. Rakha, K., Beladi, H., Timokhina, I., Xiong, X., Kabra, S., Liss, K. D., & Hodgson, P. (2014). On low temperature bainite transformation characteristics using in-situ neutron diffraction and atom probe tomography. *Materials Science and Engineering: A, 589*, 303–309.
7. Yoozbashi, M. N., Yazdani, S., & Wang, T. S. (2011). Design of a new nanostructured, high-Si bainitic steel with lower cost production. *Materials & Design, 32*(6), 3248–3253.
8. Solano-Alvarez, W., Abreu, H. F. G. D., Da Silva, M. R., & Peet, M. J. (2015). Phase quantification in nanobainite via magnetic measurements and X-ray diffraction. *Journal of Magnetism and Magnetic Materials, 378*, 200–205.
9. Huang, H., Sherif, M. Y., & Rivera-Díaz-del-Castillo, P. E. J. (2013). Combinatorial optimization of carbide-free bainitic nanostructures. *Acta Materialia, 61*(5), 1639–1647.
10. Avishan, B., Garcia-Mateo, C., Morales-Rivas, L., Yazdani, S., & Caballero, F. G. (2013). Strengthening and mechanical stability mechanisms in nanostructured bainite. *Journal of Materials Science, 48*(18), 6121–6132.
11. Garcia-Mateo, C., Caballero, F. G., Chao, J., Capdevila, C., & De Andres, C. G. (2009). Mechanical stability of retained austenite during plastic deformation of super high strength carbide free bainitic steels. *Journal of Materials Science, 44*(17), 4617–4624.
12. Bhadeshia, H. K. D. H. (2010). Nanostructured bainite. *Proceedings of the Royal Society A: Mathematical, Physical and Engineering Sciences, 466* (2113), 3–18.
13. Amel-Farzad, H., Faridi, H. R., Rajabpour, F., Abolhasani, A., Kazemi, S., & Khaledzadeh, Y. (2013). Developing very hard nanostructured bainitic steel. *Materials Science and Engineering: A, 559*, 68–73.
14. Garcia-Mateo, C., Caballero, F. G., & Bhadeshia, H. K. (2005). Mechanical properties of low-temperature bainite. In *Materials Science Forum* (Vol. 500, pp. 495–502). Trans Tech Publications Ltd.
15. Garbarz, B., & Ni-nik-Harańczyk, B. (2015). Modification of microstructure to increase impact toughness of nanostructured bainite—austenite steel. *Materials Science and Technology, 31*(7), 773–780.
16. Lan, H., Du, L., Zhou, N., & Liu, X. (2014). Effect of austempering route on microstructural characterization of nanobainitic steel. *Acta MetallurgicaSinica (English Letters), 27*(1), 19–26.

17. Garcia-Mateo, C., & Caballero, F. G. (2005). Ultra-high-strength bainitic steels. *ISIJ International*, *45*(11), 1736–1740.
18. Han, B., Chen, L., & Wu, S. J. (2015). Effect of austempering—partitioning on the bainitic transformation and mechanical properties of a high-carbon steel. *Acta Metallurgica Sinica (English Letters)*, *28*(5), 614–618.
19. Sandvik, B. P. J., & Nevalainen, H. P. (1981). Structure-property relationships in commercial low-alloy bainitic-austenitic steel with high strength, ductility, and toughness. *Metals Technology*, *8*(1), 213–220.
20. Davenport, E.S., & Bain, E.C. (1930). Transformation of austenite at constant subcritical temperatures. *Transactions of the American Institute of Mining, Metallurgical*, 1920, *90*, 117–154.
21. Clayton, P., Sawley, K. J., Bolton, P. J., & Pell, G. M. (1987). Wear behavior of bainitic steels. *Wear*, *120*(2), 199–220.
22. Guoa, Y., & Lia, Z. (2014). Effect of isothermal heat treatment on nanostructured bainitic microstructures and properties in laser cladded coatings. *Conference: Visual-JW2014At*, Osaka University.
23. Kumar, R., Dwivedi, R. K., & Ahmed, S. (2021). Stability of retained austenite in carbide free bainite during the austempering temperature and its influence on sliding wear of high silicon steel. *Silicon*, *13*, 1249–1259.
24. Hehemann, R. F., Kinsman, K. R., & Aaronson, H. I. (1972). A debate on the bainite reaction. *Metallurgical Transactions*, *3*(5), 1077–1094.
25. Aaronson, H.I. (1968). *The mechanism of phase transformation in crystalline solids, institute of metals monograph*. Proceedings of an International Symposium Organized by the Institute of Metals and Held in the University of Manchester from 3 to 5 July 1968, Issue 33; Issue 1969, 270.
26. 1st Progress Report of Sub-Committee XI, ASTM Committee E-4; Proc. ASTM, 1950, vol. 50, p. 444.
27. 2nd Progress Report of Sub-Committee XI, ASTM Committee E-4; Ibid., 1952, vol. 52, p. 543.
28. 4th Progress Report of Sub-Committee XI, ASTM Committee E~; Ibid., 1954, vol. 54, p. 568.
29. Tiabraken, L. (1958). Proc. 4th Int. Conf. on Electron Microscopy, Springer-Verlag, Berlin, p. 621.
30. Fisher, R.M., Ibid., p. 579.
31. Picketing, F. B., Ibid., p. 626.
32. Shimizu, K., Nishiyama, Z. (1963). Mere. Inst. ScL Ind. Res. Osaka Univ., vo.20, p. 43.
33. Furuhara, T., Kawata, H., Morito, S., Maki, T. (2006). Crystallography of upper bainite in Fe–Ni–C alloys. *Materials Science and Engineering: A*, *431* (2006), 228–236.
34. Katsamas, A. I. (2006). A method for the calculation of retained austenite evolution during heat-treatment of low-alloy TRIP-assisted steels. *Steel Research International*, *77*(3), 210–217.
35. Smith, G. V., & Mehl, R. F. (1942). Lattice relationships in decomposition of austenite to pearlite, bainite and martensite. *Transactions of the AIME*, *150*(211–226), 14.
36. Aaronson, H. I., & Wells, C. (1956). Sympathetic nucleation of ferrite. *JOM*, *8*(10), 1216–1223.
37. Heheman, R.F. (1970). *Phase Transformations*, ASM, Metals Park, OH, pp. 397–432.
38. Spanos, G., Fang, H. S., & Aaronson, H. I. (1990). A mechanism for the formation of lower bainite. *Metallurgical Transactions A*, *21*(6), 1381–1390.
39. Garcia-Mateo, C., FG, C., & Hkdh, B. (2003). Acceleration of low-temperature bainite. *ISIJ International*, *43*(11), 1821–1825.

40. Girsch, G., & Heyder, R. (2006, June). Advanced pearlitic and bainitic high strength rails promise to improve rolling contact fatigue resistance. In *7th World Congress on Railway Research (WCRR2006)*, Montreal, Canada.
41. Leiro, A., Vuorinen, E., Sundin, K. G., Prakash, B., Sourmail, T., Smanio, V., . . . & Elvira, R. (2013). Wear of nano-structured carbide-free bainitic steels under dry rolling—sliding conditions. *Wear, 298*, 42–47.
42. Bakshi, S. D., Leiro, A., Prakash, B., & Bhadeshia, H. K. D. H. (2014). Dry rolling/sliding wear of nanostructured bainite. *Wear, 316*(1–2), 70–78.
43. Rementeria, R., García, I., Aranda, M. M., & Caballero, F. G. (2015). Reciprocating-sliding wear behavior of nanostructured and ultra-fine high-silicon bainitic steels. *Wear, 338*, 202–209.
44. Hu, F., Wu, K. M., & Hodgson, P. D. (2016). Effect of retained austenite on wear resistance of nanostructured dual phase steels. *Materials Science and Technology, 32*(1), 40–48.
45. Hasan, S. M., Chakrabarti, D., & Singh, S. B. (2018). Dry rolling/sliding wear behaviour of pearlitic rail and newly developed carbide-free bainitic rail steels. *Wear, 408*, 151–159.
46. Zhang, C., Fu, H., Lin, J., & Lei, Y. (2019). The effect of silicon on microstructure and wear resistance in bainitic steel. *Transactions of the Indian Institute of Metals, 72*(5), 1231–1244.
47. Kumar, R., Dwivedi, R. K., & Ahmed, S. (2020). Influence of multiphase high silicon steel (Retained Austenite-RA, Ferrite-F, Bainite-B and Pearlite-P) and carbon content of RA-C γ on rolling/sliding wear. *Silicon*, 1–14.
48. Gola, A. M., Ghadamgahi, M., & Ooi, S. W. (2017). Microstructure evolution of carbide-free bainitic steels under abrasive wear conditions. *Wear, 376*, 975–982.
49. Varshney, A., Sangal, S., & Mondal, K. (2017). Exceptional work-hardening behavior of medium-carbon high-silicon low-alloy steels. *Metallurgical and Materials Transactions A, 48*(2), 589–593.

14 Impact of Nanoparticles in Lube Oil Performance
A Review

Anoop Pratap Singh, Ravi Kumar Dwivedi, and Amit Suhane

CONTENTS

14.1 Introduction 213
14.2 Review 214
14.3 Conclusion 223
References 223

14.1 INTRODUCTION

Lubrication is the process of maintaining a third medium between two tribological surfaces to reduce friction, wear, and their progressive loss. To prevent the losses in the form of frictional energy, lubrication plays a vital role in keeping the machine performing efficiently. Current progress in all areas is needed for robust lubricants. A lubricant can operate with high loads, speeds, temperatures, and adverse conditions is the basic requirement of an advanced mechanical system.

Nanotechnology presents solutions on the way to lubricant challenges. Adding nanoparticles to lubricating oil can improve its ability to reduce friction and wear as well as make lubricants work in adverse conditions. Nano-lubricant is a homogeneous mixture of a base oil, nanoparticles and a surfactant/dispersant. Nanoparticles improve the performance of lubricating oil [1].

Much research has been done on the evaluation of the performance of lubricating oils (500 SN, single-grade and multigrade oil, paraffin, etc.), with a number of nanoparticles, such asgold, bismuth, cobalt, copper, aluminum oxide (Al_2O_3), copper oxide (CuO), iron oxide (Fe_3O_4), silicon dioxide (SiO_2), polytetrafluoroethylene (PTFE), boron nitrade (BN), serpentine and others, with surfactants/dispersants such as KH-560, oleic acid, stearic acid, sodium dodecyl sulfate (SDS) and many more. Therefore, a summary of the results of past evaluations is needed for future use and understanding. So, in this review, information about nanoparticles, base oil, surfactant/dispersant, and performance parameters (minimum value of coefficient of friction) is listed in a very precise manner. Analysis of tribological test methods with the material of tribological test member is also covered in this review. All this information collectively gives a clear picture about nano-lubrication.

DOI: 10.1201/9781003093213-14

14.2 REVIEW

Several articles are available on the evaluation of the performance of nanoparticles in lubricating oil. Some researchers conducted experiments looking at one nano particle; others have looked at two or more nanoparticles and some nanocomposites. Researchers evaluate the performance of lubricating oil by using single nanoparticles [2–5]. Researchers provide a comparative angle by using two or more nanoparticles [6–8]. The nanocomposite section provides the opportunity to utilize the capacity of two or more nanoparticles at the same time [9, 10].

This review summarizes the findings of previous research, along with a listing all the elemental information about nano-lubricants, as well as tribal test methods along their surface of the material. The elemental information of nano-lubricant includes information about nanoparticle type, nanoparticle size, base oil and surfactant/dispersant. This review also demonstrates the best-performing characteristics of nano-lube oil with optimal amounts of nanoparticles.

The preparation of a nano-lubricant includes a homogeneous mix of nanoparticles in a base oil with a surfactant/dispersant. The surfactant/dispersant is an important ingredient for the proper dispersion of the nanoparticles. Sometimes, the surface of the nanoparticle is modified for a seamless mixing of nanoparticles and to prevent agglomeration. Researchers adopt a process of preparing nano-lubricants according to their convenience. The time for homogenization and sonication varies depending on the researcher. The optimum amount of nanoparticles (carboxyl [COOH]–functionalized multiwalled carbon nanotubes [MWCNTs]) are first magnetically stirred for 15 minutes, and then the mixture is stirred in a sonication bath for 1 hour [11]. Another process includes homogenizing the mixture for 10 minutes and then sonicating it for 3 hours for CuO and Al_2O_3 [8]. Some are stirred for 30 minutes and ultrasonicated for 1 hour for CuO [12].

Table 14.1 surmises the findings of various researchers in an appropriate way. Within this table, the size of nanoparticles is mentioned along with the type of nanoparticles. Base oil information is also listed in the table. Without a surfactant/dispersant, a nanoparticle gets aggregated after some time. So information about the surfactant/dispersant is important in the context of a nano-lubricant. Therefore, this study also includes information about surfactants/dispersants. Information about the type of nanoparticle, size of nanoparticle, base oil and surfactant/dispersant provide a complete information about nano-lubricants. After nano-lubricants are made, their performance should be evaluated by various tribological test methods. Therefore, the methods of tribological testing and the material of the tribological member have been critically investigated in this review. Some of the cells in the column "Material of Tribological Test Members" in Table 14.1 have a single material instead of pairs, which means that both tribological members are of the same material. The second last column of Table 14.1 provides information about the minimum COF value with optimal concentrations of nanoparticles.

Impact of Nanoparticles in Lube Oil 215

TABLE 14.1
Summary of Literature Review

S. No.	Nanoparticles	Nanoparticle Size	Base Oil	Surfactant/Dispersant	Tribological Testing Method	Material of Tribological Member	Minimum COF Obtained with % of Nanoparticles	References
1.	Al_2O_3	78 nm	VI 95	KH-560 (3-glycidoxypropyltrimethoxysilane)	Four-ball and thrust ring friction test	Steel ball GB/308–89, GCr15	0.055 in four ball tester and 0.047 in thrust ring with 0.1% of Al_2O_3	[2]
2.	Al_2O_3/SiO_2	Al_2O_3: 13 nm SiO_2: 30 nm	Polyalkylene Glycol (PAG 46)	Not mentioned	Tribology test rig	Aluminum Al 2024	0.0652 at 0.06% of Al_2O_3/SiO_2	[9]
3.	Al_2O_3 and TiO_2	Al_2O_3: 8–12 nm TiO_2: 10 nm	Engine oil (5W-30)	Oleic acid	Piston ring tester	Cylinder material: Cast iron piston ring material: Nitrided steel	0.036 (Lube Oil + 0.25% of Al_2O_3) 0.038 (Lube Oil + 0.25% of TiO_2)	[6]
4.	Al_2O_3/TiO_2	75 nm	VI 95	KH-560 (3-glycidoxypropyltrimethoxysilane)	Four-ball and thrust ring friction test	Steel ball GB/308–89, GCr15	0.051 in four-ball tester and 0.047 in thrust ring friction test with 0.1% Al_2O_3/TiO_2	[10]
5.	Bi	7 and 65 nm	BS900 (heavy base oil) and BS6500 (light base oil)	Not mentioned	Four ball tester	AISI 52100 steel balls	In BS6500 COF from 0.091 to 0.052 (for 900 mg/L) and in BS900 from 0.074 to 0.047 (for 310 mg/L)	[3]

(*Continued*)

TABLE 14.1
Summary of Literature Review (Continued)

S. No.	Nanoparticles	Nanoparticle Size	Base Oil	Surfactant/Dispersant	Tribological Testing Method	Material of Tribological Member	Minimum COF Obtained with % of Nanoparticles	References
6.	BN	100–120 nm	Epoxy	Not mentioned	Ball on disk	100Cr6 steel	0.06 with 0.5% of BN	[13]
		150 nm	Poly alpha olphaline (PAO) oil	No additive used	Sliding contact linear reciprocating rig	Surface-treated 9310 steel, carburized and borided gear steel surfaces	0.12 for borided surface	[4]
		DIA: 120 nm THK: 30 nm	SE 15W-40	Oleic acid	Friction tester	No.45 stainless	0.01 with 0.1% of BN	[14]
7.	CNT	LEN: 3–15 nm DIA: 10–20 nm	Castor oil	Oleic acid	Four-ball tester	GCr15 steel ball	0.039 with 0.02% of CNTs	[15]
8.	Cu	Not mentioned	Diesel engine oil SAE 30 and 40	Nanol®M[10.0-T]	Pin on disk	Pin—Chromium-plated steel 100Cr6 Disk—Grey cast iron (GG 25, lamellar graphite) pin material	0.075 with 0.3% of Cu	[5]
		20 nm	Mineral oil SN 650	Not mentioned	Test rig	1045 steel	0.055 with 1% of Cu	[16]
		20–130 nm	500SN	Polyisobutylenebutadimide	Pin on disk	Disk—20CrMnTi steel Pin—H62 bronze	Mending effect shown	[17]

S. No.	Nanoparticles	Nanoparticle Size	Base Oil	Surfactant/Dispersant	Tribological Testing Method	Material of Tribological Member	Minimum COF Obtained with % of Nanoparticles	References
		Not mentioned	5W-40, 5W-40 and 5W-20	DDP (di-n-octodecyldithiophosphate)	Four-ball tester	ASTMD 2783, GCr15 steel balls	CAL-1 (5W-40)—0.085 at 0.6% CAL-2 (5W-40)—0.055at 0.8% CAL-3 (5W-20) 0.05 at 1.6%	[18]
		25 nm	Avocado oil	Not mentioned	Pin-on-disk tribometer	Pin—Aluminum alloy 6061 Disc—EN-31 steel	0.022 at 1% of Cu	[19]
9.	Cu/graphene oxide	5–10 nm	Liquid paraffin	Stearic acid	Four-ball tester	GCr15 bearing steel	Sc-Cu/GO composites show great tribological properties as compared to base oil, nanoparticles with Cu and GO	[20]
10.	CuO	50 nm	Liquid paraffin	Oleic acid	Four-ball tester	AISI 52100 steel	0.2 with 0.2% of CuO	[21]
		50 nm	Castor oil and paraffin oil	Sodium dodecyl sulfate (SDS)	Four-ball tester	AISI 52100 steel	0.04 (CO + 0.1% CuO) 0.06 (PO + 1% CuO)	[12]
11.	CuO and ZnO	ZnO: 11.71 nm CuO: 4.35 nm	Soybean oil Sunflower oil Mineral oil Synthetic oil (polyalphaoleifin)	Not mentioned	High-frequency reciprocating testrig (HFRR)	A hard steel ball (570–750 HV)	0.05 (Sunflower oil and Soybean oil show inverse effects on adding nanoparticles) 0.095 (Mineral oil + Zno and synthetic oil + CuO)	[7]

(Continued)

TABLE 14.1
Summary of Literature Review (Continued)

S. No.	Nanoparticles	Nanoparticle Size	Base Oil	Surfactant/Dispersant	Tribological Testing Method	Material of Tribological Member	Minimum COF Obtained with % of Nanoparticles	References
12.	CuO and Al$_2$O$_3$	≤50 nm	GL-4 (SAE75W-85), PAO8	Not mentioned	Optimol SRV 4 tester and four-ball tribotester	AISI 52100 steel balls	0.125 (PAO8 + 1% Al2O3) 0.111 (PAO8 + 2% CuO) 0.125 (GL4 + 0.5% Al2O3) 0.115 (GL4 + 0.5% CuO)	[8]
13.	CuO, TiO$_2$ and Nano-Diamond	CuO: 5 nm TiO$_2$: 80 nm Nano-Diamond: 10 nm	API-SF (VI100) engine oil and a base oil (VI107)	Glycol	Reciprocating sliding tribotester	Chromium-coated steel ball and plate	0.084 for API-SF (VI100) engine oil 0.096 for Base oil (VI107)	[22]
14.	Fe, Cu and Co	50–80 nm	SAE 10	Converse emulsions of water in lubricant solution (CEWLS)	Four-ball tribotester	100 Cr6-bearing steel	Friction torque is minimum for SAE10+Cu, SAE10+fe+Cu, and SAE10+FeCu in its own groups	[23]
15.	Graphene and MoS$_2$	2 µm	Hydraulic oil	Not mentioned	Ball on disk	Bearing steels and brass	MoS$_2$ shows better tribological performance	[24]
16.	Graphene oxide nanoplates (GOPs)	1.53 nm	4010 AL base oil	Not mentioned	Four-ball tester	Si3N4 ceramic/GCr15 steel pairs	0.081 with 0.5% of GOPs	[25]

Impact of Nanoparticles in Lube Oil

S. No.	Nanoparticles	Nanoparticle Size	Base Oil	Surfactant/Dispersant	Tribological Testing Method	Material of Tribological Member	Minimum COF Obtained with % of Nanoparticles	References
17.	Graphene sheets	10 μm	PAO 2	No surfactant is used	Four-ball tester	Four-ball tester steel balls	COF lowest with 0.05% of graphene sheets	[26]
18.	hBN	70 nm 0.5 μm 1.5 μm and 5.0 μm	Avocado oil	Not mentioned	Pin on disk	Pin—Oxygen-free electronic copper (C101) Disk—2024 aluminum	0.015 with 70 nm	[27]
		Not mentioned	API Group III and 150N	Oleic acid PIBSI Lubrizol 6412TM Oloa 11000	Four-ball tester	Four-ball tester steel balls	Polyisobutylene succinimide (PIBSI) is most efficient	[28]
		70 nm	SAE 15W-40 diesel engine oil	Oleic acid	Four-ball friction tester	Carbon-chromium steel balls	Maximum reduction of WSD at 800N with 0.5%	[29]
19.	hBN/calcium borate	BN: 100–200 nm BCBN: 12 nm	Mineral-based oil (saturated cycloparaffin and paraffin hydrocarbon)	No external bonding element used	Four-ball tester	Steel balls GCr15	Load-carrying capacity increased COF and wear decrease as compared to BN particle	[30]
20.	MoS$_2$	MoS$_2$: 50–100 nm	Dioctyl sebacate (DOS)	Not mentioned	High-frequency reciprocating ball-on-disc tribometer	GCr15 steel ball	0.06 with 0.5% of nano MoS$_2$	[31]
		62–84 nm	Pentaerythrityl tetracaprylate/ caprate ester	Caproic, lauric, stearic, oleic acid	Four-ball friction tester	Not mentioned	0.103 with 0.5% of SCMS LA	[32]

(*Continued*)

TABLE 14.1
Summary of Literature Review (Continued)

S. No.	Nanoparticles	Nanoparticle Size	Base Oil	Surfactant/Dispersant	Tribological Testing Method	Material of Tribological Member	Minimum COF Obtained with % of Nanoparticles	References
		250 nm	Plyolester and naphthenic oil	Without any dispersant agent addition	Pin-on-disk tester	Pin—AISI 52100 steel Disk—grey cast iron	COF with naphthenic-based nano-oil with 1% concentration of nanoparticles presented a remarkable 86% reduction	[33]
		≤2 μm	Mineral base oil N-150	Trichlorooctadecylsilane	Four-ball tester, pin-on-roller tribo-tester	GCr15 steel ball and EN-31 steel pin	0.079 with 0.25% in four-ball tester 0.0226 with 0.25% in pin on roller	[34]
21.	MoS$_2$ and SiO$_2$	MoS$_2$: 90 nm SiO$_2$: 30 nm	EOT5 Engine oil	No dispersant used	Ball on flat tribometer	Magnesium alloy/ steel contacts	0.045 with 0.7% of SiO$_2$	[35]
22.	MWCNTs/ZnO	MWCNTs OD: 20–30 nm ID 8–10 nm ZnO 100 nm	SAE 20W40	Natural surfactant called gum arabic	Pin-on-disk tribometer	Not mentioned	Lube oil with nanoparticle having better tribological properties	[36]
23.	Nanodiamond and SiO$_2$	Nano diamond: 110 nm SiO$_2$: 92 nm	Liquid paraffin	Oleic acid	Ball on ring wear tester	Bearing steel GCr15	0.074 for nano-diamond (0.2%) 0.075 for SiO$_2$ (0.2%)	[37]

S. No.	Nanoparticles	Nanoparticle Size	Base Oil	Surfactant/Dispersant	Tribological Testing Method	Material of Tribological Member	Minimum COF Obtained with % of Nanoparticles	References
24.	Oxidized graphite flakes	Not mentioned	10W40	Chemically modification of flakes	Ball-on-disk tribometer	Ball—steel 100Cr6 Disk—steel 316 LN	0.128 with 0.05 mg/ml with chemically modified	[38]
25.	SiO_2	102 ± 33 nm	Rust and oxidation lubricant (R&O)	Not mentioned	Four-ball tester	GCr15 bearing steel balls	0.06 with BO68	[39]
		58, 281, 684 nm	Liquid paraffin	Oleic acid	Ball on ring	Ball—bearing steel GCr15 Ring—AISI 52100 steel	0.065 with 0.2% of SiO_2 at 58 nm particle size	[40]
26.	Stearic acid-capped cerium borate composite nanoparticles ($SA/CeBO_3$)	5–30 nm	Rapeseed oil	Stearic acid	Four-ball tester	GCr15 bearing steel balls	RP+SA/$CeBO_3$ has the least COF (0.05)	[41]
27.	TiO_2	10–25 nm	Engine oil servo 4T Synth 10W-30	Not mentioned	Pin on disk	Pin—Aluminum alloy (Al-Si7Mg)	0.015 with 0.3% of TiO_2	[42]
		50–100 nm	API-1509	Oleic acid	Four-ball tester and pin on disk	Steel balls	0.06 with 0.5% of TiO_2	[43]
		20 nm	Water	KH-570 and OP-10	Four-ball tribotester, a bench drilling machine	GGr15-bearing steel Q235 steel in drill test	0.04 with 1.6% of TiO_2	[44]

(*Continued*)

TABLE 14.1
Summary of Literature Review (Continued)

S. No.	Nanoparticles	Nanoparticle Size	Base Oil	Surfactant/Dispersant	Tribological Testing Method	Material of Tribological Member	Minimum COF Obtained with % of Nanoparticles	References
28.	WS_2	Not mentioned	Polyalphaolefin—4 (PAO 4) oil	Humin-like shell-coated	Ball on flat test rig	Stainless steel plate (AISI 316)	0.075 ± 0.005 with 1% of WS_2 coated nanoparticles	[45]
		200 nm	SN 90 and SN 150	Not mentioned	Flat-on-flat tester, roller-on-rib tester, ball-on-flat tester	AISI 2510 & AISI 2510 Brass & Steel Bearing steel & AISI 2510	The effect of IF-WS_2 as a potential additive to base Oils were investigated over a wide range of operating conditions	[46]
29.	WS_2/MoS_2	100 nm	MACs (Multialkylated cyclopentanes) and CPSO (chlorinated-phenyl with methyl-terminated silicone oil)	Not mentioned	Ball on disk	AISI 440C steel balls	0.05 with WS_2/MoS_2-MACs 0.06 with WS_2—MoS_2/CPSO	[47]
30.	ZnO	53 nm	Epoxy resin (RE)	Ionic liquid (IL) 1-octyl-3-methylimidazolium tetrafluoroborate ([OMIM]BF4)	Pin on disk	AISI 316L stainless steel pins	Resign + nanoparticle + IL shows better tribological properties	[48]

14.3 CONCLUSION

Nanoparticles show a positive effect on the performance of lubricating oil. The size of the nanoparticles and the base oil have a clear effect on the performance of lubricating oil. Therefore, this review summarized the findings of previous researchers. Nearly all types of nanoparticles are studied equally, including one, two, or more nanoparticles or nanocomposite segments. Tribological testing methods were also covered with the material of tribological members. In this review, the optimum composition of nanoparticles is mentioned along with base oil and surfactant/dispersant. The minimum value of COF with the optimal concentration of nanoparticles provides a guideline for the selection of nanoparticles according to the requirements. This review cites important observations that provide a clear picture about nanolubricants and their future.

List of Abbreviations

Abbreviations	Full Forms
API	American Petroleum Institute
COF	Coefficient of Friction
COOH	Carboxylic Acid (R—COOH)
KH-560	(3-Glycidyloxypropyl) trimethoxysilane
MWCNTs	Multi wall Carbon Nanotubes
PAO	Poly-Alpha-Olefin
PTFE	Polytetrafluoroethylene
SAE	Society of Automotive Engineers
SDS	Sodium Dodecyl Sulfate
VI	Viscosity Index

REFERENCES

1. A. P. Singh, R. K. Dwivedi, and A. Suhane, "Impact of nano particles morphology and composition in lube oil performance considering environmental issues—A review," *J. Green Eng.*, vol. 10, no. 9, pp. 4609–4625, 2020.
2. T. Luo, X. Wei, X. Huang, L. Huang, and F. Yang, "Tribological properties of Al2O3 nanoparticles as lubricating oil additives," *Ceram. Int.*, vol. 40, no. 5, pp. 7143–7149, 2014.
3. M. Flores-Castañeda, E. Camps, M. Camacho-López, S. Muhl, E. García, and M. Figueroa, "Bismuth nanoparticles synthesized by laser ablation in lubricant oils for tribological tests," *J. Alloys Compd.*, vol. 643, no. S1, pp. S67—S70, 2015.
4. A. Greco, K. Mistry, V. Sista, O. Eryilmaz, and A. Erdemir, "Friction and wear behaviour of boron based surface treatment and nano-particle lubricant additives for wind turbine gearbox applications," *Wear*, vol. 271, no. 9–10, pp. 1754–1760, 2011.
5. M. Scherge, R. Böttcher, D. Kürten, and D. Linsler, "Multi-phase friction and wear reduction by copper nanopartices," *Lubricants*, vol. 4, no. 4, p. 36, 2016.
6. M. K. A. Ali, H. Xianjun, L. Mai, C. Qingping, R. F. Turkson, and C. Bicheng, "Improving the tribological characteristics of piston ring assembly in automotive engines using Al2O3 and TiO2 nanomaterials as nano-lubricant additives," *Tribol. Int.*, vol. 103, pp. 540–554, 2016.

7. S. M. Alves, B. S. Barros, M. F. Trajano, K. S. B. Ribeiro, and E. Moura, "Tribological behavior of vegetable oil-based lubricants with nanoparticles of oxides in boundary lubrication conditions," *Tribol. Int.*, vol. 65, pp. 28–36, 2013.
8. L. Peña-Parás, J. Taha-Tijerina, L. Garza, D. Maldonado-Cortés, R. Michalczewski, and C. Lapray, "Effect of CuO and Al2O3 nanoparticle additives on the tribological behavior of fully formulated oils," *Wear*, vol. 332–333, pp. 1256–1261, 2015.
9. N. N. M. Zawawi, W. H. Azmi, A. A. M. Redhwan, and M. Z. Sharif, "Coefficient of friction and wear rate effects of different composite nanolubricant concentrations on Aluminium 2024 plate," *IOP Conf. Ser. Mater. Sci. Eng.*, vol. 257, no. 1, 2017.
10. T. Luo, X. Wei, H. Zhao, G. Cai, and X. Zheng, "Tribology properties of Al2O3/TiO2 nanocomposites as lubricant additives," *Ceram. Int.*, vol. 40, no. 7, Part A, pp. 10103–10109, 2014.
11. H. Kumar and A. P. Harsha, "Taguchi optimization of various parameters for tribological performance of polyalphaolefins based nanolubricants," *Proc. Inst. Mech. Eng. Part J J. Eng. Tribol.*, vol. 235, pp. 1262–1280, 2020.
12. R. N. Gupta and A. P. Harsha, "Tribological study of castor oil with surface-modified CuO nanoparticles in boundary lubrication," *Ind. Lubr. Tribol.*, vol. 70, no. 4, pp. 700–710, 2018.
13. M. Ekrem, H. Düzcükoglu, M. Ali Senyurt, Ö. Sinan Sahin, and A. Avci, "Friction and wear performance of epoxy resin reinforced with boron nitride nanoplatelets," *J. Tribol.*, vol. 140, no. 2, p. 022001, 2017.
14. Q. Wan, Y. Jin, P. Sun, and Y. Ding, "Tribological behaviour of a lubricant oil containing boron nitride nanoparticles," *Procedia Eng.*, vol. 102, pp. 1038–1045, 2015.
15. S. Qian, H. Wang, C. Huang, and Y. Zhao, "Experimental investigation on the tribological properties of modified carbon nanotubes as the additive in castor oil," *Ind. Lubr. Tribol.*, vol. 70, no. 3, pp. 499–505, 2018.
16. H. L. Yu, *et al.*, "Characterization and nano-mechanical properties of tribofilms using Cu nanoparticles as additives," *Surf. Coatings Technol.*, vol. 203, no. 1–2, pp. 28–34, 2008.
17. G. Liu, X. Li, B. Qin, D. Xing, Y. Guo, and R. Fan, "Investigation of the mending effect and mechanism of copper nano-particles on a tribologically stressed surface," *Tribol. Lett.*, vol. 17, no. 4, pp. 961–966, 2004.
18. Y. Li, T. T. Liu, Y. Zhang, P. Zhang, and S. Zhang, "Study on the tribological behaviors of copper nanoparticles in three kinds of commercially available lubricants," *Ind. Lubr. Tribol.*, vol. 70, no. 3, pp. 519–526, 2018.
19. W. K. Shafi, A. Raina, and M. I. Ul Haq, "Tribological performance of avocado oil containing copper nanoparticles in mixed and boundary lubrication regime," *Ind. Lubr. Tribol.*, vol. 70, no. 5, pp. 865–871, 2018.
20. Y. Meng, F. Su, and Y. Chen, "Synthesis of nano-Cu/graphene oxide composites by supercritical CO2-assisted deposition as a novel material for reducing friction and wear," *Chem. Eng. J.*, vol. 281. pp. 11–19, 2015.
21. M. Asrul, N. W. M. Zulkifli, H. H. Masjuki, and M. A. Kalam, "Tribological properties and lubricant mechanism of nanoparticle in engine oil," *Procedia Eng.*, vol. 68, pp. 320–325, 2013.
22. Y. Y. Wu, W. C. Tsui, and T. C. Liu, "Experimental analysis of tribological properties of lubricating oils with nanoparticle additives," *Wear*, vol. 262, no. 7–8, pp. 819–825, 2007.
23. J. Padgurskas, R. Rukuiza, I. Prosyčevas, and R. Kreivaitis, "Tribological properties of lubricant additives of Fe, Cu and Co nanoparticles," *Tribol. Int.*, vol. 60, pp. 224–232, 2013.

24. J. Zhao, Y. He, Y. Wang, W. Wang, L. Yan, and J. Luo, "An investigation on the tribological properties of multilayer graphene and MoS2 nanosheets as additives used in hydraulic applications," *Tribol. Int.*, vol. 97, 2016.
25. L. Wu, Z. Xie, L. Gu, B. Song, and L. Wang, "Investigation of the tribological behavior of graphene oxide nanoplates as lubricant additives for ceramic/steel contact," *Tribol. Int.*, vol. 128, no. February, pp. 113–120, 2018.
26. Y.-B. Guo and S.-W. Zhang, "The tribological properties of multi-layered graphene as additives of PAO2 oil in steel—steel contacts," *Lubricants*, vol. 4, no. 3, p. 30, 2016.
27. C. J. Reeves and P. L. Menezes, "Evaluation of boron nitride particles on the tribological performance of avocado and canola oil for energy conservation and sustainability," *Int. J. Adv. Manuf. Technol.*, vol. 89, no. 9–12, pp. 3475–3486, 2017.
28. M. K. Gupta, J. Bijwe, and A. K. Kadiyala, "Tribo-investigations on oils with dispersants and hexagonal boron nitride particles," *J. Tribol.*, vol. 140, no. 3, p. 031801, 2017.
29. M. I. H. Chua Abdullah, M. F. Bin Abdollah, N. Tamaldin, H. Amiruddin, and N. R. M. Nuri, "Effect of hexagonal boron nitride nanoparticles as an additive on the extreme pressure properties of engine oil," *Ind. Lubr. Tribol.*, vol. 68, no. 4, pp. 441–445, 2016.
30. Y. Yang, et al., "Preparation and tribological properties of BN/calcium borate nanocomposites as additive in lubricating oil," *Ind. Lubr. Tribol.*, vol. 70, no. 1, pp. 105–114, 2018.
31. X. Xu, Yong Hu, Enzhu Hu, Kunhong Xu, and Yufu Hu, "Formation of an adsorption film of MoS2 nanoparticles and dioctyl sebacate on a steel surface for alleviating friction and wear," *Tribol. Int.*, vol. 92, pp. 172–183, 2015.
32. S. H. Roslan, S. B. A. Hamid, and N. W. M. Zulkifli, "Synthesis, characterisation and tribological evaluation of surface-capped molybdenum sulphide nanoparticles as efficient antiwear bio-based lubricant additives," *Ind. Lubr. Tribol.*, vol. 69, no. 3, pp. 378–386, 2017.
33. G. Tontini, G. Dalla Lana Semione, C. Bernardi, R. Binder, J. D. B. De Mello, and V. Drago, "Synthesis of nanostructured flower-like MoS2 and its friction properties as additive in lubricating oils," *Ind. Lubr. Tribol.*, vol. 68, no. 6, pp. 658–664, 2016.
34. A. Kumar, et al., "Experimental study on the efficacy of MoS2 microfluids for improved tribological performance," *Proc. Inst. Mech. Eng. Part J J. Eng. Tribol.*, vol. 231, no. 1, pp. 107–124, 2017.
35. H. Xie, B. Jiang, J. He, X. Xia, and F. Pan, "Lubrication performance of MoS2 and SiO2 nanoparticles as lubricant additives in magnesium alloy-steel contacts," *Tribology International*, vol. 93, pp. 63–70, 2016.
36. R. Dinesh, M. J. Giri Prasad, R. Rishi Kumar, N. Jerome Santharaj, J. Santhip, and A. S. Abhishek Raaj, "Investigation of tribological and thermophysical properties of engine oil containing nano additives," *Mater. Today Proc.*, vol. 3, no. 1, pp. 45–53, 2016.
37. D. X. Peng, Y. Kang, R. M. Hwang, S. S. Shyr, and Y. P. Chang, "Tribological properties of diamond and SiO2 nanoparticles added in paraffin," *Tribol. Int.*, vol. 42, no. 6, pp. 911–917, 2009.
38. N. Kumar, A. T. Kozakov, V. I. Kolesnikov, and A. V. Sidashov, "Improving the lubricating properties of 10W40 oil using oxidized graphite additives," *J. Frict. Wear*, vol. 38, no. 5, pp. 349–354, 2017.
39. T. D. F. López, A. F. González, Á. Del Reguero, M. Matos, M. E. Díaz-García, and R. Badía-Laíño, "Engineered silica nanoparticles as additives in lubricant oils," *Sci. Technol. Adv. Mater.*, vol. 16, no. 5, p. 55005, 2015.
40. D. X. Peng, C. H. Chen, Y. Kang, Y. P. Chang, and S. Y. Chang, "Size effects of SiO2 nanoparticles as oil additives on tribology of lubricant," *Ind. Lubr. Tribol.*, vol. 62, no. 2, pp. 111–120, 2010.

41. B. Chen, K. Gu, J. Fang, W. Jiang, W. Jiu, and Z. Nan, "Tribological characteristics of monodispersed cerium borate nanospheres in biodegradable rapeseed oil lubricant," *Appl. Surf. Sci.*, vol. 353, pp. 326–332, 2015.
42. M. Laad and V. K. S. Jatti, "Titanium oxide nanoparticles as additives in engine oil," *J. King Saud Univ.—Eng. Sci.*, vol. 30, no. 2, pp. 116–122, 2018.
43. F. Ilie and C. Covaliu, "Tribological properties of the lubricant containing titanium dioxide nanoparticles as an additive," *Lubricants*, vol. 4, no. 2, p. 12, 2016.
44. Y. Gu, X. Zhao, Y. Liu, and Y. Lv, "Preparation and tribological properties of dual-coated TiO2 nanoparticles as water-based lubricant additives," *J. Nanomater.*, vol. 2014, 2014.
45. H. Sade, A. Moshkovich, J.-P. Lellouche, and L. Rapoport, "Testing of WS2 nanoparticles functionalized by a humin-like shell as lubricant additives," *Lubricants*, vol. 6, no. 1, p. 3, 2018.
46. R. Greenberg, G. Halperin, I. Etsion, and R. Tenne, "The effect of WS2 nanoparticles on friction reduction in various lubrication regimes," *Tribol. Lett.*, vol. 17, no. 2, pp. 179–186, 2004.
47. X. Quan, *et al.*, "Friction and wear performance of dual lubrication systems combining WS2-MoS2 composite film and low volatility oils under vacuum condition," *Tribol. Int.*, vol. 99, pp. 57–66, 2016.
48. J. Sanes, F. J. Carrión, and M. D. Bermúdez, "Effect of the addition of room temperature ionic liquid and ZnO nanoparticles on the wear and scratch resistance of epoxy resin," *Wear*, vol. 268, no. 11–12, pp. 1295–1302, 2010.

15 Non-Asbestos Organic Brake Pad Friction Composite Materials
A Review

Mukesh Kumar and Ashiwani Kumar

CONTENTS

15.1 Introduction 227
15.2 Characteristics Features of Binders, Fillers, Fibers, and Friction Additives/Modifiers 232
15.3 Fabrication Procedure of Brake Pad Friction Composite Materials 239
15.4 Assessment of Brake Pad Friction Composite Materials 240
15.5 Metrics for the Performance Evaluation of Brake Pad Friction Composite Materials 242
15.6 Development of Friction Composites Formulations Using Analytical Tools 246
15.7 Scope for Future Research Work 247
15.8 Conclusion 248
Acknowledgment 249
References 249

15.1 INTRODUCTION

In the 18th century, the industrial revolution changed the way mankind lives. Since then, humankind has actively engaged in developing sophisticated machinery for the production of quality goods and better means of transportation with the objective of making life easy. In such machines, friction composites were extensively used, and their usage increased exponentially with the booming of the automotive, aircraft, and shipping industries. These composites find applications in clutches and brake linings. The function of the clutch is to transmit power from one point to another, and the brake pads have to decelerate the rotating counter-rotor/disc surface. In each case, the rubbing surfaces undergo excessive frictional action, causing an instant increase in the interfacial flash temperature, leading to detrimental frictional performance. Consequently, designing and developing such components require the utmost attention to avoid performance failure [1–5].

The automotive braking system as shown in Figures 15.1A and 15.1B is certainly one of the prominent applications of friction composites [6–14]. It comprises of disc/drum rotor, brake pad lining (i.e., metallic plate with an overlay of friction

DOI: 10.1201/9781003093213-15

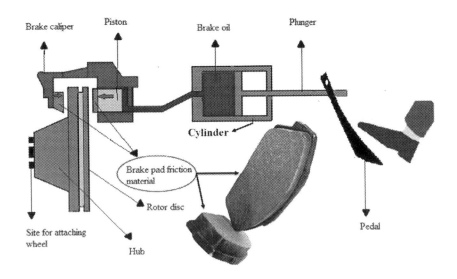

FIGURE 15.1A Working of a Rotor Disc Brake System

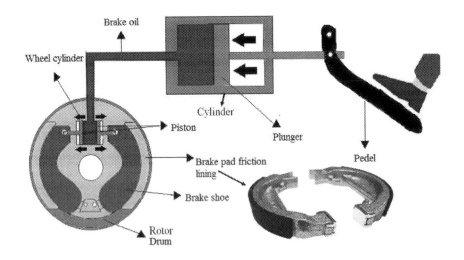

FIGURE 15.1B Working of a Rotor Drum Brake System

material), brake caliper, and hydraulic actuating system of cylinder–piston arrangement. When the brake pedal is pressed with the foot, braking oil/fluid, pumped by cylinder–piston arrangement, passes through a brake hose or lines to the brake piston. The piston, in turn, either pushes the two brake shoes/lining to expand against the brake drum rotor or presses against the brake disc rotor through the caliper mechanism. Consequently, the rotational motion/revolution of rotating wheels attached to the hub and disc/drum becomes retarded, thereby stopping or slowing the vehicle. It has been reported that during relative soft braking, the force of pushing the pads against the disk is about 5 kN, culminating in nominal contact pressure of about 1.2 MPa on the pad surface. In a severe case, this pressure could reach up to 10 MPa. In braking, via the frictional action on the interface between the pad–rotor/disk, the kinetic energy of moving vehicles can be transformed into thermal energy. The deformation of actual contact points due to plastic deformation, hysteresis, dispersion, and viscous flow [2, 4, 15–19] can result in the generation of a significant amount of heat energy. To prevent material or braking failure, this heat must be dissipated quickly, primarily through conduction through the drum/rotor or convection to surrounding components or radiations to the environment and, secondarily, through the adsorption leading to chemical, metallurgical, and wear mechanisms/processes at the interface [20].

The braking system consists of two tribo elements, that is, drum/disc rotor and brake lining/pad. In general, pearlitic grey cast iron disc of type A (having ~10 wt.% of carbon, silicon, manganese, sulfur, chromium, phosphorous, copper, titanium, niobium, and vanadium and a Brinell hardness range of 170–280 HB) [20–25], having abundant flake graphite uniformly distributed and randomly orientated in the matrix and having the lower presence of ferrite/carbide content, is used for drum/rotor/disc. The presence of graphite flakes readily initiates the formation of friction film across the pad–disc/drum interface thereby aid in the optimization of braking performance [26–28]. In recent decades, scholars have also explored other materials (shown in Figure 15.2A) for disc/rotor, and investigations are going on [17, 29–32]. On the other hand, brake lining has been reported as having a complex mix of multi-ingredients (primarily classified as the fillers, binder, fibers, and friction additives/modifiers as shown in Figure 15.2B) as they have to fulfill a certain preferred set of frictional performance attributes for effective braking, safety, reliability, and comfortably, along with appropriate reasonable cost [1, 21, 33–35]. Bijwe [34] and Kumar [20] discussed various classifications of brake pad friction composites in detail (Figure 15.3). Here, we focused on non-asbestos organic friction composites for passenger motors. Such friction composites comprise organic/inorganic fillers/fibers but no/least metallic fillers/fibers in the matrix of the organic phenolic thermosetting polymer. Across braking industries, there is tremendous research and development

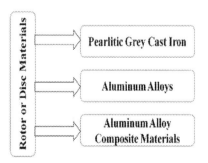

FIGURE 15.2A Materials Often Used for Rotor or Disc Fabrication

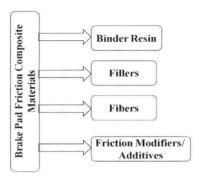

FIGURE 15.2B Classification of Brake Pad Friction Ingredients

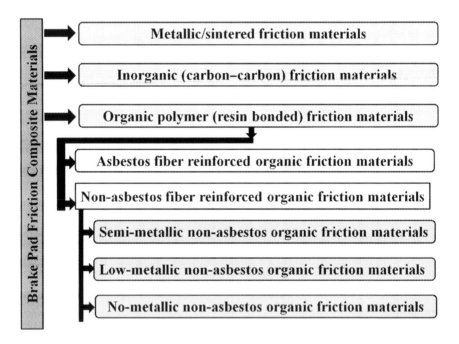

FIGURE 15.3 Classification of Brake Pad Friction Materials

going on to design, develop, and fabricate durable/efficient friction compositions that adhere to the stringent performance/environmental norms/regulations, such as better friction coefficient, negligible fade, faster recovery, improved wear resistance, low load-speed sensitivity, lower noise, and vibration tendency for safe and efficient operation over a wider range of braking applications [20, 36, 37].

The mechanistic aspect significantly affects real-time interfacial tribology, thereby affecting the friction and wear aspects of brake friction composite materials. Many scholars have reported on this theme. Serverin et al. [18] found that the effective friction mechanism probably results from the interaction between the temperature of the local friction surfaces, the strength of the local friction, the heat depends on the thermal contact resistance, and, in large brake systems, the deformation of the metal friction partner. According to Myshkin et al. [36], (1) interfacial bonds, their forms, and strength; (2) shearing and rupture of rubbing materials inside and (3) around the touch area are three fundamental elements involved in friction. Abrasion, adhesion, fatigue, and deformation components of friction/wear are affected by experimental conditions of load, sliding velocity, and temperature. Ostermeyer [37] observed that during braking, characteristic hard structures buildup on the contact patches repeatedly from the action of the flow of wear particles and friction power at a rubbing interface. The equilibrium of complex activity (birth–death–rebirth of contact patches) and the interfacial temperature decide the friction coefficient. Additionally, the brake fade and complex hysteresis are addressed in the plots of friction coefficient versus velocity.

Eriksson et al. [38–40] noted that the most apparent features of the brake pad topography are the contact plateaus/primary plateaus, that is, small, flat islands that usually lift a few microns above the rest of the surface and are constituents of the pad. A significant amount of small particles (wear debris) are moved through the thin maze between the surfaces as the disk rolls against the plateaus, milling down the weaker constituents. This small-sized wear debris forms a persistent friction film that compacts frequently in front of the initial contact plateaus and creates a few microns of larger, yet weaker, secondary plateaus. Without mechanical support from the primary plateau, these secondary plateaus cannot survive. At high nominal pressures and temperatures, the mechanism was more effective. The investigation elucidates the dramatic changes taking place on a micro-scale in the contact situation. The larger the secondary plateau thickness, the lower the wear rate, the stable response of friction. The thickness can decrease with an increase in contact pressure/temperature, thereby impacting friction and wear performance. Due to the design of the pad material, the nature of tribological interaction in automotive brakes is diverse and complicated. In a related finding, Roubicek et al. [5] stated that as the interfacial flash temperature approaches 350 °C, because of the physical and chemical changes occurring in the friction layer, the friction coefficient decreases with increasing temperature.

Myshkin [41] noted that the formation and transfer of debris were based on two key mechanical concepts: (1) adhesion and interface junction formation that was further subjected to shear and (2) the concentration of sub-surface damage (fatigue) and separation of failure fragments results from crack development and propagation. Cho et al. [42] observed that the dynamic nature of transfer layer (TL) formation and its thickness at the interface are highly dependent on the mechanical attrition (temperature <250 °C) and chemical adhesion of wear debris to the disc, relative amount and nature of ingredients, and the interface temperature. Thermo-gravimetric analysis (TGA) and differential thermo-gravimetric (DTG) plots implied that major degradation of pad ingredients takes place between 300 °C and 500 °C. The degradation of material causes a loss of adhesion between debris particles, thereby lowering the thickness of the friction film that causes higher amplitudes in coefficient of friction and wear rates. Abrasive, adhesive, fatigue, and thermal degradation control the friction and wear performance of the composite. Similar findings are reported by Wang et al. [43].

Blau [26, 27] investigates the microstructure and detachment mechanism of friction layers formed during braking, changes occurring in the micro-geometric roughness, and microstructure of the contact surfaces due to combined mechanical stresses (compression and shear) and friction-induced heating. The mechanical stresses and friction-induced heating causes pad material detachment in the form of soft-fine (due to friction modifiers and fillers) or hard-coarse grains (due to angular or rounded abrasives); these heterogeneously mechanically mixed with a binder material and forms friction film (~1–5 μm thickness) that adheres nicely with tribo-pairs. This wear debris forms a complex mixture of different phases and agglomerated iron oxide particles. The behavior of formed friction film governs friction and wear characteristics under dynamic experimental conditions. A similar observation was also made by Osterle et al. [10–14], Filip et al. [44], and Bulthe et al. [25]. Starczewski et al. [45] observed that frictional heat activates the physicochemical process across the tribo-pair interface that promotes the transfer of certain elements (like hydrogen) to the outer layer of the metal. The main source of the hydrogen is the thermal and mechanical destruction of the composite organic compound. Hydrogenization of the outer layer of the metal when it exceeds the critical level causes an increase in frictional wear, resulting in deterioration of the surface roughness. This could be avoided by using ingredients hydrogen lower in content or by nitrogenization that presents its diffusions into the metal.

15.2 CHARACTERISTICS FEATURES OF BINDERS, FILLERS, FIBERS, AND FRICTION ADDITIVES/MODIFIERS

The organic friction composite material consists of binders, fillers, fibers, and friction modifiers/additives. The binders provide physical integrity by keeping the ingredients firmly together at all times during the process of braking and under mechanical and thermal stresses so that friction composites can contribute to their performance. The binder resins like benzoxazines, cyanate esters, epoxies, polyesters, vinyl esters, pentaerythritol, polyacetal resin butanediol, and others are also used for brake pads. However, most frequently, thermosetting phenolic resins of the novolak/straight type and its variants, like silicone modified phenolics (phenolic siloxane), COPNA (condensed polynuclear aromatic)–modified phenolics, cyanate ester phenolics, epoxy-modified phenolics, thermoplastic polyimide phenolics, CNSL (cashew nut shell liquid)–modified phenolics, linseed oil–modified phenolics, nitrile butadiene rubber (NBR)–modified phenolic, alkyl benzene–modified phenolic, phosphorous-modified phenolics, boron-modified bi-phenol phenolics, alkyl-ether-modified phenolics, acryl-modified phenolics, and others [46, 47], are routinely used per the literature [20].

The straight phenolic resin of novolak type is the most widely used binder across the friction industry [20, 48]. Novolak resin is supplied with a hexamethylene-tetramine mix; hence, it needs dry conditions; otherwise, it undergoes a slow curing that limits its shelf life [49]. The peculiar characteristics that make

Brake Pad Friction Composite Materials

its huge commercial usage specifically across friction material applications are as follows:

1. Apart from better mechanical (like higher hardness, compressive strength, mild thermal strength, and resistance to creep), tribological, and thermal properties (like high T_g and low brittle–ductile transition temperature); it has superior binding properties and dimensional stability also at a lower cost. As it gives the favorable cost-to-performance ratio characteristics, it does not have any substitutes to date.
2. For several of the ingredients, it has very high wettability and hence is capable of holding the multiple ingredients intact.
3. It has extraordinary fire, smoke, and toxicity (FST) characteristics, because of the presence of an aromatic ring, and the hydroxyl group attached to the ring gives rise to excellent flame retardation characteristics (see Figure 15.4); this could be related to the safety of the component as well as a person going to use it. No other material with the same FST performance at a comparable price level is available.
4. It has delayed ignitability with low heat release or low smoke evolution with little or no toxic gas emission by nature. Additionally, it aids in significant strength retention (70%) at a serviceable temperature of 300 °C over 1–2 h, thus making the components exceptionally fire-resistant. This adds to the safety of the passengers.
5. Owing to its lower relative density, there is a weight reduction of about 60%, ensuring high fuel economy for automobiles and comparably lower manufacturing costs.

The general specifications of the phenolic resin are listed in Table 15.1.

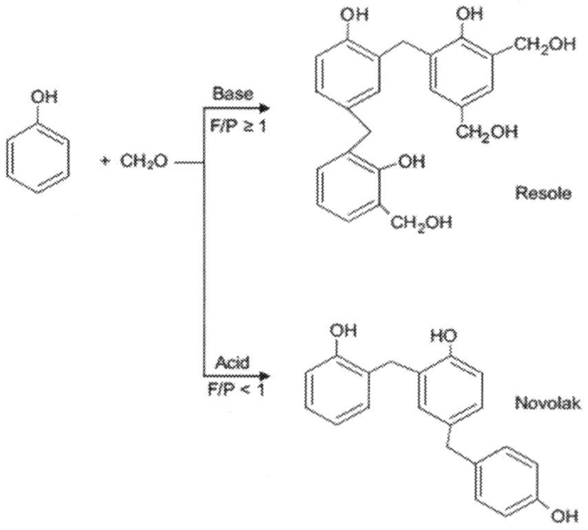

FIGURE 15.4 Structure of Phenolic Resin [49]

TABLE 15.1
General Specifications of Phenolic Resin [20]

S. No.	Property	Specifications
1.	Color (at room temperature)	White
2.	Specific density (kg/m^3)	1.28×10^3
3.	Ultimate tensile strength (MPa)	50 to 83
4.	Young's Modulus (MPa)	7600 to 18000
5.	Flexural strength (MPa)	62 to 193
6.	Compressive strength (MPa)	150 to 276
7.	Thermal expansion (20 °C)	48×10^{-6} °C^{-1}
8.	Elongation to failure (%)	0.4 to 1
9.	Hexa-content (wt.%) potentiometer method	10.78
10.	Flow distance in cm (ISO 8619)	2.4
11.	B-time (reactivity time) in seconds	41
12.	Melting point (°C)	89
13.	Curing temperature from DSC (°C)	154
14.	Initial degradation temperature from TGA (°C)	330
15.	Carbonizing temperature from TGA (°C)	420

Some demerits are also associated with phenolic resin that make us wary of its usage in certain conditions. As the curing of phenolic resin is a condensation process, there may be the liberation of volatiles like ammonia, formaldehyde, moisture, and others; thus, there may be a possibility of voids, cracks, and shrinkages in the final product. To avoid this, intermittent breathings may be planned to expel the moisture and other gaseous volatiles. The components require additional post-curing for 4–5 h at 170 °C. This additional post-curing requirement raises the expense due to the need for additional infrastructure, space, power usage, time, and manpower.

The literature study reveals that contributions of fibrous constituents are significant in enhancing various characteristics aspects of brake friction composite materials. Comprehensively, fibrous reinforcement enhances manufacturability; mechanical properties like strength, stiffness, toughness, hardness, and the like; friction and wear performance aspects; and thermal characteristics like thermal stability, resistance to oxidation, thermal degradation. This, in turn, improves the wear resistance and raises the friction coefficient in order to have stable friction properties. The degree of improvement depends on the nature and weight fraction of the fibers [50, 51, 60–65, 52–59]. The types of reinforcing fibers generally used are aramid/Kevlar pulp, ceramic fibers, kenaf fibers [66], banana fiber [67, 68], periwinkle shell particles [69], potassium titanate whiskers [70, 71], polyacrylonitrile fiber, carbon fiber [72–75], acrylic fiber, glass fiber, cellulosic fiber, rock wool fiber, lapinus fiber, and metallic fiber such as steel wool, bronze, copper, brass, and iron [76–78], among others. These high-performance fibers generally have a higher tenacity (3–6 GPa) and higher modulus (50–600 GPa), thereby having higher energy to break or work of rupture. Hence, they improve the recovery properties of friction composites and the ability to withstand repeated high energy shocks. Aramid/Kevlar is para-aramid, that is, polyphenylene-terephthal-amide (PPTA) developed by "Akzo" via condensation reaction between 1–4 phenylene-diamine or para-phenylenediamine and terephthaloylchloride in the presence of hydrogenchloride [20, 53]. It consists of an alternate benzene ring and CO-NH amide group that provides strong hydrogen bonding and hence high tenacity, modulus, thermal resistance, and an inert molecular group giving chemical resistance. Additional strength is derived from aromatic-stacking interaction between adjacent strands. Aramid fibers have a crystalline structure with strong fiber-direction covalent bonding and weak transverse hydrogen bonding and decomposition at 427 °C. Thus, it has a strong reinforcing capability that enables higher thermal and frictional stability. These interactions have a greater influence on aramid than do Van der Waals interactions. The amide segments that are regularly placed make a relatively strong hydrogen bond that enables the proper transfer of load between the chains. In chains, the hydrogen bond forms sheets that are stacked parallel into crystallites. The interaction is mainly carried out by Van der Waals forces between adjacent hydrogen-bonded planes, with some pi-bond overlaps of the phenylene fragments. This allows the hydrogen-bonded planes to behave in a way similar to close-packed planes in metals as slip planes. The presence of both aromatic rings and the double bond nature of the amide group resulting from resonance effects avoids bond rotation and hence molecular flexibility [79–83]. The important characteristics of aramid fibers are as follows:

1. Aramid fibers are soft, light, and reasonably thermally stable, with a superior stiffness-to-weight ratio and improved anti-fade properties. It is stronger than steel wire and stiffer than glass. These possess low creep and expansion coefficient, lower density, good fatigue, and abrasion resistance.
2. These prove intrinsic resilience to organic solvents, fades, lubricants, and flame exposure.
3. These are sensitive to ultraviolet (UV) radiation between 300–450 nm.
4. Aramid fiber has a remarkable application in preparing brake lining composites because it shows good processing assistance and good filler retention. In pulp form, it serves better by evenly mixing with other ingredients. It offers good wear resistance and a stable friction response. In general, 10–25 vol.% of resin is blended with 2–10 vol.% of aramid pulp to provide adequate green strength to pre-form and achieve the appropriate mechanical properties. It improves the operating lifetime of the brakes by providing low sequel, smooth friction, durability, and frictional stability. It forms a viscous, glassy coating over the counterface during sliding under high pressure (P)-velocity (V) conditions, which this viscous layer improves coefficient of friction performance. It reduces wear by modifying the consistency of transfer film and its adherence to the rotor disc. Better the transfer film and its adherence to the counterface, which leads to lower wear; also, the transfer film efficiency is better established, which delays fading.

5. Aramid fiber had no reduction in compressive strength with modulus, so it blended well with resin and offered great structural features that demonstrate great damage tolerance. When exposed to compressive strain (0.5%), para-aramid molecules tend to buckle instead of fracturing. This may also be due to the molecular rotation of the amide carbon to nitrogen bonds, which makes configurationally modification without ensuing in bond cleavage. As a result, the structure's load-bearing capacity is increased.
6. Absence of unstable linkage, as it is a wholly aromatic polyamide and due to the formation of the rigid molecular symmetric chain, it exhibits a higher degree of crystallinity that brings dimensional stability and thermal resistance to the product.

The general specification of aramid pulp is listed in Table 15.2.

The filler's ingredients [85, 86] are added to improve the manufacturability and adjust the relative weight fraction of the constituents, thereby lowering the net cost of the brake lining. It is accepted that these are inert materials and do not have any impact on the tribological properties of the friction composites. Commonly used organic fillers are cashew dust and rubber while inorganic fillers are barium sulfate/barites ($BaSO_4$), mica (aluminum silicate), vermiculite (hydrated calcium aluminum silicate), calcium carbonate ($CaCO_3$), calcium hydroxide ($Ca(OH)_2$), wollastonite (calcium silicate), molybdenum trioxide, alkali metal-titanate, for example, sodium titanate, kolin, fly ash, bottom ash, cenosphere, sulfate-rich scrubber sludge, clay, and others. Jeganmohan et al. [87] found improvement in mechanical as well as frictional performance specifically with 10% $CaSO_4$. They also reported similar results with NBR and styrene butadiene rubber (SBR) [88]. $BaSO_4$ is a white crystalline powder widely used as an inert filler in friction materials. The main source is barite mineral and can be easily extracted from the blast furnace.

TABLE 15.2
General Specifications of Aramid/Kevlar Fiber [20, 53]

S. No.	Property	Specifications
1.	Specific density (g/cm³)	1.45
2.	Ultimate tensile strength (MPa)	2790
3.	Young's modulus (MPa)	124
4.	Specific strength (MPa)	2000
5.	Specific modulus (GPa)	85
6.	Thermal degradation range (°C)	425–482
7.	Chemical resistance (PH range)	3–11
8.	Break elongation	3%
9.	Fiber length (μm)	1410 to 3170
10.	Fiber diameter (μm)	25.4 to 50.8
11.	Mohs hardness	7
12.	Elongation to break (%)	3.3

This enhances its red-hot properties. The important characteristics of barium sulfate are as follow [20, 89]:

1. It has a higher bulk density (4.5 g/cc) and higher inertness to both acids as well as alkaline. It is insoluble in water (max 0.5%) and has low oil absorption (10–12%).
2. It poses a strong crystalline structure and high refractive index (~1.64) that imparts high strength.
3. It lowers the cracking tendency of friction composites.
4. It has lower abrasive action (hardness ~3–3.5 Mohs) during the braking operation.
5. It improves the shock resistance capability of the friction composite system.
6. It enhances the surface quality of the finished pads.
7. It improves processability, low mill sticking, and uniform cure rates.
8. It is easily dispersible and highly stable at high temperatures (melting point ~1580 °C).

Frictional additives primarily adjust the value of the friction coefficient, stabilize it, and reduce wear rates [28]. These are split into two major categories: abrasives and lubricants. The main purpose of the lubricant is to stabilize the coefficient of friction during braking, particularly at high temperatures [90–92]. Generally used lubricants [93, 94] are graphite (synthetic and natural), metal sulfides, for example, antimony sulfide, tin sulfide, copper sulfide, lead sulfide, antimony trisulfide (Sb_2S_3) [95], molybdenum disulfide, CNSL powder, metal powder, coke powder; copper, brass, and iron [96, 97]; hexagonal boron nitride [98]; CPC, H-BN, TP (talcum powder) [99]. Abrasive particulate destroyed the pyrolyzed/carbonaceous glazed friction film formed on the rotor disc during sliding. This film is made of particles worn out from the surfaces of rotor disc and pad during braking and may be of some other undesirable film. Through the destruction of this film, it improves the grip and enables reestablishing it, thereby controlling the friction level during braking; this, in turn, enhances the frictional performance to the desired level [41, 42, 44, 90]. Abrasive particulates of various sizes and concentrations are generally optimized in the frictional material. In general, their hardness range is 7–10 Mohs. Commonly used abrasives are zirconium oxide (ZrO_2); chromium oxide; carbonaceous material (including different types of graphite, coke, and carbon black); iron oxide; and ceramic materials, for example, alumina (Al_2O_3; Mohs hardness: 7–8), silicon carbide (SiC>9 Mohs hardness), zirconium silicate/zirconia ($ZrSiO_4$; Mohs hardness: 7–8), quartz (SiO_2), and silicon carbide [100] alumina [101, 102]. Therefore, the relative amounts of solid lubricants and abrasives should be carefully controlled for desirable brake performance [103]. Graphite (a solid lubricant) structure possesses a hexagonal planar network of carbon atoms, with a strong bonding of atoms within the basal plane relative to the in-between planes that easily promotes the shearing of planes thereby accounts for its lubricant nature. There is a wide gap between the basal planes (0.334), and the Van der Waals connections between them are weak (7 to 1 kJ mol). Lattice defects, interstitial foreign atoms, and other anomalies further expand the interplanar gap. Because of this, the basal planes will slide on top of each other without losing their coherence, and graphite is able to yield plastically. The synthetic graphite is derived from organic precursors like petroleum coke, coal tar, and pyrolytic graphite; from methane and other gaseous hydrocarbons; vitreous carbon; and fiber from the polymers, carbon black from natural gas, charcoal from wood, coal from plants, and so on [104–107]. The important characteristics of graphite are as follows:

1. The graphite crystalline lamellar structure allows the crystal planes to easily slip over each other without disintegration, making it an excellent solid lubricant. The lubricity may be due to the strong bonding forces between the individual carbon atoms on the same plane and the comparatively weak bonding forces between the plane, and the presence of condensable vapors such as water. Graphite lubricity is considered to be moisture-dependent, so the same graphite shows a dramatic decrease to high temperatures and to act as a moderate abrasive. Graphite in friction composites is generally added to smooth unwanted frictional response variations with operating parameters. For a stable frictional coefficient of reaction, it easily forms a self-sustaining lubricant layer on the opposite counter-surface.
2. It remains chemically inert.
3. It serves as an exceptionally strong lubricant with lower intrinsic friction, which can be attributed to the ease of basal plane slippage with low shear strength. When rubbed against the metallic surface, a thin transfer film is formed that reduces the frictional coefficient.
4. Graphite may enhance the overall heat conductivity of the friction material, thereby assisting in the faster removal of heat across the brake pad without increasing too much temperature of the material.

The general specifications of the graphite are listed in Table 15.3.

The analysis of the effects of wear debris particulates on the health of organisms of ecology is elaborately studied by Kukutschova et al. [108]. The ball-milled pad shows potential mutagenicity, that is, a change in bacterial cell DNA, and induces

TABLE 15.3
General Specifications of Graphite [20]

S. No.	Property	Specifications
1.	Specific density (g/cc)	1.3–1.9
2.	Young's modulus (MPa)	5–10
3.	Flexural strength	10–100 MPa
4.	Compressive strength	65–89 MPa
5.	Vickers hardness (H_{VI})	40–100
6.	Porosity (%)	0.7–53
7.	Melting point (°C)	~800
8.	Thermal conductivity	25–470 W/mK
9.	Hardness	1–2 Mohs

an acute inflammatory reaction to bronchi and a translocation to the rats' lymph tissue for a particle fraction of less than 5 µm and doses of 3 mg/l after metabolic activation. The chemical debris compound responsible for the findings reported is still undetectable [109]. Research on new friction materials for vehicle brake pad use to mitigate emissions and health problems was documented by Jadhav et al. [110]. According to them, the toxin produced as a result of wear must be investigated as aquatic life could be at risk.

15.3 FABRICATION PROCEDURE OF BRAKE PAD FRICTION COMPOSITE MATERIALS

The designed formulations of friction composite materials are effectively developed through the addition/deletion of ingredients or varying their proportions, depending on a scholar's understanding of the literature and practical experience. The finalized formulation is fabricated in line with commercial norms, and the procedure as mentioned in the following:

1. The finalized ingredients are measured as per designed proportion amounting to 100% by weight for uniform mixing.
2. The plough shear–type mixer is used with a feeder at 150 rpm and a chopper at 3000 rpm to ensure mechanical isotropy of the mix. A proper mixing sequence of the ingredients needs to be maintained. For example, fibrous and fillers should be mixed first for few minutes, say, 10 minutes; this ensures the proper opening and dispersion of the fibers within the filler. Thereafter, other powdery ingredients, like additives and binders having a catalyst, must be added and mixed for another 5–10 minutes.
3. The green mix is then put in the mold cavity of specimen size and with the help of a compression molding machine, a pre-form is prepared. Depending on the curing isotherm of the matrix novolak resin, the molding conditions are followed according to the standard commercial procedure (i.e., keeping 10 °C higher within the plates than the actual curing temperature of 153 °C as determined from differential scanning calorimetry [DSC] to compensate any heat/thermal losses through the mold to the atmosphere).
4. In order to prevent composite cracking, moisture and other gaseous by-products released (due to the poly-condensation of the phenolic resin) were permitted to be expelled through multiple periodic breathings during the curing process.
5. Finally, the brake pads are post-cured (in the muffle furnace or autoclave for 5 hours at 170 °C) so that frozen-in stresses get relieved and cure the residual resin, if any.
6. In the same conditions, the resulting composite pads are subjected to surface polishing using a grinding wheel to facilitate uniform/maximum contact during the bedding run and to retain comparable levels of surface roughness. Next, physical, mechanical, thermal, and tribological properties of the samples are determined [4, 20, 111–113].

15.4 ASSESSMENT OF BRAKE PAD FRICTION COMPOSITE MATERIALS

The newly formed brake pad friction composites need to characterize for their various physical, mechanical, and braking tribology. This established its specifications and applications. Hence, it may relatively be compared with a benchmark to establish its commercial acceptability. Such common characterizations areas follow [4, 20, 112–114]:

1. *Theoretical density (g/cc):* The theoretical density (ρ_{ct}) of the composite materials in terms of weight fraction may be obtained from Equation 15.1:

$$\rho_{ct} = \frac{1}{\dfrac{w_r}{\rho_r} + \dfrac{w_f}{\rho_f} + \dfrac{w_{fl}}{\rho_{fl}} + \dfrac{w_a}{\rho_a} + \cdots}, \quad (15.1)$$

where w and ρ represent the weight fraction and density, respectively, and the subscripts, f, fi, and a correspond, respectively, to resin binder, fibrous ingredients, fillers, additives, and so on.

2. *Actual density (g/cc):* The actual density (ρ_{ca}) of the composites may be determined by the standard water immersion method following ASTM D792.

3. *Voids content/fraction (%):* The fraction of voids presence within the composites may be calculated using Equation 15.2:

$$V_o = \frac{\rho_{ct} - \rho_{ca}}{\rho_{ct}} \quad (15.2)$$

4. *Ash content (%):* The ash content may be measured by roasting the samples at very high temperatures (~840 °C) for approximately 4 hours inside the muffle furnace following gravimetric methods. It is calculated as the ratio of difference in weights of the sample before and after roasting to the preliminary weight of the sample.

5. *Acetone extractable (%):* The acetone extractable test determines the extent of curing of resin or any other organic fraction inside the composite or it measures the amount (%) of residual/uncured resin/organic fraction within the composite. A powdered composite pad sample of about 2 g is wrapped in filter paper and placed inside the upper extraction thimble (Whatman no.- 44), and the lower flask has adequate acetone placed inside the Soxhlet extraction apparatus. The acetone in the lower flask is made to boil by heating it at 90 °C, consequently, acetone evaporates and strikes the water flowing above the sample, resulting in its cooling. The condensed acetone vapors fall back to the lower flask by passing through the sample. This acetone falling through the sample extracts the uncured resin by dissolving it. The process is allowed to continue for about 2–3 hours.

Thereafter, the sample was removed and transferred to a drying oven set at 80 °C for 1 hour, followed by cooling in desiccators, and is reweighed. The acetone extract (%) is calculated using Equation 15.3:

$$\text{Acetone extractable } (\%) = \frac{(W_2 - W_1)}{W_1} \times 100, \qquad (15.3)$$

where W_2 is the initial weight of thimble and powdered friction material and W_1 is the weight of dried thimble and pad material after completion of the test.

6. *Hardness:* The hardness measures the degree of cross-linking/curing and the surface resistance to indentation under load as an indirect measure of the mechanical isotropy achieved when mixing the ingredients. According to ASTM D785–65, it can be measured using Rockwell hardness testers on the R-scale.

7. *Shear strength:* It evaluates composite integrity throughout the bulk and adhesion with the back plates. It may be measured using a universal testing machine.

8. *Compressibility:* The compressibility measures the change in the thickness of the pad under elastic or standard load during the compressibility test performed on the compressibility testing machine. At first, the specimen thickness is measured using vernier and placing it in the machine. Thereafter, the load of varying magnitude is applied, and the reduction in the thickness of the pad is measured. The ratio of the difference between reductions in thickness at a particular load to the mean thickness gives the compressibility of the specimen. The average of such values is taken as the compressibility of the samples.

9. *Tensile and flexural strength:* The strength of composite sample (specimen dimension of 140 × 12 × 10 mm³; the crosshead speed of 2 mm/min.; span length of 65 mm as per ASTM D 3039–76) under tensile loading gives the tensile strength, whereas the strength of sample (specimen dimension of 127 × 12.5 × 4 mm³; the crosshead speed of 2 mm/min.; span length of 70 mm as per ASTM D 2344–84) under flexural loading gives flexural strength. Both may be measured using a universal testing machine following the standards.

10. *Impact strength:* Izod impact test may be performed on (standard specimen size = 70 mm × 10 mm × 10 mm, 45° notch and 2 mm deep at 47 mm clamped height in the vice) pendulum impact tester following ASTM D-256 to measure impact strength.

11. *Surface morphology:* Scanning electron microscopy (SEM) is used to examine and analyze the wear mechanisms and structural integrity of multiphase hybrid composite friction materials.

12. *Dynamic mechanical analysis (DMA):* Temperature-dependent dynamic mechanical characteristics, such as stiffness and damping properties of polymeric composite materials (visco-elastic material), vary significantly in temperature near the glass transition region (Tg) and are recognized by

means of a DMA method. It aids in defining the storage modulus (E′), loss modulus (E″), loss tangent (Tan δ), and Tg for qualitatively investigating the reinforcement effects and quantitatively ascertaining a shift (if any) in the nature of the moduli decay/build-up of the composites within the experimental temperature range.

13. *Thermo-gravimetric analysis (TGA) and differential thermo-gravimetric (DTG) analysis:* TGA and DTG are vital thermal analytical techniques often employed to assess thermal stability/quantities, ash content, and so on of materials. It measures the difference between the flow of heat from the sample (which may occur when heat is absorbed or emitted by the sample due to thermal effects such as melting, crystallization, chemical reactions, polymorphic transitions, vaporization, etc.) and the sensor's reference side as a function of temperature or time.

14. *DSC:* DSC is an instrument for thermal analysis to measure changes in material physical characteristics along with temperature and time. As a function of time and temperature, it decides the temperature and heat flow associated with material transitions. Regardless of the temperature difference between the sample and the reference material, DSC measures a heat quantity radiated or absorbed in excessive amounts by the sample. The use of this equipment helps in determining glass transitions, curing temperature, and other features of polymers.

15. *Tribological performance evaluation method:* The braking tribological performance may be evaluated using a brake dynamometer. There is a variety of such dynamometers available, for example, full-scale inertia brake dynamometer (Dyno) [88], pin on disk [115–116], and others.

15.5 METRICS FOR THE PERFORMANCE EVALUATION OF BRAKE PAD FRICTION COMPOSITE MATERIALS

Friction composite systems are evidently a multi-ingredient system to have required performance parameters. The braking performance is extremely complex dynamical processes at a rotor–pad interface that depends on heterogeneous composition; operating parameters like temperature, rubbing/sliding speed, and contact pressure; and surface characteristics of counterface [38–40, 45]. These must have adequate strength/compressibility to withstand high contact pressure. These must have adequate shear strength to transfer the friction forces to the structure. These have-to-have superb thermal properties, such as high heat capacity, excellent thermal conductivity, and stability to oxidation. This makes it possible to tolerate high interfacial frictional heat without degrading the composite lining. High resistance to thermal cracking/fatigue is supported by good thermal resilience [4, 20]. The two mating surfaces (rotor/drum surface and the lining) should be able to generate a sufficiently high friction coefficient in the range of 0.3 to 0.5, say, for a middle-range vehicle or, as per Figure 15.5, with at least fading and variations or oscillations in the friction coefficient that should approach zero over a wider range of operating parameters, namely, temperature, pressure, sliding speed, and braking

Brake Pad Friction Composite Materials

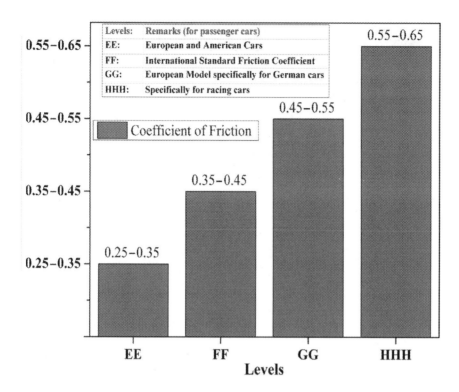

FIGURE 15.5 Levels of the Coefficient of Friction as per SAE Standards [117]

durations regardless of their age and serving environment [37, 117]. These materials should resist the fade phenomenon and possess a higher recovery property. Also, these should have minimum sensitivity to the load-speed variations, noise, shudder, and vibration; that is, they must work smoothly [76]. As these materials repeatedly remain in the contact with the counterface metallic drum/rotor, these should have higher wear resistance to avoid the wear effects such as scoring, galling, and ablation that bring higher wear volumes or rapid eating of the lining. Therefore, both parts should be tribological compatible with each other. The wearing of linings should be little as possible for their longer life, and it should generate a small amount of wear debris, which, in turn, proves environmentally safer. These should be at least sensitive or resistant to the environmental conditions in which they operate like water, oil, or corrosive environments that are salty, muddy, dusty, and so on. These should be safer to use and acceptable under environmental standards norms. The fabrication of these composites should be cost-effective and reliable [28, 34, 51, 118–120]. The stability coefficient should be high whereas the variability coefficient should be as low as possible. The physical, mechanical, and thermal properties should fulfill industrial standards for the commercial viability of the composite. The wear of the pad/lining and rise in disc temperature should be as low as possible.

The detailed description of various attributes often used for braking performance is as follows [20, 111–113]:

1. *μ-fade (μ_f):* The μ-fade is the minimum friction coefficient developed after 270 °C for the fade period. Due to the loss of friction between the braking surfaces as a result of frictional heat, the temporary drop in braking efficiency (typically in the range of 300–400 °C) is known as fade. Fade occurs when there is a decrease in the kinetic friction coefficient (μ) due to frequent and intense braking. A high amount of fade is undesirable as it impairs performance, utility, and reliability, especially during severe/frequent braking while maneuvering the vehicle's speed and during intermittent conditions of stop-and-go city driving [81–83, 121–124]. The potential causes of fade are high interfacial flash temperatures that decrease the shear strength of the pad due to resin degradation, which ultimately decreases the frictional force leading to fade, and the evolution of gas at the braking interfaces due to pyrolysis and thermal degradation of the materials (specifically organic) results in a decrease in the forces applied at elevated temperatures. It may also be due to the formation of a load-carrying friction film that is prone to exhibit shear-thinning interfacial rheology at elevated temperatures across the braking junction. As the film's deformability increases, the contact area increases, and it may diminish the effect of the applied normal force, eventually causing fade [20, 22, 56, 125–128].
2. *μ-recovery (μ_r):* μ-recovery is the maximum friction coefficient taken after 100 °C for the recovery cycle. As brake lining cools with the release of brakes, the restoration of friction coefficient (μ) to its original levels is termed as recovery. At lower temperatures, the friction layer made up of loosely attached wear-debris particles will deteriorate, and the underlying surface will harden. Such wear debris is stuck in the mating zone in the recovery run and acts as hard abrasives in the form of third bodies contributing to third-body abrasion (TBA). Therefore, the abrasive action through a rolling abrasion mechanism/TBA leads to a higher degree of friction. As a consequence, the characteristic frictional response of the composites is restored [20, 111–113].
3. *μ-performance (μ_p):* μ-performance is the average coefficient of friction for fade and recovery cycles at a temperature above 100 °C. (Range is typically 0.4 to 0.45 for small-sized cars, 0.5 to 0.6 for racing cars, and 0.3 to 0.4 for heavier automobiles [Figure 15.5]).
4. *μ-maximum (μ_{max}):* μ_{max} is the highest coefficient of friction for cold, fade, and recovery cycles.
5. *μ-minimum (μ_{min}):* μ_{min} is the lowest coefficient of friction for cold, fade, and recovery cycles.
6. *Frictional fluctuations ($\Delta\mu$):* $\Delta\mu = \mu_{max} - \mu_{min}$. It should be low as possible and should be constant over a range of operating conditions.

Brake Pad Friction Composite Materials

7. *Fading tendency (%):* The fading tendency of friction composites may be computed using Equation 15.4:

$$\% \text{ Fade} = \frac{\mu_{performance} - \mu_{fade}}{\mu_{performance}} \times 100. \quad (15.4)$$

A value of about 30% or lower is desirable for brake pad friction composite materials (see Figure 15.6).

8. *Recovery (%):* Recovery performance of friction composites may be computed using Equation 15.5:

$$\% \text{ Recovery} = \frac{\mu_{recovery}}{\mu_{performance}} \times 100. \quad (15.5)$$

A value of about 100–120% of recovery is desirable for brake pad friction composite materials (see Figure 15.6).

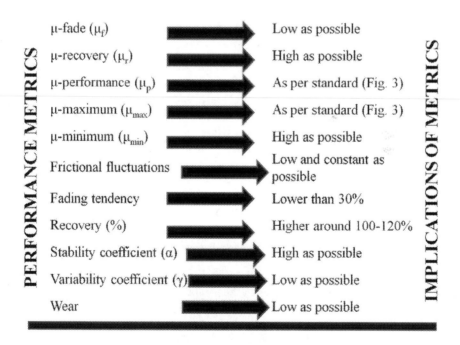

FIGURE 15.6 Performance Metrics and Their Implications for Assessing Friction Composite Materials [20]

9. *Stability coefficient (α):* The stability coefficient of friction composites may be computed using Equation 15.6:

$$\text{Stability coefficient } (\alpha) = \frac{\mu_{performance}}{\mu_{maximum}}. \quad (15.6)$$

A higher value for the stability coefficient is desirable for brake pad friction composite materials (see Figure 15.6).

10. *Variability coefficient (γ):* The variability coefficient of friction composites may be computed using Equation 15.7:

$$\text{Variability coefficient } (\gamma) = \frac{\mu_{minimum}}{\mu_{maximum}}. \quad (15.7)$$

A lower value for the variability coefficient is desirable for brake pad friction composite materials (see Figure 15.6).

11. *Wear (weight loss):* $\Delta W = W_2 - W_1$, where W_2 is the weight of the brake pad after the test and W_1 is the weight of the brake pad before the test (see Figure 15.6).

15.6 DEVELOPMENT OF FRICTION COMPOSITES FORMULATIONS USING ANALYTICAL TOOLS

Brake friction composite components need to have an amalgam of multi-ingredients broadly categorized as the resin, fibers, fillers, and friction additives. The real challenge faced by industry and academics scholars is the selection of appropriate ingredients, the design of formulations, the understanding of prevalent braking interface tribology, and the development of formulations based on the experimental findings. There are three main problems in the development of friction material, namely, how to select the raw material, how to find interaction among the components, and how to optimize the formulation. Sometimes researchers make use of simulation tools to predict the behavior of such materials [129–143].

The literature analysis shows that the brake friction material should meet strict norms (e.g., eco-friendly, higher friction coefficient, low fading, quick recovery, high wear resistance, low sensitivity toward load-speed alterations, low noise and vibration propensity, etc.) over a wide variety of braking variables. Thus, the selection of an acceptable formulation among many that satisfies a set of performance criteria actually becomes a multi-norm/criteria/attribute decision-making exercise. A wider review of literature on decision-making techniques under contradictory multi-criteria, that is, multi-criteria decision-making (MCDM) techniques is briefly outlined by Jahan et al. [144]. They reported numerous state-of-the-art material quantitative screening methods (e.g., cost per unit property method, chart method, product selection data, knowledge-based systems, neural network, etc.) and resource rankings (e.g., technique for order of preference by similarity to ideal solution (TOPSIS),

Elimination ET Choix Traduisant la REalite (ELECTRE), analytic hierarchy process (AHP), simple additive weighting (SAW), fuzzy MCDM (multi criteria decision making), goal programming, preference ranking organization method for enrichment evaluation (PROMETHEF), etc. Satapathy et al. [82] use balancing and ranking approaches to determine friction materials ranking. Maniya et al. [145] and Jeganmohan et al. [116] use the preference selection index (PSI) approach for material ranking. The PSI approach is considered to be an innovative technique for choosing the best alternative among multi-alternatives. Thus, it considers usability where there are conflicts in determining the relative value of attributes. The alternative with a higher PSI value is picked as the best alternative. Hu et al. [146] apply the principles of a hybrid AHP-PROMETHEF to determine the optimal formulation consisting of aramid and $CaSO_4$ whiskers.

15.7 SCOPE FOR FUTURE RESEARCH WORK

The detailed study of literature for understanding of braking tribology, braking performance metrics, design/development/fabrication of friction formulations, usage of various analytical, and decision-making tools aids scholars or engineers to have the following scope for future research work:

1. There is less research on other variants or grades of phenolic resin; subsequently it's quite interesting to explore/extent the research based on a different modified form of phenolic resins or other grades of the binder resin. A comparative analysis of the performance data, then, clearly reveals the applicability/significance of resin to a particular brake lining for commercial application.
2. There is ample research reporting on the influence of single-fiber reinforcement relative to binary/ternary combinations. It is interesting to explore/extend research in this area; this aid material scientists in demonstrating that the proper selection of fibrous combination helps in replacing expensive fiber by inexpensive fiber keeping the same functional level. Thus, research based on various combinations like organic–inorganic, organic–ceramic, inorganic–ceramic, organic–inorganic–ceramic, and others have ample scope for exploration [54].
3. It is rendered that the incorporation of combinations of high-performance fibrous ingredients having complementary nature leads to improvement in the overall friction performance. Such friction composites show enhanced physical, mechanical, and thermal characteristics that could be verified experimentally and via different characterizations. The composition must have an agent like aramid fiber that enables ease in pre-form and hence enhanced manufacturability of such composites along with controlling lubricity and stabilization of overall friction and wear response.
4. The research could also further explore using natural plant fibers like banana fiber, among others.
5. The experimental investigation observed that interfacial temperature during dynamic rubbing reaches about >300 °C, to keep the frictional

response effective the presence of thermally resilient ingredients like ceramic fibers, inorganic fibers, abrasives, and the like must be there in composition.

6. Literature based on metallic fibers and thermally resistant ceramic fibers as reinforcements is scarcely available. Thus, research based on their single, binary, or ternary combination may prove significant for high-temperature applications.
7. Careful constituent selection and weight fraction determination may assist formulation designers in adjusting/replacing one fiber in conjunction with others while maintaining the same functional level of performance for cost optimization and commercial use.
8. Research on the filler ingredients, for example, comparative analysis of different fillers, binary or ternary combinations of complementary filler, fillers of same or varying size, and the like, on braking performance is also lacking in the literature. The incorporation of natural hazardous fillers like fly ash, marble dust, red mud, and the like further would widen the area.
9. The effect of friction modifiers (abrasive/lubricants) particulates and their relative size variations in the friction composite is another potential area that has received the least research attention. The effects of binary or ternary combinations of such constituents are scarcely available.
10. During braking, certain things are undesirable like noise propensity, squeal, judder, creep groan phenomenon, and so on, and literature targeting the sensitivity of friction composite materials to these undesirable phenomena is rarely available [124, 147, 148]. The use of simulation tools could strengthen the scope of work.
11. The braking performance of such friction composites under wet conditions could be carried out (scarcely available), and then it could be compared with parameters obtained under dry conditions [149–151].
12. The study of thermal conductivity, diffusivity, specific heat capacity, and other aspects would further enhance the understanding of co-relation between such parameters and braking tribology/performance.
13. Ascertaining the toxicity of such friction composites and the wear debris could be done to fulfill the legislative norms coined by environmental agencies.
14. The use of analytical and decision-making tools in formulation design, as well as ranking of compositions, would further widen the scope of research.

15.8 CONCLUSION

The discussion on organic brake pad friction composite materials reveals the following salient observations:

1. As newer and newer modern-age lightweight, high-speed, and fuel-efficient vehicles are continuously evolving, the requirement for efficient and safe braking systems continues to rise rapidly.

2. The brake pad friction composite materials must satisfy the stringent performance norms in modern automobiles at all ranges of braking variables. This means that brake pad composition should include a variety of different ingredients such as resin, fibers, fillers, and friction additives. The vital areas of current research are the design and development of non-asbestos organic matrix-based friction composite materials for industrial use.
3. The research focuses on the development of appropriate formulation design by selecting appropriate ingredients that take care of environmental norms. The development of such formulation must be based on experimental findings of dynamic braking tribological performance fulfilling brake standards.
4. The use of analytical tools in the development of friction formulations and decision-making techniques in ranking the formulations are the additional thrust areas where material scholars and engineers are working continuously.

ACKNOWLEDGMENT

The authors express their sincere gratitude to their respective departments and institutes for their overall support.

REFERENCES

1. M. G. Jacko, P. H. S. Tsang, and S. K. Rhee, Automotive friction materials evolution during the past decade, Wear **100**, 503 (1984).
2. G. P. Ostermeyer and M. Müller, Dynamic interaction of friction and surface topography in brake systems, Tribol. Int. **39**, 370 (2006).
3. O. M. Braun and A. G. Naumovets, Nanotribology: Microscopic mechanisms of friction, Surf. Sci. Rep. **60**, 79 (2006).
4. M. Kumar and A. Kumar, Thermomechanical analysis of hybrid friction composite material and its correlation with friction braking performance, Int. J. Polym. Anal. Charact. **0**, 1 (2020).
5. V. Roubicek, H. Raclavska, D. Juchelkova, and P. Filip, Wear and environmental aspects of composite materials for automotive braking industry, Wear **265**, 167 (2008).
6. B. N. J. Persson, On the theory of rubber friction, Surf. Sci. **401**, 445 (1998).
7. C. Y. Teoh, Z. M. Ripin, and M. N. A. Hamid, Analysis of friction excited vibration of drum brake squeal, Int. J. Mech. Sci. **67**, 59 (2013).
8. M. R. K. Vakkalagadda, D. K. Srivastava, A. Mishra, and V. Racherla, Performance analyses of brake blocks used by Indian Railways, Wear **328–329**, 64 (2015).
9. N. Langhof, M. Rabenstein, J. Rosenlöcher, R. Hackenschmidt, W. Krenkel, and F. Rieg, Full-ceramic brake systems for high performance friction applications, J. Eur. Ceram. Soc. **36**, 3823 (2016).
10. W. Österle, H. Kloß, I. Urban, and A. I. Dmitriev, Towards a better understanding of brake friction materials, Wear **263**, 1189 (2007).
11. W. Österle, I. Dörfel, C. Prietzel, H. Rooch, A. L. Cristol-Bulthé, G. Degallaix, and Y. Desplanques, A comprehensive microscopic study of third body formation at the interface between a brake pad and brake disc during the final stage of a pin-on-disc test, Wear **267**, 781 (2009).

12. W. Österle and I. Urban, Third body formation on brake pads and rotors, Tribol. Int. **39**, 401 (2006).
13. W. Österle and I. Urban, Friction layers and friction films on PMC brake pads, Wear **257**, 215 (2004).
14. W. Österle, M. Griepentrog, T. Gross, and I. Urban, Chemical and microstructural changes induced by friction and wear of brakes, Wear **250–251**, 1469 (2001).
15. H. Deng, K. Li, H. Li, P. Wang, J. Xie, and L. Zhang, Effect of brake pressure and brake speed on the tribological properties of carbon/carbon composites with different pyrocarbon textures, Wear **270**, 95 (2010).
16. C. Ferrer, M. Pascual, D. Busquets, and E. Rayón, Tribological study of Fe-Cu-Cr-graphite alloy and cast iron railway brake shoes by pin-on-disc technique, Wear **268**, 784 (2010).
17. M. Siroux, A. L. Cristol-Bulthé, Y. Desplanques, B. Desmet, and G. Degallaix, Thermal analysis of periodic sliding contact on a braking tribometer, Appl. Therm. Eng. **28**, 2194 (2008).
18. D. Severin and S. Dörsch, Friction mechanism in industrial brakes, Wear **249**, 771 (2001).
19. M. Kalin, Influence of flash temperatures on the tribological behaviour in low-speed sliding: A review, Mater. Sci. Eng. A **374**, 390 (2004).
20. M. Kumar, Performance Assessment of Hybrid Composite Friction Materials : Effect of Ceramic, Organic and Inorganic Fibre Combinations, Ph.D. Thesis, Mech. Eng. Dep. NIT Hamirpur, H.P., INDIA 1 (2015).
21. K. Kato, Wear in relation to friction – a review, Wear **241**, 151 (2000).
22. A. Sellami, M. Kchaou, R. Elleuch, A. L. Cristol, and Y. Desplanques, Study of the interaction between microstructure, mechanical and tribo-performance of a commercial brake lining material, Mater. Des. **59**, 84 (2014).
23. P. J. Blau, Fifty years of research on the wear of metals, Tribol. Int. **30**, 321 (1997).
24. M. Terheci, R. R. Manory, and J. H. Hensler, The friction and wear of automotive grey cast iron under dry sliding conditions Part 1-relationships between wear loss and testing parameters, Wear **180**, 73 (1995).
25. A. L. Cristol-Bulthé, Y. Desplanques, G. Degallaix, and Y. Berthier, Mechanical and chemical investigation of the temperature influence on the tribological mechanisms occurring in OMC/cast iron friction contact, Wear **264**, 815 (2008).
26. P. J. Blau and B. C. Jolly, Wear of truck brake lining materials using three different test methods, Wear **259**, 1022 (2005).
27. P. J. Blau, Microstructure and detachment mechanism of friction layers on the surface of brake shoes, J. Mater. Eng. Perform. **12**, 56 (2003).
28. Blau, PJ (Oct 2001). Compositions, Functions, and Testing of Friction Brake Materials and Their Additives; TOPICAL (ORNL/TM--2001/64). United States.
29. G. J. Howell and A. Ball, Dry sliding wear of particulate-reinforced aluminium alloys against automobile friction materials, Wear **181–183**, 379 (1995).
30. P. J. Blau and H. M. Meyer, Characteristics of wear particles produced during friction tests of conventional and unconventional disc brake materials, Wear **255**, 1261 (2003).
31. M. Kermc, M. Kalin, and J. Vižintin, Development and use of an apparatus for tribological evaluation of ceramic-based brake materials, Wear **259**, 1079 (2005).
32. A. Jahan, K.L. Edwards Multi-criteria decision analysis for supporting the selection of engineering materials in product design Butterworth-Heinemann, Oxford (2013).
33. J. D. Kiser, K. E. David, C. Davies, R. Andrulonis, and C. Ashforth, Updating composite materials handbook-17 volume 5—ceramic matrix composites, Ceram. Trans. **263**, 413 (2018).

34. J. Bijwe, Recent developments in non-asbestos fiber reinforced friction materials – a review, Polym. Compos. **18**, 378 (1997).
35. S. Venkatesh and K. Murugapoopathiraja, Scoping review of brake friction material for automotive, Mater. Today Proc. **16**, 927 (2019).
36. N. K. Myshkin, M. I. Petrokovets, and A. V. Kovalev, Tribology of polymers: Adhesion, friction, wear, and mass-transfer, Tribol. Int. **38**, 910 (2005).
37. G. P. Ostermeyer, On the dynamics of the friction coefficient, Wear **254**, 852 (2003).
38. M. Eriksson, J. Lord, and S. Jacobson, Wear and contact conditions of brake pads: Dynamical in situ studies of pads on glass, Wear **249**, 272 (2001).
39. M. Eriksson and S. Jacobson, Tribological surfaces of organic brake pads, Tribol. Int. **33**, 817 (2000).
40. M. Eriksson, F. Bergman, and S. Jacobson, On the nature of tribological contact in automotive brakes, Wear **252**, 26 (2002).
41. N. K. Myshkin, Friction transfer film formation in boundary lubrication, Wear **245**, 116 (2000).
42. M. H. Cho, K. H. Cho, S. J. Kim, D. H. Kim, and H. Jang, The role of transfer layers on friction characteristics in the sliding interface between friction materials against gray iron brake disks, Tribol. Lett. **20**, 101 (2005).
43. X. Wang, S. Wang, S. Zhang, and D. Wang, Wear mechanism of disc-brake block material for new type of drilling rig, Front. Mech. Eng. China **3**, 10 (2008).
44. P. Filip, Z. Weiss, and D. Rafaja, On friction layer formation in polymer matrix composite materials for brake applications, Wear **252**, 189 (2002).
45. L. Starczewski and J. Szumniak, Mechanisms of transferring the matter in a friction process in a tribology system: Polymeric composite-metal, Surf. Coatings Technol. **100**, 33 (1998).
46. A. A. Byerlin and L. K. Pakhomova, Polymeric matrices for high-strength reinforced composites, Polym. Sci. U.S.S.R. **32**, 1275 (1990).
47. R. A. Scott and N. A. Peppas, Compositional effects on network structure of highly cross-linked copolymers of PEG-containing multiacrylates with acrylic acid, Macromolecules, Macromolecules **32**, 6139 (1999).
48. A. H. R. M. Al-Sarraf, Study on adhesion wear damage done on the hybrid composite Novolac under the experimental variables, Energy Procedia **157**, 644 (2019).
49. A. P. Luz, C. G. Renda, A. A. Lucas, R. Bertholdo, C. G. Aneziris, and V. C. Pandolfelli, Graphitization of phenolic resins for carbon-based refractories, Ceram. Int. **43**, 8171 (2017).
50. J. Kim and Y.W. Mai, *Engineered Interfaces in Fiber Reinforced Composites* (1998).
51. M. L. Halberstadt, S. K. Rhee, and J. A. Mansfield, Effects of potassium titanate fiber on the wear of automotive brake linings, Wear **46**, 109 (1978).
52. J. P. Favre and M. C. Merienne, Characterization of fibre/resin bonding in composites using a pull-out test, Int. J. Adhes. Adhes. **1**, 311 (1981).
53. Gopal, P., Dharani, L.R., Blum, F.D.: Hybrid phenolic friction composites containing Kevlar® pulp part II—wear surface characteristics. Wear **193**, 180–185 (1996).
54. J. J. Lee, J. A. Lee, S. Kwon, and J. J. Kim, Effect of different reinforcement materials on the formation of secondary plateaus and friction properties in friction materials for automobiles, Tribol. Int. **120**, 70 (2018).
55. B. Z. Jang, Control of interfacial adhesion in continuous carbon and kevlar fiber reinforced polymer composites, Compos. Sci. Technol. **44**, 333 (1992).
56. S. C. Ho, J. H. C. Lin, and C. P. Ju, Effect of fiber addition on mechanical and tribological properties of a copper/phenolic-based friction material, Wear **258**, 861 (2005).
57. M. H. Cho, S. J. Kim, D. Kim, and H. Jang, Effects of ingredients on tribological characteristics of a brake lining: An experimental case study, Wear **258**, 1682 (2005).

58. K. J. Lee, S. W. Lee, J. S. C. Jang, and H. Z. Cheng, Effect of sol-gel boehmite infiltration on tribological and mechanical behavior of brake lining materials, Wear **264**, 337 (2008).
59. Z. C. Zhu, Y. X. Peng, Z. Y. Shi, and G. A. Chen, Three-dimensional transient temperature field of brake shoe during hoist's emergency braking, Appl. Therm. Eng. **29**, 932 (2009).
60. T. Singh and A. Patnaik, Performance assessment of lapinus-aramid based brake pad hybrid phenolic composites in friction braking, Arch. Civ. Mech. Eng. **15**, 151 (2015).
61. T. Singh, A. Patnaik, B. Gangil, and R. Chauhan, Optimization of tribo-performance of brake friction materials: Effect of nano filler, Wear **324–325**, 10 (2015).
62. T. Singh, A. Patnaik, and R. Chauhan, Optimization of tribological properties of cement kiln dust-filled brake pad using grey relation analysis, Mater. Des. **89**, 1335 (2016).
63. T. Singh, A. Patnaik, R. Chauhan, and A. Rishiraj, Assessment of braking performance of lapinus–wollastonite fibre reinforced friction composite materials, J. King Saud Univ.—Eng. Sci. **29**, 183 (2017).
64. M. Baklouti, A. L. Cristol, Y. Desplanques, and R. Elleuch, Impact of the glass fibers addition on tribological behavior and braking performances of organic matrix composites for brake lining, Wear **330–331**, 507 (2015).
65. J. H. Gweon, B. S. Joo, and H. Jang, The effect of short glass fiber dispersion on the friction and vibration of brake friction materials, Wear **362–363**, 61 (2016).
66. A. Mustafa, M. F. B. Abdollah, F. F. Shuhimi, N. Ismail, H. Amiruddin, and N. Umehara, Selection and verification of kenaf fibres as an alternative friction material using Weighted Decision Matrix method, Mater. Des. **67**, 577 (2015).
67. U. D. Idris, V. S. Aigbodion, I. J. Abubakar, and C. I. Nwoye, Eco-friendly asbestos free brake-pad: Using banana peels, J. King Saud Univ.—Eng. Sci. **27**, 185 (2015).
68. M. Amirjan, Microstructure, wear and friction behavior of nanocomposite materials with natural ingredients, Tribol. Int. **131**, 184 (2019).
69. D. S. Yawas, S. Y. Aku, and S. G. Amaren, Morphology and properties of periwinkle shell asbestos-free brake pad, J. King Saud Univ.—Eng. Sci. **28**, 103 (2016).
70. Y. C. Kim, M. H. Cho, S. J. Kim, and H. Jang, The effect of phenolic resin, potassium titanate, and CNSL on the tribological properties of brake friction materials, Wear **264**, 204 (2008).
71. K. H. Cho, M. H. Cho, S. J. Kim, and H. Jang, Tribological properties of potassium titanate in the brake friction material; morphological effects, Tribol. Lett. **32**, 59 (2008).
72. Q. F. Guan, G. Y. Li, H. Y. Wang, and J. An, Friction-wear characteristics of carbon fiber reinforced friction material, J. Mater. Sci. **39**, 641 (2004).
73. Y. Z. Wan, H. L. Luo, Y. L. Wang, Y. Huang, Q. Y. Li, F. G. Zhou, and G. C. Chen, Friction and wear behavior of three-dimensional braided carbon fiber/epoxy composites under lubricated sliding conditions, J. Mater. Sci. **40**, 4475 (2005).
74. W. Österle, A. I. Dmitriev, B. Wetzel, G. Zhang, I. Häusler, and B. C. Jim, The role of carbon fibers and silica nanoparticles on friction and wear reduction of an advanced polymer matrix composite, Mater. Des. **93**, 474 (2016).
75. M. Hao, R. Luo, Z. Hou, W. Yang, Q. Xiang, and C. Yang, Effect of fiber-types on the braking performances of carbon/carbon composites, Wear **319**, 145 (2014).
76. H. Jang, K. Ko, S. J. Kim, R. H. Basch, and J. W. Fash, The effect of metal fibers on the friction performance of automotive brake friction materials, Wear **256**, 406 (2004).
77. M. Kumar and J. Bijwe, Optimized selection of metallic fillers for best combination of performance properties of friction materials: A comprehensive study, Wear **303**, 569 (2013).
78. F. Eddoumy, H. Kasem, H. Dhieb, J. G. Buijnsters, P. Dufrenoy, J. P. Celis, and Y. Desplanques, Role of constituents of friction materials on their sliding behavior between room temperature and 400 °C, Mater. Des. **65**, 179 (2015).

79. S. J. Kim and H. Jang, Friction and wear of friction materials containing two different phenolic resins reinforced with aramid pulp, Tribol. Int. **33**, 477 (2000).
80. S. J. Kim, M. H. Cho, D. S. Lim, and H. Jang, Synergistic effects of aramid pulp and potassium titanate whiskers in the automotive friction material, Wear **250–251**, 1484 (2001).
81. B. K. Satapathy and J. Bijwe, Performance of friction materials based on variation in nature of organic fibres Part I. Fade and recovery behaviour, Wear **257**, 573 (2004).
82. B. K. Satapathy and J. Bijwe, Performance of friction materials based on variation in nature of organic fibres Part II. Optimisation by balancing and ranking using multiple criteria decision model (MCDM), Wear **257**, 585 (2004).
83. B. K. Satapathy and J. Bijwe, Composite friction materials based on organic fibres: Sensitivity of friction and wear to operating variables, Compos. Part A Appl. Sci. Manuf. **37**, 1557 (2006).
84. Y. S. Suyev, Reinforcement of polymers by finely dispersed fillers. Review, Polym. Sci. U.S.S.R. **21**, 1315 (1979).
85. H. Czichos, D. Klaffke, E. Santner, and M. Woydt, Advances in tribology: The materials point of view, Wear **190**, 155 (1995).
86. S. R. Jeganmohan, T. V. Christy, D. G. Solomon, and B. Sugozu, Influence of calcium sulfate whiskers on the tribological characteristics of automotive brake friction materials, Eng. Sci. Technol. an Int. J. (2019).
87. S. Raj Jeganmohan, D. Gnanaraj Solomon, and T. V. Christy, Effect of two different rubbers as secondary binders on the friction and wear characteristics of non-asbestos organic (NAO) brake friction materials, Tribol.—Mater. Surfaces Interfaces **12**, 71 (2018).
88. S. J. Kim, M. H. Cho, R. H. Basch, J. W. Fash, and H. Jang, Tribological properties of polymer composites containing barite ($BaSO_4$) or potassium titanate ($K_2O_6 \cdot (TiO_2)$), Tribol. Lett. **17**, 655 (2004).
89. T. Liu, S. K. Rhee, and K. L. Lawson, A study of wear rates and transfer films of friction materials, Wear **60**, 1 (1980).
90. S. K. Rhee, M. G. Jacko, and P. H. S. Tsang, The role of friction film in friction, wear and noise of automotive brakes, Wear **146**, 89 (1991).
91. J. Wang, N. Jiang, and H. Jiang, The high-temperatures bonding of graphite/ceramics by organ resin matrix adhesive, Int. J. Adhes. Adhes. **26**, 532 (2006).
92. J. Spreadborough, The frictional behaviour of graphite, Wear **5**, 18 (1962).
93. J. K. Lancaster, Instabilities in the frictional behaviour of carbons and graphites, Wear **34**, 275 (1975).
94. H. Jang and S. J. Kim, The effects of antimony trisulfide (Sb_2S_3) and zirconium silicate ($ZrSiO_4$) in the automotive brake friction material on friction characteristics, Wear **239**, 229 (2000).
95. M. Křístková, P. Filip, Z. Weiss, and R. Peter, Influence of metals on the phenol-formaldehyde resin degradation in friction composites, Polym. Degrad. Stab. **84**, 49 (2004).
96. S. C. Ho, J. H. C. Lin, and C. P. Ju, Effect of phenolic content on tribological behavior of carbonized copper-phenolic based friction material, Wear **258**, 1764 (2005).
97. G. Yi and F. Yan, Effect of hexagonal boron nitride and calcined petroleum coke on friction and wear behavior of phenolic resin-based friction composites, Mater. Sci. Eng. A **425**, 330 (2006).
98. G. Yi and F. Yan, Mechanical and tribological properties of phenolic resin-based friction composites filled with several inorganic fillers, Wear **262**, 121 (2007).
99. V. Matějka, Y. Lu, L. Jiao, L. Huang, G. S. Martynková, and V. Tomášek, Effects of silicon carbide particle sizes on friction-wear properties of friction composites designed for car brake lining applications, Tribol. Int. **43**, 144 (2010).

100. B. K. Satapathy and J. Bijwe, Analysis of simultaneous influence of operating variables on abrasive wear of phenolic composites, Wear **253**, 787 (2002).
101. B. K. Satapathy and J. Bijwe, Wear data analysis of friction materials to investigate the simultaneous influence of operating parameters and compositions, Wear **256**, 797 (2004).
102. G. Petzow, R. Telle, and R. Danzer, Microstructural defects and mechanical properties of high-performance ceramics, Mater. Charact. **26**, 289 (1991).
103. S. J. Kim, M. Hyung Cho, K. Hyung Cho, and H. Jang, Complementary effects of solid lubricants in the automotive brake lining, Tribol. Int. **40**, 15 (2007).
104. D. Kolluri, A. K. Ghosh, and J. Bijwe, Analysis of load-speed sensitivity of friction composites based on various synthetic graphites, Wear **266**, 266 (2009).
105. I. J. Antonyraj and D. L. Singaravelu, Tribological characterization of various solid lubricants based copper-free brake friction materials – a comprehensive study, Mater. Today Proc. (2019).
106. M. G. Faga, E. Casamassa, V. Iodice, A. Sin, and G. Gautier, Morphological and structural features affecting the friction properties of carbon materials for brake pads, Tribol. Int. **140**, 105889 (2019).
107. J. Kukutschová, V. Roubíček, M. Mašláň, D. Jančík, V. Slovák, K. Malachová, Z. Pavlíčková, and P. Filip, Wear performance and wear debris of semimetallic automotive brake materials, Wear **268**, 86 (2010).
108. J. Kukutschová, V. Roubíček, K. Malachová, Z. Pavlíčková, R. Holuša, J. Kubačková, V. Mička, D. MacCrimmon, and P. Filip, Wear mechanism in automotive brake materials, wear debris and its potential environmental impact, Wear **267**, 807 (2009).
109. S. P. Jadhav and S. H. Sawant, A review paper: Development of novel friction material for vehicle brake pad application to minimize environmental and health issues, Mater. Today Proc. **19**, 209 (2019).
110. A. Patnaik, M. Kumar, B. K. Satapathy, and B. S. Tomar, Performance sensitivity of hybrid phenolic composites in friction braking: Effect of ceramic and aramid fibre combination, Wear **269**, 891 (2010).
111. M. Kumar, B. K. Satapathy, A. Patnaik, D. K. Kolluri, and B. S. Tomar, Hybrid composite friction materials reinforced with combination of potassium titanate whiskers and aramid fibre: Assessment of fade and recovery performance, Tribol. Int. **44**, 359 (2011).
112. M. Kumar, B. K. Satapathy, A. Patnaik, D. K. Kolluri, and B. S. Tomar, Evaluation of fade-recovery performance of hybrid friction composites based on ternary combination of ceramic-fibers, ceramic-whiskers, and aramid-fibers, J. Appl. Polym. Sci. **124**, 3650 (2012).
113. S. Bhaskar, M. Kumar, and A. Patnaik, Silicon carbide ceramic particulate reinforced AA2024 alloy composite – Part I: Evaluation of mechanical and sliding tribology performance, Silicon **12**, 843 (2019).
114. P. Gill, T. T. Moghadam, and B. Ranjbar, Differential scanning calorimetry techniques: Applications in biology and nanoscience, J. Biomol. Tech. **21**, 167 (2010).
115. S. Pujari, S. Srikiran, Experimental investigations on wear properties of palm kernel reinforced composites for brake pad applications. Def. Technol. **15**(3), 295–299 (2019).
116. Nasi Auto Parts, "Friction coefficient of brake pads," NAP, 2018. [Online]. Available: http://www.napbrake.com/2018/02/24/friction-coefficient-brake-pads/. [Accessed 02 May 2020].
117. D. Chan and G. W. Stachowiak, Review of automotive brake friction materials, Proc. Inst. Mech. Eng. Part D J. Automob. Eng. **218**, 953 (2004).
118. A. M. Pye, A review of asbestos substitute materials in industrial applications, J. Hazard. Mater. **3**, 125 (1979).

119. P.G. Sanders, T.M. Dalka, R.H. Basch, A reduced-scale brake dynamometer for friction characterization Tribol. Int., **34** (2001), pp. 609-615120.
120. J. Bijwe and M. Kumar, Optimization of steel wool contents in non-asbestos organic (NAO) friction composites for best combination of thermal conductivity and triboperformance, Wear **263**, 1243 (2007).
121. W. K. Lee and H. Jang, Moisture effect on velocity dependence of sliding friction in brake friction materials, Wear **306**, 17 (2013).
122. M. Kchaou, A. Sellami, R. Elleuch, and H. Singh, Friction characteristics of a brake friction material under different braking conditions, Mater. Des. **52**, 533 (2013).
123. M. Federici, S. Gialanella, M. Leonardi, G. Perricone, and G. Straffelini, A preliminary investigation on the use of the pin-on-disc test to simulate off-brake friction and wear characteristics of friction materials, Wear **410–411**, 202 (2018).
124. J. Bijwe, Nidhi, N. Majumdar, and B. K. Satapathy, Influence of modified phenolic resins on the fade and recovery behavior of friction materials, Wear **259**, 1068 (2005).
125. Nidhi and J. Bijwe, NBR-modified resin in fade and recovery module in non-asbestos organic (NAO) friction materials, Tribol. Lett. **27**, 189 (2007).
126. J. Bijwe, M. Kumar, P. V. Gurunath, Y. Desplanques, and G. Degallaix, Optimization of brass contents for best combination of tribo-performance and thermal conductivity of non-asbestos organic (NAO) friction composites, Wear **265**, 699 (2008).
127. P. D. Neis, N. F. Ferreira, and F. P. da Silva, Comparison between methods for measuring wear in brake friction materials, Wear **319**, 191 (2014).
128. Y. Lu, A golden section approach to optimization of automotive friction materials, J. Mater. Sci. **38**, 1081 (2003).
129. D. Thuresson, Influence of material properties on sliding contact braking applications, Wear **257**, 451 (2004).
130. L. Han, L. Huang, J. Zhang, and Y. Lu, Optimization of ceramic friction materials, Compos. Sci. Technol. **66**, 2895 (2006).
131. Y. Lu, A combinatorial approach for automotive friction materials: Effects of ingredients on friction performance, Compos. Sci. Technol. **66**, 591 (2006).
132. Y. Zhao, Y. Lu, and M. A. Wright, Sensitivity series and friction surface analysis of non-metallic friction materials, Mater. Des. **27**, 833 (2006).
133. G. Xiao and Z. Zhu, Friction materials development by using DOE/RSM and artificial neural network, Tribol. Int. **43**, 218 (2010).
134. D. Aleksendrić, Neural network prediction of brake friction materials wear, Wear **268**, 117 (2010).
135. D. Aleksendrić and Č. Duboka, Fade performance prediction of automotive friction materials by means of artificial neural networks, Wear **262**, 778 (2007).
136. D. Aleksendrić, D. C. Barton, and B. Vasić, Prediction of brake friction materials recovery performance using artificial neural networks, Tribol. Int. **43**, 2092 (2010).
137. A. I. Dmitriev, W. Österle, and H. Kloß, Numerical simulation of typical contact situations of brake friction materials, Tribol. Int. **41**, 1 (2008).
138. A. I. Dmitriev, A. Y. Smolin, S. G. Psakhie, W. Osterle, H. Kloss, and V. L. Popov, Computer modeling of local tribological contacts by the example of the automotive brake friction pair, Phys. Mesomech. **11**, 73 (2008).
139. V. L. Popov, S. G. Psakhie, E. V Shilko, A. I. Dmitriev, K. Knothe, F. Bucher, and M. Ertz, Friction coefficient in rail-wheel contacts as a function of material and loading parameters, Phys. Mesomech. **5**, 17 (2002).
140. M. Mueller and G. P. Ostermeyer, Cellular automata method for macroscopic surface and friction dynamics in brake systems, Tribol. Int. **40**, 942 (2007).

141. M. Müller and G. P. Ostermeyer, A Cellular Automaton model to describe the three-dimensional friction and wear mechanism of brake systems, Wear **263**, 1175 (2007).
142. S. Qi, H and A. J. Day, Investigation of disc/pad interface temperature in friction braking, Wear **262**, 505 (2007).
143. A. Jahan, M. Y. Ismail, S. M. Sapuan, and F. Mustapha, Material screening and choosing methods – a review, Mater. Des. **31**, 696 (2010).
144. K. Maniya and M. G. Bhatt, A selection of material using a novel type decision-making method: Preference selection index method, Mater. Des. **31**, 1785 (2010).
145. Z. Zhu, L. Xu, G. Chen, and Y. Li, Optimization on tribological properties of aramid fibre and $CaSO_4$ whisker reinforced non-metallic friction material with analytic hierarchy process and preference ranking organization method for enrichment evaluations, Mater. Des. **31**, 551 (2010).
146. A. Lazzari, D. Tonazzi, and F. Massi, Squeal propensity characterization of brake lining materials through friction noise measurements, Mech. Syst. Signal Process. **128**, 216 (2019).
147. E. Davin, A. L. Cristol, J. F. Brunel, and Y. Desplanques, Wear mechanisms alteration at braking interface through atmosphere modification, Wear **426–427**, 1094 (2019).
148. P. J. Blau and J. C. McLaughlin, Effects of water films and sliding speed on the frictional behavior of truck disc brake materials, Tribol. Int. **36**, 709 (2003).
149. N. S. M. EL-Tayeb and K. W. Liew, On the dry and wet sliding performance of potentially new frictional brake pad materials for automotive industry, Wear **266**, 275 (2009).
150. N. S. M. El-Tayeb and K. W. Liew, Effect of water spray on friction and wear behaviour of noncommercial and commercial brake pad materials, J. Mater. Process. Technol. **208**, 135 (2008).

16 Impact of Water Particles on Fly Ash–Filled E-Glass Fiber–Reinforced Epoxy Composites

Pankaj Kumar Gupta, Rahul Sharma, and Gaurav Kumar

CONTENTS

16.1 Introduction 257
16.2 Materials and Methodology 258
16.3 Experiments...260
    16.3.1 Moisture Content Test 260
    16.3.2 Water Absorption Test 260
    16.3.3 Linear Swelling Test 261
    16.3.4 Density Measurement Test 261
    16.3.5 Cost Analysis of Composite Material...261
16.4 Results..262
    16.4.1 Moisture Content 262
    16.4.2 Water Absorption 262
    16.4.3 Linear Swelling 262
    16.4.4 Density Test 265
    16.4.5 Cost Analysis of Composite Material...265
16.5 Conclusion...266
References..266

16.1 INTRODUCTION

A composite is a combination of "a stronger and a weaker material" or at least two dissimilar macroscopically identifiable materials. These are combined together to achieve better properties [1, 2]. The composite made of polymer matrix with E-glass fiber reinforcement have wide applications in aerospace, sports, automobile engineering, and electrical applications. A polymer matrix with E-glass fiber reinforcement has several advantages such as better mechanical strength, high modulus, high strength and cost-effective, more storage modulus, more loss modulus and higher glass transition temperatures [3]. The literature reveals that tribological response of epoxy-based composites is better. Therefore, it is widely used in aircraft and

DOI: 10.1201/9781003093213-16

automobile industry applications such as seal, gear, cams shafts and bushes [4, 5], chemical processing equipment, microelectronic devices, satellites, sporting goods, submarines, ships, automobiles, medical prosthesis, and civil infrastructure [1, 6].

It was investigated that resin composites up to 54 wt.% of fly ash content have better mechanical properties like impact strength and compressive strength. A composite having a certain type of resin material as a matrix in a polymer matrix composite [7]. Fly ash particles have shown increased mechanical properties, such as compressive strength, tensile strength and hardness with an increase in the percentage of fly ash concentration, and density was decreased [8]. It had been evaluated from experiments that the microwave curing method has shown better mechanical characteristics compared to the conventional thermal curing method. It was also observed from experiments that flexural and tensile modulii of untreated fly ash–filled epoxy composites were improved with an increase in the quantity of fly ash. The results revealed that 0.5 wt.% KBM603 fly ash surface–treated epoxy composites have better mechanical characteristics [9]. S. M. Kulkarni found at 5 volume fraction (V_f)% of fiber and 25 V_f% of fly ash–reinforced epoxy composite showed compressive strengths (99.7 MPa) and modulii (2.5 GPa) [10]. Fly ash can be hybridized easily with other filler materials to increase suitability in structural applications [11]. S. M. Kulkarni studied the values of strengths and modulii of epoxy (LAPOX L-12) composites with filler fly ash. Silane coupling agent (3-Triethoxysilyl-propylamine [3 TESPA]) was used for the treatment of the surface of particles. It was investigated that the magnitude of strengths and modulii for all volume fractions of treated filler particles were improved [12]. S. M. Kulkarni studied better compressive strengths and modulus for the fly ash–reinforced epoxy composites fabricated by open-mold casting technique [13]. It was investigated that composites having 10% volume of fly ash as a filler showed high impact strength as compared tooth filler materials such as 10% volume of aluminum oxide (Al_2O_3) or magnesium oxide ($Mg(OH)_2$) [14]. A small amount of silane-treated fly ash content was preferable for the overall improvement of composite properties [15]. Literature reveals that impact of water on fly ash–filled composite is not investigated by any of the researcher. The impact of water plays a major role on composite material particularly in outdoor applications. Therefore, this research is carried out to observe the impact of water with varying the percentage of the concentration of fly ash–filled E-glass fiber–reinforced epoxy composite (E-GFREC).

16.2 MATERIALS AND METHODOLOGY

An epoxy resin (LY 556) and a curing agent or hardener (HY951), as shown in Figure 16.1a, were mixed 10:1 for composite manufacturing. A release gel was spread on the inside surface of the sheet of the bottom mold. The fly ash and epoxy mix was spread uniformly on the mold sheet using a brush. E-glass fiber, as shown in Figure 16.1b, was placed on it. A hand roller was used to remove the air traps present on the mold surface and to remove the excess polymer present on the surface. Mild pressure was applied on the roller while moving it on the surface. The same process was repeated until the desired layers were stacked. The release gel was spread on inside surface of the top mold sheet before placing it on the mold. Pressure was applied after placing it on the stacked layers. Curing was carried out at ambient

FIGURE 16.1 Composite Raw Materials: (a) Epoxy Resin (LY 556) and Hardener (HY951) and (b) E-Glass Fiber

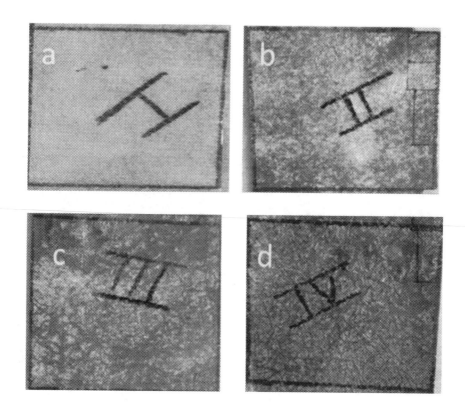

FIGURE 16.2 Composite Specimens with Fly Ash (a) 0 wt.%, (b) 4 wt.%, (c) 8 wt.%, and (d) 12 wt.% Concentration

temperature in 24 hours; subsequently, the mold was opened, and the composite part was taken out. The fly ash particulates were reinforced into the epoxy resin at varying concentrations of 0 wt.%, 4 wt.%, 8 wt.% and 12 wt.% as shown in Figure 16.2a–d. The E-glass fiber–reinforced composite (70 wt.% resin and 20 wt.% fiber) with 10 wt.% fly ash had been shown better compressive strength, tensile strength and hardness. Therefore, the higher limit of fly ash concentration in this research was kept closed to 10 wt.% and confined to 12 wt.% [16]. Three specimens for each concentration were prepared, and average results were taken into consideration.

16.3 EXPERIMENTS

The impact of water affects the dimensional stability of a composite material, which affects the application of the product. The addition of fly ash also affects the density and cost of the material. Therefore, the impact of water was observed by a moisture content test, a water absorption test, and a linear swelling test. A cost analysis and a density test are carried out in succeeding sections.

16.3.1 MOISTURE CONTENT TEST

The moisture content test is necessary to find out the water particles uptake by the composite material. Fibers are treated as more sensitive to water and humidity. It affects the outdoor application of the composites. This test is based on the mass fraction method. In this test, the specimen's weight was measured, and then it is kept in an oven for drying until a constant weight was reached. The drying time to reach constant weight depends on the types of material, the quantity of the material and the condition of the material. In most cases, a 16-hour to 24-hour drying period was enough to gain a constant weight. The weight of the specimen was again taken when the constant weight was reached. The size of the specimen for moisture content test was taken according to ASTM D570, which was 20 mm × 20 mm with a 5-mm average thickness. The moisture content was calculated [13] using Equation 16.1:

$$Moisture\ Content = \frac{W_{mci} - W_{mcf}}{W_{mcf}} \times 100, \quad (16.1)$$

where W_{mci} is the initial weight of the specimen before heating it in the oven and W_{mcf} is the final weight of the specimen after heating it in an oven.

16.3.2 WATER ABSORPTION TEST

Water absorption can be found by the weight percent method. The size of the specimen for the moisture content test was also taken according to ASTM D570, which was similar to the previous test. All specimens were kept in an electric oven at 80 °C for 24 hours to remove the moisture. Then the weight of each dried specimen was taken; then they were dipped in water for 24 hours. The weight of the specimen was taken after dipping again. The water absorption was calculated [12] by Equation 16.2:

$$Water\ Absorption = \frac{W_{wbf} - W_{wbi}}{W_{wbi}} \times 100, \quad (16.2)$$

where W_{waf} is the final weight of the sample after dropping it in water and W_{wai} is the initial weight of the sample before dropping it in water.

16.3.3 LINEAR SWELLING TEST

Linear swelling gives an idea of change in length, width and thickness. Linear swelling can be measured by length or width or thickness percent method. The size of the specimen for the linear swelling test was also the same as that taken in the previous test according to ASTM D570. All specimens kept in an electric oven at 80 °C for 24 hours to eliminate the moisture. In order to observe the change in length or width or thickness of each dried specimen was dipped in water for 24 hours and measured again. The linear swelling was calculated [17] by Equation 16.3:

$$Linear\ Swelling = \frac{L_{lsf} - L_{lsi}}{L_{lsi}} \times 100, \quad (16.3)$$

where L_{lsf} is the final length/width/thickness of the specimen after dipping it in water and L_{lsi} is the initial length/width/thickness of the specimen before dipping it in water.

16.3.4 DENSITY MEASUREMENT TEST

Density is an important physical property of the composites. Low-density composites are favorable due to their low weight cost. The density of composites depends on quantities of fibers, filler particulates and epoxy resin. The specimens for the density test were prepared by cutting a piece of composite product in 30 mm in length and 15 mm in width. The density was calculated [17] by Equation 16.4:

$$Density = \frac{W_a \times \sigma_w}{W_a - W_w}, \quad (16.4)$$

where W_w is the mass of the sample weighed in distilled water,
W_a is the mass of the sample weighed in air and
σ_w is the density of distilled water at normal temperature and pressure (0.998 gm/cm³)

16.3.5 COST ANALYSIS OF COMPOSITE MATERIAL

The cost of fly ash–filled E-GFREC is calculated by Equation 16.5:

$$\text{Total cost of composite material} = \text{Fixed cost} + \text{Variable cost.} \quad (16.5)$$

The fixed cost includes the cost of the E-glass fiber, which is constant irrespective of the type of matrix material. The cost of the E-glass fiber was 450 rupees per square meter. The variable cost includes the cost of the fly ash and the epoxy. The cost of the epoxy was 450 rupees per liter, and the cost of the fly ash was negligible. There was only transportation cost per ton for the fly ash. The epoxy was replaced by fly ash in order to maintain the total weight of the material constant. Thus, with an increase in fly ash, the cost of composite material was reduced.

16.4 RESULTS

16.4.1 MOISTURE CONTENT

The variation in moisture content with the percentage weight of fly ash particulate in E-GFREC is shown in Figure 16.3. The E-GFREC without the particulate filler has a minimum magnitude of percentage of moisture content. This is attributed to free hydroxyl group inside the resin matrix as well as fiber [18] and the absence of fly ash. The nature of fly ash is to absorb water particles. Therefore, an increase in the moisture content showed the presence of fly ash according to concentration.

16.4.2 WATER ABSORPTION

The variation of water absorption with percentage weight of fly ash particulate in E-GFREC is shown in Figure 16.4. The E-GFREC without the fly ash particulate filler has a minimum magnitude of percentage of water absorption due to the good compatibility between epoxy resin and glass fiber [19]. The results show that fly ash particles are more prone to water absorption.

16.4.3 LINEAR SWELLING

The variation in linear swelling with the percentage weight of fly ash particulate in E-GFREC is shown in Figures 16.5 through 16.7. E-GFREC without the particulate filler has a minimum magnitude of percentage of swelling in any one linear

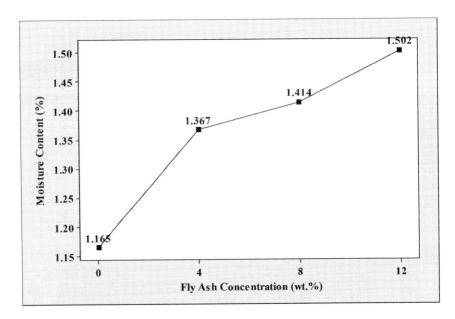

FIGURE 16.3 Effect of Percentage Weight of Fly Ash on Moisture Content of E-GFREC

Fly Ash–Filled E-GFREC 263

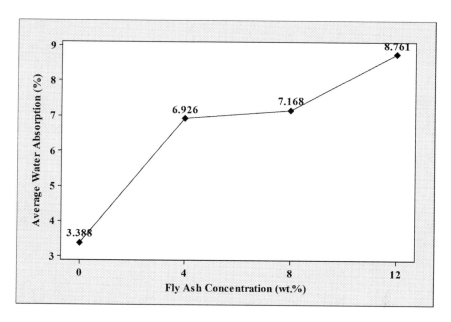

FIGURE 16.4 Effect of Percentage Weight of Fly Ash on Water Absorption of E-GFREC

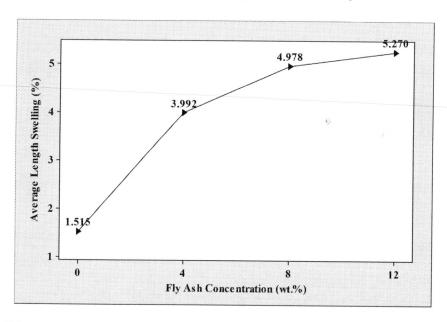

FIGURE 16.5 Effect of Percentage Weight of the Fly Ash on Length Swelling of E-GFREC

dimension due to good compatibility between epoxy resin and glass fiber as discussed in the previous section [19]. All linear dimensions are in good correlation with water absorption and moisture content. Therefore, the same reasons are responsible for the variation in length, width and thickness.

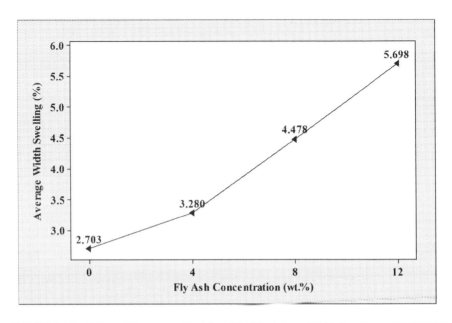

FIGURE 16.6 Effect of Percentage Weight of the Fly Ash on Width Swelling of E-GFREC

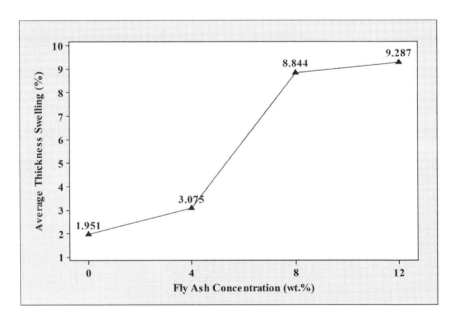

FIGURE 16.7 Effect of Percentage Weight of the Fly Ash on the Thickness Swelling of E-GFREC

16.4.4 Density Test

The variation of density with the percentage weight of fly ash particulate in E-GFREC is shown in Figure 16.8. The E-GFREC without the fly ash particulate filler has a maximum magnitude of density. It is attributed to better bonding between the epoxy resin and the glass fiber as well as minimum void formation [19]. It is shown in Figure 16.8 that the magnitude of density of the fly ash particulate–filled epoxy composite from 0 wt.% to 4 wt.% suddenly decreased due to the formation of voids. The void formation in the composite is caused by air bubbles entrapped within the composite. Vapors were developed during the curing of the epoxy, hardener, and residual solvents. The resin has difficulty wetting the fiber completely after mixing in of fly ash particles, which also leads to void formation [20].

16.4.5 Cost Analysis of Composite Material

The variation of composite material cost with the percentage weight of fly ash particulate in E-GFREC is shown in Figure 16.9. The E-GFREC without the particulate filler has a maximum value of cost due to both epoxy resin and glass fibers having the highest cost contribution. The total material cost decreases by the addition of fly ash while keeping the total weight of the composite constant. The cost is inversely proportional to the weight of fly ash and is reduced by the cost of resin material.

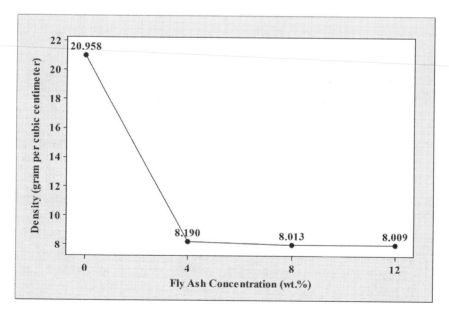

FIGURE 16.8 Effect of Percentage of Weight of Fly Ash on the Density of E-GFREC

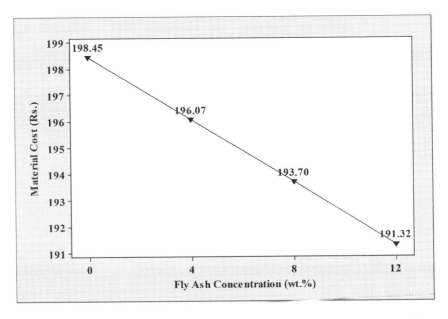

FIGURE 16.9 Effect of Percentage Weight of Fly Ash on Material Cost of E-GFREC

16.5 CONCLUSION

The effect of water particles was observed with different concentrations of fly ash in the composite material. Water particles affect the dimensional stability of the composite. For a low-cost criterion, a higher quantity of fly ash is preferable whereas for better dimensional stability, a low concentration is desirable. However, a small percentage of fly ash content can be used for the overall enhancement of the polymer composite.

The results of experimental investigation and analytical evaluation of physical properties on fly ash–filled E-GFRECs have led to the following specific conclusions:

- The specimen with the maximum 12 wt.% has a maximum magnitude of percentage of moisture content (1.165%), water absorption (3.388%) and thickness swelling (1.951%).
- The E-GFREC without particulate filler has a maximum magnitude of density and cost.

REFERENCES

1. A. Patnaik, "Development, Characterization and Solid Particle Erosion Response of Polyester Based Hybrid Composites," ethesis, nitrkl.ac.in, 2008, http://ethesis.nitrkl.ac.in/166/1/thesis.doc.pdf.
2. A. K. Tanwer, "Mechanical Properties Testing of Unidirectional and Bi-Directional Glass Fiber Reinforced Epoxy Based Composites," International Journal of Research in Advent Technology, Vol. 2, Pp. 34–39, 2014.
3. P. K. Mallick, Fiber Reinforced Composites Materials Manufacturing and Design, Third Ed., Taylor & Francis Group, 2008.

4. S. Vashishtha and K. Gupta, "Mechanical and Abrasive Wear Characterization of Bidirectional and Chopped E-Glass Fiber Reinforced Composite Materials," Materials and Design, Vol. 35, Pp. 467–479, 2012.
5. S. Agrawal, K. K. Singh and P. K. Sarkar, "A Comparative Study of Wear and Friction Characteristics of Glass Fibre Reinforced Epoxy Resin, Silding under Dry, Oil Lubricated and Inert Gas Environments," Tribology International, Vol. 96, Pp. 217–224, 2016.
6. P.-S. Shin, Z.-J. Wang, D.-J. Kwon and J.-Y. Choi, "Optimum Mixing Ratio of Epoxy for Glass Fiber Reinforced Composites with High Thermal Stability," Composites Part B, Vol. 79, Pp. 132–137, 2015.
7. M. Singla and V. Chawla, "Mechanical Properties of Epoxy Resin–Fly Ash Composite," Journal of Minerals and Materials Characterization and Engineering, Vol. 3, Pp. 199–210, 2010.
8. R. Manimaran, I. Jayakumar, R. M. Giyahudeen and L. Narayanan, "Mechanical Properties of Fly Ash Composites—A Review," Energy Sources, Part A: Recovery, Utilization, and Environmental Effects, Vol. 40, No. 8, Pp. 887–893, 2018.
9. T. Chaowasakoo and N. Sombatsompop, "Mechanical and Morphological Properties of Fly ash/Epoxy Composites using Conventional Thermal and Microwave Curing Methods," Composites Science and Technology, Vol. 67, Pp. 2282–2291, 2007.
10. S. M. Kulkarni and Kishore, "Effect of Filler-Fiber Interactions on Compressive Strength of Fly Ash and Short-Fiber Epoxy Composites," Journal of Applied Polymer Science, Vol. 87, Pp. 836–841, 2003.
11. S. Mutalikdesai, A. Hadapad, S. Patole and G. Hatti, "Fabrication and Mechanical Characterization of Glass Fibre Reinforced Epoxy Hybrid Composites Using Fly Ash/ Nano Clay/ Zinc Oxide as Filler," IOP Conference Series: Materials Science and Engineering, Vol. 376, P. 012061, 2018.
12. S. M. Kulkarniand and Kishore, "Effects of Surface Treatments and Size of Fly Ash Particles on the Compressive Properties of Epoxy Based Particulate Composites," Journal of Materials Science, Vol. 37, Pp. 4321–4326, 2002.
13. S. M. Kulkarni and Kishore, "Studieson Fly Ash–Filled Epoxy-Castslabsunder Compression," Journal of Applied Polymer Science, Vol. 84, Pp. 2404–2410, 2002.
14. K. Devendra and T. Rangaswamy, "Strength Characterization of E-Glass Fiber Reinforced Epoxy Composites with Filler Materials," Journal of Minerals and Materials Characterization and Engineering, Vol. 1, Pp. 353–357, 2013.
15. P. Shubham and S. K. Tiwari, "Effect of Fly Ash Concentration and Its Surface Modification on Fiber Reinforced Epoxy Composite's Mechanical Properties," International Journal of Scientific and Engineering Research, Vol. 4, No. 8, 2013.
16. R. S. Raja, K. Manisekar and V. Manikandan, "Study on Mechanical Properties of Fly Ash Impregnated Glass Fiber Reinforced Polymer Composites Using Mixture Design Analysis," Materials and Design, Vol. 55, Pp. 499–508, 2014.
17. A. Atiqah, M. Jawaid, M. R. Ishak and S. M. Sapuan, "Moisture Absorption and Thickness Swelling Behaviour of Sugar Palm Fibre Reinforced Thermoplastic Polyurethane," Advances in Material & Processing Technologies Conference, No. 184, Pp. 581–586, 2017.
18. J. Sahari, S. M. Sapuan, E. S. Zainudin and M. A. Maleque, "Mechanical and Thermal Properties of Environmentally Friendly Composites Derived From Sugar Palm Tree," Materials and Design, Vol. 49, Pp. 285–289, 2013.
19. M. Jawaid, H. P. S. Abdul Khalil, A. Hassan and E. Abdallah, "Bi-Layer Hybrid Biocomposites: Chemical Resistant and Physical Properties," Bioresources, Vol. 7, No. 2, Pp. 2344–2355, 2012.
20. M. L. Sánchez, L. Y. Morales and J. D. Caicedo, "Physical and Mechanical Properties of Agglomerated Panels Made from Bamboo Fiber and Vegetable Resin," Construction and Building Materials, Vol. 156, Pp. 330–339, 2017.

17 A Study on the Silane Chemistry and Sorption/Solubility Characteristics of Dental Composites in a Wet Oral Environment

Sukriti Yadav and Swati Gangwar

CONTENTS

17.1 Introduction: Background and Driving Forces 269
17.2 Silane Coupling Agent in Dentistry 270
17.3 How Does a Silane Coupling Agent Modify the Surfaces of Inorganic Fillers? 271
17.4 Selection of Silane for Surface Modification of Inorganic Substrate 273
17.5 Sorption and Solubility Characteristics in Dental Composites 274
17.6 Conclusion 276
References 277

17.1 INTRODUCTION: BACKGROUND AND DRIVING FORCES

The interconnection of aesthetic features with mechanical properties and reasonable cost makes the resin composites remarkably significant for dentistry. Resin-based composite materials currently are mostly favored as the effective material for occlusal loads as posterior teeth [1, 2]. Dental resins ideally should be highly stable and moisture resistant. The structure and chemical constituents of the base monomer change the solubility and water sorption, the degree of conversion, and the color stability of dental resins. The solubility and sorption of dental restoratives depend on the materials and are considerably manipulated by the hydrophilicity of resin amount, along with the properties of the fillers [3, 4].

In dentistry, silane coupling agents are more effective as adhesion boosters to chemically consolidate the resin and silica/non-silica-based fillers to improve the resin bonding. The surface functionalization of inorganic fillers enhances the chemical interaction between functionalized inorganic fillers and organic polymer matrix, which improves the interfacial strength between them and changes the hydrophilic nature of fillers into the hydrophobic nature [5, 6]. Water sorption is well thought of as the important aspect for deciding the clinical triumph of materials because it

DOI: 10.1201/9781003093213-17

adversely affects the hydrolytic stability of resin composites that contribute to the discoloration, poor wear resistance, inferior mechanical properties, and hydrolytic deterioration of the matrix–filler interface bonding in dental composites [7, 8].

It also promotes hygroscopic stresses due to the hygroscopic expansion of the materials that could lead to micro-cracks or cracked cups in dental restorations [9, 10]. It has been found in the literature that fracture, as well as secondary caries, is the major reason for the deterioration of composites used for posterior teeth restorations, whereas the restorations of anterior teeth are usually carried out for aesthetics aspects such as color characteristics, uneven teeth, and dental crowns, among others. The discoloration is vital in the lifespan of dental restorations [11–13]. The formulation of composites for dental applications shows a vital role in analyzing the degradation resistance and material properties [14, 15]. The sorption of liquids into a polymer-based material is governed by the diffusion-controlled process, that is, by the interaction and the free volume theory [10]. In the interaction concept, water links to the specific ionic groups of the polymeric network and based on the water harmony of ionic groups, whereas in the free volume theory, solvent or liquid incorporation takes place across voids in the polymers [9, 16]. Usually, the extent of water sorption is affected by the crosslinking density, the hydrophilicity of the polymeric structure, and the degree of conversion. Depending on the degree of the crosslinking density of the polymers, the soaked water may lead to the bulging and broadening of positions between polymer chains. This concedes unreacted monomers captured in the polymeric structure to disperse out into the liquid solution based on the affinity to the liquid solution and its molecular size [10, 17].

Although many investigations focus on the solubility and sorption of resin-based dental restoratives for short-term physio-mechanical characteristics in water and some other studies focuses on different storage media, for example, artificial saliva, ethanol, different pH solutions, and so on, but still there are inadequate data regarding the long-term stability of dental composites in an aqueous storage medium compared to traditional dental materials. This study focuses on gaining and improving knowledge about the silane chemistry procedure for the surface functionalization of silica/non-silica-based inorganic fillers and clinical studies regarding solubility and sorption of several resin-based materials in different oral environments such as distilled water, artificial saliva, citric acid solution, and others that improve the application performance of materials in biomedical and dental applications.

17.2 SILANE COUPLING AGENT IN DENTISTRY

The proficiency and durability of resin composites greatly depend on the type of organic polymer, inorganic fillers and interphase bonding along with the efficiency of the polymerization reaction [18–19]. The synergy endures at the interphase, between the matrix and the reinforcing phase, and is dominantly mediated by the interfacial bonding. The interface associate with these materials approaches a vibrant field of chemistry that modifies the surfaces to develop the required heterogeneous atmosphere or to assimilate the gross properties of several stages into a homogeneous material structure [20]. Surface functionalization of inorganic fillers alters the surface free energy associated with it and impedes the clustering of particles by reducing the intensity of synergy between fillers. This leads to enhancement through the good distribution of inorganic fillers during fabrication, which, in turn, positively affects the mechanical characteristics of polymer composite [21, 22]. The surfaces of inorganic filler particles generally consist of OH group, owing to its reaction with atmospheric water, and make it easy to accrue at the matrix–filler interface by diffusion through the matrix. Therefore, interfacial bonds between matrix and fillers either become weak or degraded over time when composites are exposed to a moist atmosphere.

Hence, most of the time, these inorganic filler surfaces are treated/coated with the help of coupling agents that changes the hydrophilic characteristics of fillers to hydrophobic and significantly reduces the moisture absorption characteristics of particulate-filled dental composites [23–26]. In dentistry, silane coupling agents are more effective as adhesion boosters to chemically consolidate the resin and silica-based fillers/material having a surface chemistry with siliceous properties, for example, aluminates, silicates, borates, and others. For non-silica-based restorative materials, a pre–surface treatment is required to boost the silica amount that improves the resin bonding [27–29]. Generally, silane coupling agents are acknowledged for many purposes, such as enhancing the wettability of resin with filler particles/fibers and modifying the compatibility of two dissimilar materials, which, in turn, strengthens the physical, chemical, and mechanical bonding between polymers and filler particles/fibers. It also provides the preeminent durability and water resistance property at the interphase. Despite the major problem of interfacial bonding through coupling agent, silane, among others, is the loosening or degradation of the bond in the soaked oral atmosphere throughout a period [30–32].

The chemistry of organo-silanes becomes quite complex due to the involvement of hydrolytically originated self-condensation reactions that end up in polymeric silsesquioxane patterns, interchange reactions with carboxylated or hydroxylated monomers to produce silyl esters along with ethers, along with the creation of silane procured configuration by adhesive coupling agents as well as siliceous mineral facade [33–35]. Silane creates an extensive group of organic mixtures that typically consist of silicon atoms. It may be monofunctional (i.e., molecule consist of one silicone atom along three alkoxy groups) or bifunctional (i.e., two silicon atoms along three alkoxy groups having dual reactivity). An organic polymer matrix can react with organic functional constituents of silane (e.g., allyl—CH_2 $CH=CH_2$, vinyl—$CH=CH_2$; allyl, isocyanate—$N=C=O$; amino—NH_2). The inorganic substance can react with alkoxy group (e.g., ethoxy—O—CH_2 CH_3; methoxy—O—CH_3) by establishing covalent bonds between matrices [36–38].

17.3 HOW DOES A SILANE COUPLING AGENT MODIFY THE SURFACES OF INORGANIC FILLERS?

A silane agent, that is, a trialkoxysilane, consists of two functional units which join an unpolymerized polymer matrix and inorganic filler particles (surfaces). The general formula of silane represented by R-$(CH_2)_n$—Si—X_3, where "X" represents a hydrolysable group, that is, acryloxy, alkoxy, amine, or halogen, and the "R" group represents a non-hydrolyzable organic radical having the functionality to convey favored characteristics (remains accessible for the physical interaction with different

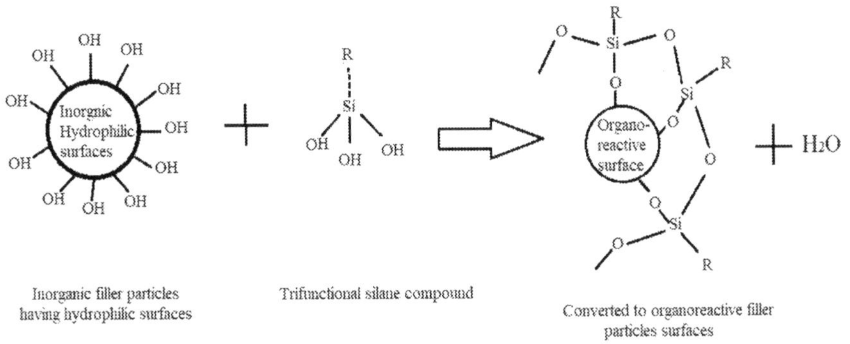

FIGURE 17.1 Schematic Representation of the Surface Functionalization of Inorganic Filler Surfaces with a Silane Coupling Agent

phases and covalent bond formation). During a majority of the surface functionalization procedure, the hydrolysis of the alkoxy group's results into a formation of a silanol-consisting species that can condense with several silanol groups; for example, a siloxane linkage is formed on the siliceous fillers surface as represented in Figure 17.1.

However, some of the oxides such as those of nickel, aluminum, tin, zirconium, and titanium has formed stable condensation products. The oxides of iron, boron, and carbon form fewer stable bonds, whereas carbonates and alkali metal oxides do not form stable bonds with Si-O- [39–41]. Silane coupling agents can be used for the silanization of inorganic surfaces by various techniques such as the following: [42]

- Deposition of silane through an aqueous alcohol
- Deposition of silane through an aqueous solution
- Bulk deposition of silane onto inorganic powder by spraying
- Addition of silane through integral blend method during composite fabrication
- Addition of silane as a primer

The minimum amount of silane (binding agent) recommended for accessing the minimum multilayer coverage can be obtained by Arkles formula [42, 43] as shown in Equation 17.1. The values of the relative surface area of fillers and wetting surface area of silane are different for all inorganic fillers and each group of silanes, respectively, that can be provided by the manufacturer or theoretical values can be used.

$$\text{Amount of silane / binding agent}(g) = \frac{\text{Amount of filler}(g) \times \text{surface area of filler}(m^2/g)}{\text{wetting surface of silane}(m^2/g)} \quad (17.1)$$

The surface functionalization reaction of silane through aqueous solution is executed in four steps. The schematic representation of working mechanism of all four steps is as follows [44]:

I. Hydrolysis of the alkoxy groups
II. Condensation of oligomers
III. Formation of hydrogen bond of oligomers with OH groups of filler particles surfaces
IV. Finally, covalent bond formation during curing or drying

After the hydrolysis step, condensation and hydrogen bonding reactions can happen simultaneously. Surface modification with silane can be done under anhydrous situation persistence with vapor phase deposition and monolayer requirements. The typical extended reaction time is 4–12 hours at an elevated temperature of 50–120 °C. The reversible nature of the hydrolysis of the chemical bond is one of the significant benefits of silane coupling agents, as it may decrease the effect of internal stresses in the composite material. The final outcome of the surface functionalization process varies with the adhesion or wetting characteristics of the substrate, the utilization of the substrate to initiate chemical conversions at the different interfaces, the regulation of the interfacial region, and the improvement of its partition characteristics. Importantly, it comprises the ability to affect the covalent bonds between inorganic and organic materials [45–48].

17.4 SELECTION OF SILANE FOR SURFACE MODIFICATION OF INORGANIC SUBSTRATE

There are many factors that need to be known before the selection of silane for inorganic substrate modification [49–50]:

I. What is the concentration of substrate hydroxyl groups?
II. What type of substrate hydroxyl groups present?
III. Hydrolytic stability of the chemical bond formed.
IV. What are the physical attributes of the inorganic substrate?

To fulfill all these requirements, the optimum silane concentration and thorough study of inorganic substrate and silane chemistry are helpful. The maximum surface functionalization can be achieved when silane and the substrate surface react with each other and allows the prompting of the maximum number of available sites with a relevant surface energy. Additionally, physical and chemical attributes (i.e., modulus, water/hydroxyl content, etc.) of the interphase region significantly affect the same. The types and concentrations of the hydroxyl groups present deviate extensively in hydroxyl-consisting substrate. A minimum number of hydroxyl groups is present in freshly fused substrate reserved under neutral circumstances. Silane coupling agents prompt a quicker response with hydrogen-bonded silanols, whereas free or isolated hydroxyls give respond to them less.

In trialkoxysilane, the three alkoxy groups of silanes are a prevalent starting point for substrate functionalization. This silane frequently contributes to covering up organic polymeric films, enhancing the addition of organic functionality and total coverage. It is widely utilized in adhesive sealants, composites, and coatings. There are certain inherent limitations in the utilization of multilayer deposition on nanoparticles/nanocomposites in which multilayer deposition on the substrate produces interphase dimensions. Mono-alkoxy silane furnishes a widely used substitute for nano-featured substrate, considering deposition is controlled to a monolayer. The poor hydrolytic stability of the oxane bond (i.e., between silane and substrate) or utilization is in a contentious aqueous atmosphere leads to generous performance enhancement of dipole silane. These materials are suitable for primer applications because it develops a stronger network and may provide 10 times higher hydrolysis resistance [51–54].

17.5 SORPTION AND SOLUBILITY CHARACTERISTICS IN DENTAL COMPOSITES

Dental restorative materials are invariably exposed to the wet surrounding and their bioactivity, sorption and solubility are of greater interest for clinical use. The sorption of the restorative material is a diffusion-controlled phenomenon that leads to chemical deterioration of materials due to the matrix–filler interface debonding and release of residual monomers. The assessment of the diffusion coefficient is not mentioned in the ISO standard. It denotes the water diffusion rate in the polymeric structure that could be linked to the deterioration of restorative composites. It is presumed that the greater diffusion coefficient leads to rapid degradation action in the restorative composites [55].

Presently, some literature focuses on the in vivo conditions; the test specimen was immersed in the aqueous solutions without desiccation. However, some studies use ISO 4049: 2009 to access the volumetric expansion, solubility, and sorption of restoratives in which test samples are desiccated to get the initial stable mass after clearing excess water from the specimen. As per ISO 4049: 2009, the highest value of water sorption is 40 µg/mm^3, and the solubility is 7.5 µg/mm^3. Since the soaking period of 7 days (ISO 4049: 2009 standard) is inadequate to many restorative materials, which was usually saturated for 7–60 days [56, 57]. The significance of the immersion period has been clearly demonstrated by Ferracane et al. [57]; they have reported that the immersion period has much significance in defining the accurate features of restorative composites. The decreased mechanical properties of restorative composites were predominantly allied with the water sorption of the restoratives. The mechanical attributes of restoratives were continuing to decrease until the polymer structure was fully saturated; after that, it was stabilized. A similar observation was also made by Asaoka and Hirano [58]. They have suggested that for clinical use, the diffusion coefficient plays a significant role in evaluating the time-dependent hygroscopic expansion and time-dependent mechanical attributes of restorative composites. Mustafa et al. [59] had investigated the consequences of desiccation on the volumetric expansion, sorption, and solubility of dental composites for Biodentine TM (BD), Ionolux (IO), and GC Fuji IX GP FAST (FJ; i.e., a calcium silicate–based composite, a resin-modified glass ionomer, and a conventional glass ionomer, respectively) at distinct time intervals (1 hr, 24 hr, 3 days, 7 days, and 30 days). They have found that more water absorbed by the materials without desiccation. The lower sorption value was observed with desiccation of FJ and IO, i.e., 79.97 µg/mm^3 and 124.33 µg/mm^3 sequentially whereas, without desiccation, water sorption was found 130.35 µg/mm^3 for IO, 122.07 µg/mm^3 for BD, and 107.21 µg/mm^3 for FJ (IO > BD > FJ). With desiccation, solubility was −20.19 µg/mm^3 for FJ and −12.36 µg/mm^3 for IO, and volumetric expansion was −2.35% for FJ along with 3.01% for IO.

The solubility was found to be decreased without desiccation, that is, 154.83 µg/mm^3 for BD, 88.82 µg/mm^3 for FJ, and 25.67 µg/mm^3 for IO (BD > FJ > IO). FJ and IO exhibit a considerable variation in solubility and water sorption with and without desiccation. The hygroscopic expansion of restorative composites was reported as IO > BD > FJ. The calcium silicate–based material (BD) shows the maximum solubility while resin-modified glass ionomer shows the minimum solubility. The resin-modified glass ionomer (IO) exhibits lower solubility but maximum hygroscopic expansion. Hence, desiccation significantly influences the solubility, water sorption, and volumetric expansion of polymeric restorative composites. The stability of restorative composites in water and its interfacial properties could be significantly enhanced by enhancing the surface attributes of the fillers.

Liu et al. [60] have reported that grafting of poly-bisphenol A-glycidyl methacrylate (poly Bis-GMA) on the hydroxyapatite whiskers (HWs) enhances the solubility, the water sorption of the restorative material without negotiating the bioactivity. They have grafted the HWs by poly Bis-GMA (PGHW) with a different grafting ratio as PGHW-1h (8.5 wt.% grafting ratio and PGHW-18h (32.8 wt.% grafting ratio). They had reported that PGHW-1h filled restorative composite absorbs minimum water (27.16 µg/mm^3 in 7 days), while the untreated HW-filled restorative composites absorbs the maximum amount of water, and PGHW-18h filled restorative composites shows the highest solubility due to decreased monomer conversion in this composite.

The fracture surface of the resulting restorative composites also demonstrates improved PGHW resin–matrix interfacial bonding and compatibility that facilitates the homogeneous distribution of PGHW-1h in the restorative composites. From in vitro bioactivity study of these dental restoratives in simulated body fluid (SBF), it has been found that the surface grafting of HW did not prominently influence the ability of apatite formation in the restorative composites. As we know that fruit juices and carbonated soft drinks are the frequently consumed beverages by most of the population worldwide. Statistics reveal that over the past 50 years, the consumption of these beverages has increased dramatically (25% of daily fluid intake), with the majority of groups consuming these drinks being children and adolescents. The chemical activities of these drinks could lead to surface erosion, andthe consumption of these acidic beverages may deteriorate the teeth and restorative composites [61, 62].

McKenzie et al. [62] investigated the properties of dental restoratives after immersing the material in cola and orange juice, and they have found a significant decrease in the hardness of dental restorative composites. Orange juice and cola consist of higher quantities of citric acid and phosphorous acid, respectively, that may cause increased sorption and decrease the important properties of the restoratives. Similarly, Rahim et al. [63] emphasized in evaluating the influence of acidic drinks (cola and orange juice) on solubility, diffusion coefficient, and water sorption of distinct restorative composites (Filtek TM Z350 Durafill® VS, and Spectrum® TPH®3), where Filtek TM Z350 was a nano-zirconia- and silica-filled composite, Durafill VS was pre-polymerized silica-filled composite, and Spectrum TPH3 was micro-silica filled composite. The experimental data of diffusion coefficient nearly fitted to the theoretical data points obtained by Fick's second law.

The experimental results demonstrate the maximum value of the diffusion coefficient (32.23–45.25 × 10^{-13} m^2/s) for Durafill VS, whereas Filtek TM Z350 and Spectrum TPH3 showed the maximum value of diffusion coefficient when dipped in cola and then decreased in distilled water followed by orange juice. Filtek TM Z350 showed the maximum sorption characteristics when dipped in cola and distilled water (18.22 μg/mm^3 and 16.13 μg/mm^3, respectively), whereas Durafill VS showed maximum solubility (7.20–9.27 μg/mm^3). Usually, the immersion of most dental restoratives in cola increases the water sorption, but Spectrum TPH3 reveals a rise in the solubility for cola. It can be concluded from this study that acidic drinks influence prominently on the sorption, solubility, and diffusion properties of restorative composites. However, there are still many other factors that are responsible for regulating the water sorption and solubility characteristics, for example, filler type and size, filler loading, and the polymer matrix.

17.6 CONCLUSION

It can be demonstrated from the study that the clinical success of a restorative compositeis not only associated with its mechanical or tribological properties since there are some other factors that affect the clinical success of dental materials such as water sorption, solubility, and diffusion coefficient.

- For clinical use, diffusion coefficient plays a significant role in evaluating the time-dependent hygroscopic expansion and time-dependent mechanical properties of restorative materials. Ideally, restorative composites should be able to withstand occlusal loading and should not degrade in the harsh oral environment.
- The exposure of restorative composites in acidic media may lead to increased solubility, diffusion coefficient, and water sorption and encourage the degradation process of composites and thus decreases the longevity of composite restorations. The stability of restorative composites in water and its interfacial properties could be significantly enhanced by improving the surface properties of the filler materials, desiccation, and the like and thus increases the lifespan of the restorative composites.
- One of the main advantages of trialkoxysilyl group's silane over dialkoxysilyl is greater crosslinking density with substrate that leads to higher adhesion property. This three-dimensional crosslinking formed due to the hydrolytic condensation of trialkoxysilyl groups whereas two-dimensional crosslinking was formed due to hydrolytic condensation of dialkoxysilyl.

- Among the alkoxysilyl groups and methoxysilyl groups, hydrolyze quickly than that of ethoxysilyl groups. Hydrolysis of the methoxysilyl groups produces methanol whereas ethanol was formed during the hydrolysis of the ethoxysilyl groups. Whenever you are fretful about the absolution of methanol, ethoxysilyl group silane should be used.

REFERENCES

1. Alrahlah A., Silikas N., Watts D.C., 2014, Hygroscopic expansion kinetics of dental resin-composites, Dent. Mater., 30, 143–148.
2. Lynch C.D., Opdam N.J., Hickel R., Brunton P.A., Gurgan S., Kakaboura A., 2014, Guidance on posterior resin composites: Academy of operative dentistry – European section, J. Dent., 42, 377–383.
3. Bertolo M.V.L., Moraes R.C.M., Peifer C., Salgado V.E., Correr A.R.C., Schneider L.F.J., 2017, Influence of photo initiator system on physical-chemical properties of experimental self-adhesive composites, Braz. Dent. J., 28, 35–39.
4. Um C.M., Ruyter I.E., 1991, Staining of resin-based veneering materials with coffee and tea, Quintessence Int., 22, 377–386.
5. Ferracane J.L., Hilton T.J., Stansbury J.W., Watts D.C., Silikas N., Ilie N., Heintze S., Cadenaro M., Hickel R., 2017, Academy of Dental Materials guidance—Resin composites: Part II—Technique sensitivity (handling, polymerization, dimensional changes), Dent. Mater, 33(11), 1–56.
6. Sarrett D.C., 2005, Clinical challenges and the relevance of materials testing for posterior composite restorations, Dent. Mater., 21(1), 9–20.
7. Kupiec A.S., Pluta K., Drabczyk A., Włos M., Tyliszczak B., 2018, Synthesis and characterization of ceramic polymer composites containing bioactive synthetic hydroxyapatite for biomedical applications, Ceram. Int., 44(12), 13630–13638.
8. Ito S., Hashimoto M., Wadgaonkar B., Svizero N., Carvalho R.M., You C., 2005, Effects of resin hydrophilicity on water sorption and changes in modulus of elasticity, Biomaterials, 26, 6449–6459.
9. Sideridou I.D., Karabela M.M., Vouvoudi E.C., 2008, Volumetric dimensional changes of dental light-cured dimethacrylate resins after sorption of water or ethanol, Dent. Mater., 24, 1131–1136.
10. Alshali R.Z., Salim N.A., Satterthwaite J.D., Silikas N., 2015, Long-term sorption and solubility of bulk-fill and conventional resin-composites in water and artificial saliva, J. Dent., 43, 1511–1518.
11. Yadav S., Gangwar S., 2018, An overview on Recent progresses and future perspective of biomaterials, IOP Conf. Ser. Mater. Sci. Eng., 404, 012013.
12. Correa M.B., Peres M.A., Peres K.G., Horta B.L., Barros A.D., Demarco F.F., 2012, Amalgam or composite resin? Factors influencing the choice of restorative material, J. Dent., 40, 703–710.
13. Ferracane J.L., Pfeifer C.S., Hilton T.J., 2014, Microstructural features of current resin composite materials, Curr. Oral. Health. Rep., 1, 205–212.
14. Hashem D.F., Foxton R., Manoharan A., Watson T.F., Banerjee A., 2014, The physical characteristics of resin composite-calcium silicate interface as part of a layered/laminate adhesive restoration, Dent. Mater., 30, 343–349.
15. Fonsecaa A.S.Q.F., Moreirad A.D.L., Paulo A.C.P., Pfeiferg C.S., Schneidera L.F.J., 2017, Effect of monomer type on the C C degree of conversion, water sorption and solubility, and color stability of model dental composites, Dent. Mater., 33(4), 394–401.

16. Chaves L.P., Graciano F.M.O., Júnior, Vale Pedreira A.P.R., Manso A.P., Wang L., 2013, Water interaction with dental luting cements by means of sorption and solubility, Braz. Dent. Sci., 15, 29–35.
17. Wei Y., Silikas N., Zhang Z., Watts D.C., 2011, Diffusion and concurrent solubility of self-adhering and new resin-matrix composites during water sorption/ desorption cycles, Dent. Mater., 27, 197–205.
18. Correa M.B., Peres M.A., Peres K.G., Horta B.L., Barros A.D., Demarco F.F., 2012, Amalgam or composite resin? Factors influencing the choice of restorative material, J. Dent., 40(9), 703–710.
19. Ilie N., Hilton T.J., Heintze S.D., Hickel R., Watts D.C., Silikas N., 2017, Academy of Dental Materials guidance—resin composites: Part I—mechanical properties, Dent. Mater., 33(8), 880–894.
20. Moszner N.H.T., 2012, New polymer-chemical developments in clinical dental polymer materials: Enamel-dentin adhesives and restorative composites, J. Polym. Sci. Part A: Polym. Chem., 50(21), 4369–4402.
21. Hornsby P.R., Watson C.L., 1995, Interfacial modification of polypropylene composites filled with magnesium hydroxide, J. Mater. Sci., 30(21), 5347–5355.
22. Ilie N., Hickel R., 2006, Silorane-based dental composite: Behavior and abilities, Dent. Mater. J., 25(3), 445–454.
23. Vaz C.M., Reis R.L., Cunha A.M., 2002, Use of coupling agents to enhance interfacial in starch-EVOH/HA composites, Biomaterials, 23(2), 629–635.
24. Halvorson R.H., Erickson R.L., Davidson C.L., 2003, The effect of filler and silane content on conversion of resin-based composite, Dent. Mater., 19(4), 327–333.
25. Shah P.K., Stansbury J.W., 2014, Role of filler and functional group conversion in the evolution of properties in polymeric dental restoratives, Dent. Mater., 30(5), 586–593.
26. Leal F.B., Pereira C.M., Ogliari F.A., 2015, Synthesis, characterization, and photocuring of siloxane-oxirane monomers, J. Polym. Sci. Part A: Polym. Chem., 53(14), 1728–1733.
27. Sharpe L.H., 1992, Recent Advances in Adhesion, Lee L.H., Ed., Gordon and Breach, New York.
28. Miller J.D., Ishida H., 1991, Adhesive-adherend interface and interphase, in: Fundamentals of Adhesion, Lee L.H., Ed., Plenum Press, New York, 291–324.
29. Antonucci J.M., Fowler B.O., Dickens S.H., Richards N.D., 2002, Novel dental resins from trialkoxysilanes and dental monomers by in-situ formation of oligomeric silsesquioxanes, Polymer Preprints, 43(2), 633–634.
30. Arkles B., Larson G.L., 2013, Silicon compounds: Silane and silicones: A survey of properties and chemistry, 3rd Edition Pennsylvania: Gelest, Inc., 55(1), 24–25.
31. Deschler U., Kleinschmit P., Panster P., 1986, 3-Chloropropyltrialkoxysilanes—key intermediates for the commercial production of organo functionalized silanes and polysiloxanes, Angew. Chem. Int. Ed. Engl., 25, 236–252.
32. Wilson K.S., Zhang K., Antonucci J.M., 2005, Systematic variation of interfacial phase reactivity in dental nanocomposites, Biomaterials, 26(25), 5095–5103.
33. Blatz M.B., Sadan A., Kern M., 2003, Resin ceramic bonding: A review of the literature, J. Prosthet. Dent., 89, 268–274.
34. Matinlinna J.P., et al., 2017, Silane adhesion mechanism in dental applications and surface treatments: A review, Dent. Mater., 34(1), 13–28.
35. Suzuki N., IsidaH., 1996, A review on the structure and characterization techniques of silane/Matrix interphases, Mucromol, 108, 19–53.
36. Wilson K.S., Antonucci J.M., 2006, Interphase structure—property relationships in thermoset dimethacrylate nanocomposites, Dent. Mater., 22(11), 995–1001.

37. Karmaker A., Prasad A., Sarkar N.K., 2007, Characterization of adsorbed silane on fillers used in dental composite restoratives and its effect on composite properties, J. Mater. Sci.: Mater. Med., 18, 1157–1162.
38. Antonucci J.M., Dickens S.H., Fowler B.O., Xu H.H.K., McDonough W.G., 2005, Chemistry of silanes: Interfaces in dental polymers and composites, J. Res. Natl. Inst. Stand. Technol., 110(5), 541–558.
39. Arklese B., Maddox A., Singh M., Zazyczny J., Maatisons J., 2014, Silane Coupling Agents Connecting Across Boundaries, Gelest, Inc., Morrisville, PA., 2–72.
40. Lung C.Y.K., Matinlinna J.P., 2012, Aspects of silane coupling agents and surface conditioning in dentistry: An overview, Dent. Mater., 28(5), 467–477.
41. Lung C.Y.K., Matinlinna J.P., 2013, Silanes for adhesion promotion and surface modification, in: Silane: chemistry, applications and performance, Moriguchi K., Utagawa S., Eds., New York: Nova Science Publishers, 87–109.
42. Debnath S., Ranade R., Wunder S., McCool J., Boberick K., Baran G., 2004, Interface effects on mechanical properties of particle-reinforced composites, Dent. Mater., 20(7), 677–686.
43. Aydinoglu A., Yoruç A.B.H., 2017, Effects of silane modified fillers on properties of dental composite resin, Mater. Sci. Eng., 79(1), 382–389.
44. Shin-Etsu chemical Corporation Limited, Japan, Silane Coupling Agents: Combination of Organic and Inorganic Materials Report, 2017.
45. Matinlinna J.P., Lassila L.V.J., Vallittu P.K., 2007, The effect of five silane coupling agents on the bond strength of a luting cement to a silica-coated titanium, Dent. Mater., 23(9), 1173–1180.
46. Rosen M.R., 1978, From treating solution to filler surface and beyond. The life history of a silane coupling agent, J. Coat. Technol., 50, 70–82.
47. Antonucci J.M., Dickens S.H., Fowler B.O., Xu H.H.K., McDonough W.G., 2005, Chemistry of silanes: Interfaces in dental polymers and composites, J. Res. Natl. Inst. Stand. Technol., 110(5), 541–558.
48. McDonough W.G., Antonucci J.M., Dunkers J.P., 2001, Interfacial shear strengths of dental resin-glass fibers by the microbond test, Dent. Mater., 17(6), 492–498.
49. Arklese B., Maddox A., Singh M., Zazyczny J., Maatisons J., 2014, Silane Coupling Agents Connecting Across Boundaries, Gelest, Inc. Morrisville, PA, 2–72.
50. Joseph M.A., Sabine D., Bruce O.F., Hockin H.K.X., Walter G.M.D., 2005, Chemistry of silanes: Interfaces in dental polymers and composites, J. Res.Nist., 110(5), 541–558.
51. Arklese B., 2015, Hydrophobicity, Hydrophilicity and Silanes, Gelest Inc., Morrisville, PA, 114–135.
52. Haider A., Haider S., Han S.S., Kang I.K., 2017, Recent advances in the synthesis, functionalization and biomedical applications of hydroxyapatite: A review, R. Soc. Chem., 7(13), 7442.
53. Oral O., Lassila L.V., Kumbuloglu O., Vallittu P.K., 2014, Bioactive glass particulate filler composite: Effect of coupling of fillers and filler loading on some physical properties, Dent. Mater., 30(7), 570–577.
54. Hirano H., Kadota J., Yamashita T., Agari Y., 2012, Treatment of inorganic filler surface by silane-coupling agent: Investigation of treatment condition and analysis of bonding state of reacted agent, J. Miner. Metall. Mater. Engin., 6(1), 1–5.
55. Da Silva E., Gonçalves L., Guimarães J., Poskus L., Fellows C., 2010, The diffusion kinetics of a nanofilled and a midifilled resin composite immersed in distilled water, artificial saliva, and lactic acid, Clin. Oral. Invest., 15, 393–401.
56. International Organization for Standardization, 2000, BS EN ISO 4049: 2000 Polymer-based Filling, Restorative and Luting Materials, Brussels.

57. Ferracane J.L., Berge H.X., Condon J.R., 1998, In vitro aging of dental composites in water—effect of degree of conversion, filler volume, and filler/matrix coupling, J. Biomed. Mater. Res., 42, 465–472.
58. Asaoka K., Hirano S., 2003, Diffusion coefficient of water through dental composite resin, Biomaterials, 24, 975–979.
59. Mustafaa R., Alshali R.Z., Silikas N., 2018, The effect of desiccation on water sorption, solubility and hygroscopic volumetric expansion of dentine replacement materials, Dent. Mater., 34(8), 205–213.
60. Liu F., Jiang X., Bao S., Wang R., Sun B., Zhu M., 2015, Effect of hydroxyapatite whisker surface graft polymerization on water sorption, solubility and bioactivity of the dental resin composite, Mater. Sci. Eng. C, 53, 150–155.
61. Kitchens M., Owens B.M., 2007, Effect of carbonated beverages, coffee, sports and high energy drinks, and bottled water on the in vitro erosion characteristics of dental enamel, J. Clin. Pediatr. Dent., 31, 153–159.
62. McKenzie A.M., Linden R.W.A., Nicholson J.W., 2004, The effect of Coca-Cola and fruit juices on the surface hardness of glass-ionomers and 'compomers', J. Oral. Rehabil., 31, 1046–1052.
63. Rahima T.N.A., Mohamada D., Akilb H.M., Rahman I.A., 2012, Water sorption characteristics of restorative dental composites immersed in acidic drinks, Dent. Mater., 28, e63–e70, http://doi.org/10.1016/j.dental.2012.03.011.

Index

A

abrasive machining, 36
additive manufacturing, 135, 136, 149
air-jet erosion, 3, 6
AL6061 alloy, 1–7
alginate, 99, 100
ANOVA analysis, 87
artificial, 220, 303
artificial neural network (ANN), 120
atomic force microscopy (AFM), 179
Auger electron spectroscopy (AES), 185

B

binders, 65, 332, 239
biofunctionality, 96
boron nitrade (BN), 213
brake pad friction composite materials, 229

C

carbide-free bainite (CBFs), 199, 209
carbon allotropes, 2
cellulose, 99
chitin, 99
chitosan, 99
composites fabricated, 110, 116, 258
computer-aided design (CAD), 24, 135, 149
continuous liquid interface production (CLIP), 136
corrosion resistance, 96, 101, 102, 174, 187
current AFM, 182

D

degrees of freedom, 25, 148
density measurement test, 261
dental composites, 270, 271, 274
digital light processing (DLP), 136, 140
dimensional accuracy, 39, 41
directed energy deposition (DED), 136
direct metal laser sintering (DMLS), 136
drug delivery devices (DDD), 145

E

electrical discharge machining (EDM), 36
electron beam melting (EBM), 136
electron energy loss spectroscopy (EELS), 185
electrostatic force microscopy (EFM), 181
Energy-Dispersive X-Ray Spectroscopy (EDX, EDS), 177, 184

F

fatigue behaviour, 159, 160, 167
fatigue properties, 156–159, 162–164
flexibility, 2, 38, 60, 235
Food and Drug Administration (FDA), 145

G

geometry, 25, 36, 136, 141, 146
GO preparation, 1, 2
grey relational analysis, 12, 14

H

hardness testing, 3, 4, 109
hybrid alloy composites, 79
hydrophilicity, 269, 270

I

inorganic fillers, 271, 272
inorganic substrate, 273

L

laser engineered net shaping (LENS), 136
linear swelling test, 261
lower bainite (LB), 205

M

material removal rate, 40, 67, 120
matrix interface, 136
metallic biomaterials, 96, 101
metal matrix composite, 1, 108
microstructural characterization, 110
minimum quantity lubrication (MQL), 11, 20
moisture content test, 260, 262
mono-alkoxy silane, 274

N

nanofluid preparation, 11, 13

P

piezo force microscopy (PFM), 182
polymer composites, 156–159, 163, 164
polytetrafluoroethylene (PTFE), 213
powder bed fusion (PBF), 143

R

reinforcements, 53, 78, 248
reliable, 243
retained austenite (RA), 191, 199, 209
rolling contact fatigue, 191, 206

S

scanning capacitance microscope (SCM), 182
scanning electron microscopy (SEM), 13, 108, 127, 126
scanning probe microscopy (SPM), 175
selective laser sintering (SLS), 136
semiconductor, 174, 182, 188
silane coupling, 258, 270, 271
sliding velocity, 81, 110, 231
sliding wear, 2, 78, 80, 81, 108
sodium dodecyl sulfate, 213
specific wear rate, 80, 81, 86, 89–92
Standard Tessellation Language (STL), 136
stereolithography, 136, 140
structured, 141
surface engineering, 173–175, 187

T

Taguchi method, 78, 87
tool wear rate, 41, 59, 61
transmission electron microscopy (TEM), 176
tribological behaviour, 112, 113, 203, 208
turbine blades, 24, 262, 450

U

ultrasonic cutting, 36
ultrasonic transducers, 38
upper bainite (UB), 205

W

water absorption test, 260
wire electric discharge machining (WEDM), 119
worn surface, 110, 114, 115

X

X-ray diffraction (XRD), 183, 184